CLASSICAL SOLUTIONS IN QUANTUM FIELD THEORY

Classical solutions play an important role in quantum field theory, high energy physics, and cosmology. Real time soliton solutions give rise to particles, such as magnetic monopoles, and extended structures, such as domain walls and cosmic strings, that have implications for the cosmology of the early universe. Imaginary time Euclidean instantons are responsible for important nonperturbative effects, while Euclidean bounce solutions govern transitions between metastable states.

Written for advanced graduate students and researchers in elementary particle physics, cosmology, and related fields, this book brings the reader up to the level of current research in the field. The first half of the book discusses the most important classes of solitons: kinks, vortices, and magnetic monopoles. The cosmological and observational constraints on these are covered, as are more formal aspects, including BPS solitons and their connection with supersymmetry. The second half is devoted to Euclidean solutions, with particular emphasis on Yang–Mills instantons and on bounce solutions.

ERICK J. WEINBERG is a Professor of Physics in the Department of Physics, Columbia University. Since 1996 he has been Editor of *Physical Review D*. His research interests include the implications of solitons and instantons for high energy physics, cosmology, and black holes, as well as a variety of other topics in quantum field theory.

CAMBRIDGE MONOGRAPHS ON MATHEMATICAL PHYSICS

General Editors: P. V. Landshoff, D. R. Nelson, S. Weinberg

Classical Solutions in Quantum Field Theory

Solitons and Instantons in High Energy Physics

ERICK J. WEINBERG
Columbia University

CAMBRIDGE
UNIVERSITY PRESS

University Printing House, Cambridge CB2 8BS, United Kingdom

Cambridge University Press is part of the University of Cambridge.

It furthers the University's mission by disseminating knowledge in the pursuit of
education, learning and research at the highest international levels of excellence.

www.cambridge.org
Information on this title: www.cambridge.org/9780521114639

First published 2012

A catalogue record for this publication is available from the British Library

Library of Congress Cataloguing in Publication data
Weinberg, Erick J.
Classical solutions in quantum field theory : solitons and instantons in high
energy physics / Erick J. Weinberg.
p. cm. – (Cambridge monographs on mathematical physics)
Includes bibliographical references and index.
ISBN 978-0-521-11463-9 (hardback)
1. Quantum theory – Mathematics. I. Title.
QC174.17.M35W45 2012
530.12–dc23
2012015503

ISBN 978-0-521-11463-9 Hardback

To Carolyn, Michael, and Cate

Contents

Preface

Semiclassical methods based on classical solutions play an important role in quantum field theory, high energy physics, and cosmology. Real-time soliton solutions give rise both to new particles, such as magnetic monopoles, and to extended structures, such as domain walls and cosmic strings. These could have been produced as topological defects in the very early universe. Confronting the consequences of such objects with observation and experiment places important constraints on grand unification and other potential theories of high energy physics beyond the standard model. Imaginary-time Euclidean instanton solutions are responsible for important nonperturbative effects. In the context of quantum chromodynamics they resolve one puzzle—the U(1) problem—while raising another—the strong CP problem—whose resolution may entail the existence of a new species of particle, the axion. The Euclidean bounce solutions govern transitions between metastable vacuum states. They determine the rates of bubble nucleation in cosmological first-order transitions and give crucial information about the evolution of these bubbles after nucleation. These bounces become of particular interest if there is a string theory landscape with a myriad of metastable vacua.

This book is intended as a survey and overview of this field. As the title indicates, there is a dual focus. On the one hand, solitons and instantons arise as solutions to classical field equations. The study of their many varieties and their mathematical properties is a fascinating subfield of mathematical physics that is of interest in its own right. Much of the book is devoted to this aspect, explaining how the solutions are discovered, their essential properties, and the topological underpinnings of many of the solutions. However, the physical significance of these classical objects can only be fully understood when they are seen in the context of the corresponding quantum field theories. To that end, there is also a discussion of quantum effects, including those arising from the interplay of fermion fields with topologically nontrivial classical solutions, and of some of the phenomenological consequences of instantons and solitons.

The first half of this book focuses on real-time classical solutions. I focus in particular on three classes of solitons—kinks, vortices, and magnetic monopoles—in one, two, and three spatial dimensions, respectively. Several chapters are devoted to their classical properties and many aspects of their quantum behavior. These are followed by a chapter that discusses the cosmological consequences of domain walls and cosmic strings—the dimensionally extended manifestations of kinks and vortices—and of magnetic monopoles, and the implications of these for proposed high energy theories. Finally, there is a chapter discussing solitons in the BPS limit, including the connections with supersymmetry and duality.

After considering solitons, I turn to Euclidean solutions. Although these are solutions of classical equations, they are associated with tunneling processes that are truly quantum mechanical phenomena. An introductory chapter presenting an overview of this connection is followed by two chapters on Yang–Mills instantons. The first of these is primarily concerned with the mathematical properties of these solutions and their interpretation in terms of vacuum tunneling. Fermions are introduced in the second chapter, which discusses the physical consequences flowing from the instantons. A final chapter describes the bounce solutions and vacuum transitions.

Of necessity, some topics had to be omitted. In particular, Q-balls, nontopological solitons whose existence is based on the possession of a conserved charge rather than on topology, are not covered, nor are skyrmions, a fascinating class of topological solitons.

My goal has been to make the book accessible to advanced graduate students and other newcomers to the field, but also useful for more experienced researchers. I assume that the reader has had an introductory course in quantum field theory and some familiarity with non-Abelian gauge theories, but only the mathematical background of a typical physics graduate student. The homotopy theory needed to understand the topological underpinnings of the solitons is presented and explained. An appendix discusses roots, weights, and other necessary properties of Lie groups and algebras, building on the familiar results associated with SU(2).

I owe much to the colleagues and students with whom I have collaborated in research in this field. I thank Claude Bernard, Xingang Chen, Norman Christ, Huidong Guo, Alan Guth, Jim Hackworth, Conor Houghton, Roman Jackiw, Tom Kibble, Alex Kusenko, Bum-Hoon Lee, Choonkyu Lee, Hakjoon Lee, Kimyeong Lee, Sang-Hoon Lee, Arthur Lue, Ali Masoumi, Dimitrios Metaxas, Chris Miller, Doug Rajaraman, Alex Ridgway, Jon Rosner, Koenraad Schalm, and Piljin Yi. I am grateful to the late Sidney Coleman, from whom I learned field theory and much more.

I am also grateful for the suggestions and comments on aspects of this book from Adam Brown, Dan Kabat, Kimyeong Lee, Eugene Lim, Andy Millis, I-Sheng Yang, and Piljin Yi. I am particularly grateful to Ali Masoumi and Xiao Xiao for carefully reading and pointing out errors in the final text.

Parts of this book were written at the Korea Institute for Advanced Study and at the Aspen Center for Physics. I am grateful to both institutions. My research over the years has been supported in part by the U.S. Department of Energy. My stays at Aspen were supported in part by the U.S. National Science Foundation.

Finally, I thank Carolyn for her support, her encouragement, and her gentle prodding.

1

Introduction

1.1 Overview

It is a familiar fact that when a field theory is treated quantum mechanically the wave solutions of the classical theory lead to elementary quanta that have a natural interpretation as particles in the quantum theory. This suggests a one-to-one correspondence between fields and particle species and is the basis for the standard applications of perturbative quantum field theory.

However, many classical field theories have solutions that are already particle-like at the classical level. These are characterized by an energy density that is localized in space and that does not dissipate over time. It is natural to ask whether these "solitons", as they are called, have counterparts in the quantum version of the theory. If so, they would presumably be a new species of particle, quite distinct from the "elementary" particle associated with the wave solutions of the free field theory.

It is instructive to compare the classical size of the soliton with the Compton wavelength that it would have in the quantum theory. If the elementary particles of the theory have masses of order m and a characteristic coupling of order g, one typically finds that the soliton has a classical energy

$$E_{\text{classical}} \sim \frac{m}{g} \tag{1.1}$$

and a characteristic spatial size $\ell_{\text{soliton}} \sim 1/m$. Hence,

$$\lambda_{\text{Compton}} \sim \frac{1}{E_{\text{classical}}} \sim g\,\ell_{\text{soliton}}. \tag{1.2}$$

(I am using units with $\hbar = 1$.) If the coupling is weak, the Compton wavelength is much less than the classical size, and so we might expect the soliton to survive, perhaps with slight modifications, after quantization.

A possible objection is the stark contrast between the smooth profile of the classical solution and the fuzziness of quantum field theory. It is certainly true that the quantum fluctuations of the field are large, even divergent, when the field is measured at very short distances. However, these fluctuations are reduced when the field is averaged over a larger smearing distance. We will see that the same weak-coupling regime that gives $\ell_{\text{soliton}} \gg \lambda_{\text{Compton}}$ also guarantees the existence of a smearing distance that is both large enough to suppress the quantum fluctuations and small enough that the classical field profile is still evident.

The inverse dependence on the coupling implies that in this weak-coupling regime the soliton mass is large, tending toward infinity as the coupling goes to zero. This explains why the effects of the soliton are not seen in perturbation theory. Nevertheless, once the classical solution is known, perturbative methods can be used to quantize the fields about the soliton and to demonstrate that there is indeed a corresponding one-particle state in the quantum theory. Furthermore, the quantum corrections to the classical energy are calculable and give a mass of the form

$$M_{\text{quantum}} = E_{\text{classical}} \left(1 + c_1 g + c_2 g^2 + \cdots\right). \tag{1.3}$$

What about the strong-coupling regime? Even though the soliton may still be a solution of the classical field equations, the perturbative analysis of the quantum theory breaks down here, and the arguments for a quantum counterpart to the soliton are no longer so clear-cut. However, a new and striking phenomenon may now come into play. There are examples of theories—a particularly well-known pair being the sine-Gordon and massive Thirring models—that are related by a duality that maps the weak-coupling regime of one onto the strong-coupling regime of the other. The sine-Gordon soliton states correspond to elementary particle states of the massive Thirring model, while the elementary particle of the sine-Gordon model becomes a massive Thirring bound state. One must conclude that there is no intrinsic difference between an elementary particle and a soliton. The distinction between them is simply that one viewpoint or the other is more convenient for calculation in a particular coupling regime.

Although we live in a world with three spatial dimensions (and perhaps some additional hidden ones), it can be instructive to consider solitons in fewer dimensions. The analysis of these toy models is often more tractable and helps elucidate issues of principle. Their solutions can also be trivially extended to higher dimensions, where they acquire new physical significance. A particle-like soliton in one dimension can be interpreted as a planar solution in three dimensions, corresponding to a domain wall. Similarly, a two-dimensional particle-like soliton becomes a line solution, or string, in three dimensions.

One can also consider solitons in more than three spatial dimensions. Of particular interest are those in four dimensions. These could be viewed as particles in a hypothetical world with four spatial dimensions. Alternatively, and more importantly, they can be interpreted as solutions in a Euclideanized version of our four-dimensional spacetime. Such Euclidean solutions, or instantons, have no obvious physical significance in a classical context. However, they become meaningful quantum mechanically because wavefunctions extend into classically forbidden regions where the potential energy is greater than the total energy. Roughly speaking, one can view this as implying a negative kinetic energy, corresponding to evolution in a Euclidean spacetime with imaginary time. A well-known consequence is that quantum systems can tunnel though potential energy barriers to effect transitions that would be classically forbidden. This leads to important and unexpected nonperturbative effects in gauge theories, with magnitudes that are determined by the action of the relevant instanton. A further result of tunneling processes in field theory is the decay of metastable vacua by bubble nucleation, a process of considerable importance for cosmology. The Euclidean solutions that govern such bubble nucleation are known as bounces.

Finally, a note on terminology. I follow the practice in high energy physics of using the term soliton for any localized classical solution that does not dissipate over time. However, the reader should be aware that some other fields use a more restrictive definition, with the term only used for solutions, arising in integrable systems, that emerge from scattering processes without deformation or loss of energy.

1.2 Conventions

Metric and indices

For the spacetime metric I use the "mostly minus" convention, with the metric $\eta_{\mu\nu} = \mathrm{diag}\,(1, -1, -1, -1)$ in flat four-dimensional spacetime. Coordinates are defined by

$$x^\mu = (t, x, y, z) = (t, \mathbf{x}) \tag{1.4}$$

so that

$$\partial_\mu = (\partial/\partial t, \boldsymbol{\nabla}). \tag{1.5}$$

Lorentzian spacetime indices are denoted by Greek letters and summation over repeated indices, one upper and one lower, is to be understood. Purely spatial indices are denoted by Latin letters, generally from the middle of the alphabet; summation over repeated indices (possibly both upper or both lower) is also to be understood. Euclidean spacetime indices are denoted by Latin letters.

The antisymmetric tensor in any dimension is defined to be unity when all of its indices are upper and in numerical order. Thus, $\epsilon^{123} = \epsilon^{0123} = \epsilon^{1234} = 1$.

Dirac matrices

The Dirac matrices in four-dimensional Lorentzian spacetime obey

$$\{\gamma^\mu, \gamma^\nu\} = 2g^{\mu\nu}. \tag{1.6}$$

Of these, γ^0 is Hermitian, while the remaining three are anti-Hermitian. The matrix

$$\gamma^5 = i\gamma^0\gamma^1\gamma^2\gamma^3 \tag{1.7}$$

is Hermitian and obeys $(\gamma^5)^2 = I$.

Units

I use natural units with c, $\hbar$, and Boltzmann's constant k_B all equal to unity.

Gauge fields

Conventions associated with gauge fields vary within the soliton and instanton literature. Those used in this book are described below.

The electromagnetic potential is

$$A^\mu = (\Phi, \mathbf{A}) \tag{1.8}$$

where Φ and $\mathbf{A}$ are the usual scalar and vector potentials. The field strength tensor is

$$F_{\mu\nu} = \partial_\mu A_\nu - \partial_\nu A_\mu \tag{1.9}$$

so that, e.g., $F_{12} = F^{12} = -B_z$ and $F_{03} = F^{30} = E_z$. The covariant derivative of a complex field carrying electromagnetic [or any other U(1)] charge q is given by

$$D_\mu\phi = (\partial_\mu + iqA_\mu)\phi. \tag{1.10}$$

The Lagrangian is then invariant under U(1) gauge transformations of the form

$$\begin{aligned} \phi &\to e^{iq\Lambda(x)}\phi, \\ A_\mu &\to A_\mu - \partial_\mu\Lambda(x). \end{aligned} \tag{1.11}$$

In non-Abelian gauge theories the gauge field is written as a Hermitian element of the Lie algebra

$$A_\mu = A_\mu^a T^a, \tag{1.12}$$

where the Hermitian generators T^a are normalized so that

$$\operatorname{tr} T^a T^b = \frac{1}{2}\delta^{ab}. \tag{1.13}$$

They obey

$$[T^a, T^b] = if_{abc}T^c, \tag{1.14}$$

with the structure constants f_{abc} being totally antisymmetric. This corresponds to the standard normalization for the fundamental representation of SU(2), with the generators being $\sigma^a/2$, where the σ^a are the Pauli matrices. The field strength is

$$F_{\mu\nu} = \partial_\mu A_\nu - \partial_\nu A_\mu - ig[A_\mu, A_\nu], \tag{1.15}$$

with components

$$F_{\mu\nu}^a = \partial_\mu A_\nu^a - \partial_\nu A_\mu^a + gf_{abc}A_\mu^b A_\nu^c. \tag{1.16}$$

A matter field ϕ can be written as a column vector transforming under an irreducible representation of the gauge group. Its covariant derivative is

$$D_\mu\phi = \partial_\mu\phi - igA_\mu\phi. \tag{1.17}$$

With components written out explicitly, this is

$$(D_\mu \phi)_j = \partial_\mu \phi_j - ig A_\mu^a (t^a)_{jk} \phi_k \tag{1.18}$$

with i, j, and k running from 1 to N and $(t^a)_{jk}$ denoting the appropriate representation of the generators.

Under a non-Abelian gauge transformation $U(x)$, the various quantities above transform as

$$\begin{aligned} A_\mu &\longrightarrow U A_\mu U^{-1} - \frac{i}{g}(\partial_\mu U)U^{-1} = U A_\mu U^{-1} + \frac{i}{g} U \, \partial_\mu U^{-1}, \\ F_{\mu\nu} &\longrightarrow U F_{\mu\nu} U^{-1}, \\ \phi &\longrightarrow \mathcal{U}\phi. \end{aligned} \tag{1.19}$$

where $\mathcal{U}$ is the transformation written in the appropriate representation of the group. For an infinitesimal gauge transformation

$$U = e^{i\Lambda} \approx I + i\Lambda + \cdots \tag{1.20}$$

the change in the gauge potential is

$$\delta A_\mu = \frac{1}{g}\partial_\mu \Lambda - i[A_\mu, \Lambda] = \frac{1}{g} D_\mu \Lambda. \tag{1.21}$$

If the matter fields transform under the adjoint representation, an alternative notation is to write them as linear combinations of the generators,

$$\phi = \phi^a T^a \tag{1.22}$$

with

$$D_\mu \phi = \partial_\mu \phi - ig[A_\mu, \phi]. \tag{1.23}$$

In the special case of a triplet field in an SU(2) gauge theory (where $f_{abc} = \epsilon_{abc}$) I sometimes adopt the standard three-dimensional vector notation and write

$$\begin{aligned} D_\mu \phi &= \partial_\mu \phi + g\mathbf{A}_\mu \times \phi, \\ \mathbf{F}_{\mu\nu} &= \partial_\mu \mathbf{A}_\nu - \partial_\nu \mathbf{A}_\mu + g\mathbf{A}_\mu \times \mathbf{A}_\nu. \end{aligned} \tag{1.24}$$

It is sometimes convenient to absorb the gauge coupling in the gauge field by a rescaling $A_\mu \to g A_\mu$. The Yang–Mills Lagrangian is then

$$\mathcal{L} = -\frac{1}{4g^2} F_{\mu\nu}^a F^{\mu\nu a} = -\frac{1}{2g^2} \text{tr}\, F_{\mu\nu} F^{\mu\nu}. \tag{1.25}$$

2

One-dimensional solitons

Field theories in one spatial dimension provide a natural starting point for the study of solitons. Because of the simplifications that result from working in one dimension, many more calculations can be carried through explicitly. In addition, the topological considerations that play an important role in all dimensions are particularly easy to visualize in one-dimensional theories. Although the primary value of these theories is as toy models, some of the results we will obtain find application in the real world. First, there are condensed matter systems that can be treated as essentially one-dimensional, some of which support solitons. Second, some of the one-dimensional solitons that we will find can be trivially extended to higher dimensions, so that a localized one-dimensional soliton can become a planar domain wall in higher dimensions.

2.1 Kinks

The classic example of a soliton in one spatial dimension [1, 2] arises in a theory with a single scalar field ϕ and Lagrangian density

$$\mathcal{L} = \frac{1}{2}(\partial_\mu \phi)(\partial^\mu \phi) - V(\phi), \tag{2.1}$$

where the scalar field potential

$$
\begin{aligned}
V(\phi) &= -\frac{1}{2}m^2\phi^2 + \frac{\lambda}{4}\phi^4 + \frac{\lambda}{4}v^4 \\
&= \frac{\lambda}{4}(\phi^2 - v^2)^2.
\end{aligned}
\tag{2.2}
$$

Here m^2 and λ are both positive and

$$v = \sqrt{\frac{m^2}{\lambda}}. \tag{2.3}$$

For later reference, note that in two spacetime dimensions the scalar field is dimensionless, so that the coupling constants are dimensionful. The dimensionless parameter that signals weak or strong coupling is the ratio λ/m^2.

The Lagrangian is invariant under the transformation $\phi \to -\phi$. However, this symmetry is spontaneously broken, with $V(\phi)$ having two degenerate minima, at $\phi = \pm v$. The constant term in Eq. (2.2), which has no effect on the dynamics, was chosen so that $V = 0$ at these minima. When the theory is quantized, these two minima correspond to two physically equivalent vacua. Choosing either one of them, say $\phi = v$, and then expanding in terms of the shifted field $\phi - v$, one finds that the theory has a single elementary scalar particle, with mass $\sqrt{2}\,m$.

The classical Euler–Lagrange equation of the theory is

$$\frac{d^2\phi}{dt^2} - \frac{d^2\phi}{dx^2} = -\lambda(\phi^2 - v^2)\phi. \tag{2.4}$$

We are particularly interested in static solutions, which satisfy

$$0 = \frac{d^2\phi}{dx^2} - \lambda(\phi^2 - v^2)\phi. \tag{2.5}$$

This is a nonlinear equation, and it may not be obvious from the outset that it has any nonsingular solutions other than the two constant vacuum solutions, $\phi(x) = v$ and $\phi(x) = -v$. To persuade ourselves that it does, note that Eq. (2.5) is also the condition for a configuration $\phi(x)$ to be a stationary point of the potential energy[1]

$$U[\phi(x)] = \int dx \left[\frac{1}{2}\left(\frac{d\phi}{dx}\right)^2 + V(\phi) \right]. \tag{2.6}$$

This is not surprising. For a system whose kinetic energy is purely quadratic in time derivatives, the static solutions of the equations of motion are just the stationary points of the potential energy; the stable solutions are given by the local minima of U.

Thus, our task is to show that there are configurations other than the vacuum solutions that are local minima of $U[\phi(x)]$. To this end, consider a configuration, such as the one shown in Fig. 2.1, in which $\phi(\infty) = v$ and $\phi(-\infty) = -v$. Unless we have made a remarkably lucky choice, this configuration will not be a solution. This means that it can be smoothly varied in such a way as to lower its potential energy. Continuing this process until a minimum of U is reached will lead us to a static solution. Because a smooth variation cannot change the values of ϕ at spatial infinity, the solution we are led to will have $\phi(\infty) \neq \phi(-\infty)$, and so

---

[1] It is important to remember that the potential energy includes not just the contribution from the scalar field potential $V(\phi)$, but also that from the spatial gradient terms.

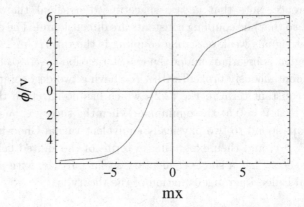

Fig. 2.1. A field configuration that cannot be smoothly deformed to a vacuum solution.

cannot be either of the vacuum solutions. It must instead be a nontrivial spatially varying solution; i.e., the soliton that we are seeking.[2]

Some important properties of this solution are revealed by a rescaling of variables. If we write $\phi = vf$ and $u = mx$, Eq. (2.5) becomes

$$0 = \frac{d^2 f}{du^2} - f(f^2 - 1),$$
(2.7)

with $f(\pm\infty) = \pm 1$, while the energy of this static solution takes the form

$$E = \frac{m^3}{\lambda} \int du \left[\frac{1}{2} \left(\frac{df}{du} \right)^2 + \frac{1}{4}(f^2 - 1)^2 \right].$$
(2.8)

It is evident from Eq. (2.7) that $f(u)$ does not contain any explicit factors of m or λ. Hence, its spatial variation is characterized by a distance that is of order unity when measured in terms of u, and thus of order m^{-1} when measured in terms of x. Because the integral on the right-hand side of Eq. (2.8) is also independent of m and λ, it must be of order unity, so the solution has an energy of order m^3/λ.

This is much greater than the mass $\sqrt{2}\,m$ of the elementary scalar when the coupling is weak, and diverges in the limit $\lambda \to 0$. This fact, which is characteristic of solitons in field theory, explains why solitons are not encountered in ordinary perturbative approaches to quantum field theory.

___

[2] This argument is not rigorous, and must be used with care. Because the space of field configurations is not compact, there need not be any configuration that minimizes the potential energy. Although this does not happen here, we will encounter such a situation when we discuss multisoliton configurations later in this section.

It is now time to tackle the field equation directly. Multiplying both sides of Eq. (2.5) by $d\phi/dx$ gives

$$0 = \frac{d}{dx}\left[\frac{1}{2}\left(\frac{d\phi}{dx}\right)^2 - \frac{\lambda}{4}(\phi^2 - v^2)^2\right]. \qquad (2.9)$$

Hence, the quantity in brackets must be independent of x. Evaluating it at $x = \infty$, we see that it actually vanishes. It follows that

$$\frac{d\phi}{dx} = \pm\sqrt{\frac{\lambda}{2}}(\phi^2 - v^2). \qquad (2.10)$$

Our boundary conditions require the upper sign. Straightforward integration then gives

$$\phi(x) = v\tanh\left[\frac{m}{\sqrt{2}}(x - x_0)\right], \qquad (2.11)$$

where x_0 is a constant of integration. This solution, which is shown in Fig. 2.2, is known as the kink solution; x_0 can be viewed as specifying the position of the kink. The solution obtained by starting with the opposite boundary conditions and taking the lower sign in Eq. (2.10),

$$\phi(x) = -v\tanh\left[\frac{m}{\sqrt{2}}(x - x_0)\right], \qquad (2.12)$$

is called the antikink.

The energy density, shown in Fig. 2.3, is the same for both solutions. It is concentrated within a region of width $\sim m^{-1}$ centered about x_0. Outside this region, the field is essentially indistinguishable from that in a vacuum. Although it is a different vacuum on opposite sides of the kink, this is not evident to a local

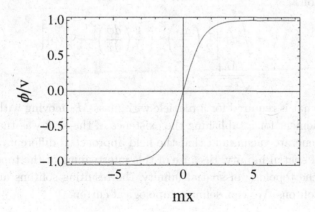

Fig. 2.2. The kink solution of Eq. (2.11), with $x_0 = 0$.

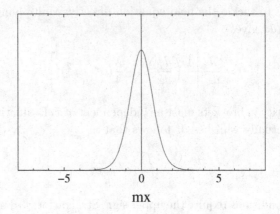

mx

Fig. 2.3. The energy density of the kink solution with $x_0 = 0$.

observer. This localization of the energy suggests that the kink be interpreted as a kind of particle, with its mass given by the total energy of the static solution,

$$M_{\text{cl}} = \frac{2\sqrt{2}}{3} \frac{m^3}{\lambda}. \tag{2.13}$$

The subscript here is intended to indicate that this is just the classical approximation to the mass. We will see in the next section that there are quantum corrections to the mass.

If the kink is to be interpreted as a particle, then there should also be solutions corresponding to moving kinks. Lorentz transforming the static solution gives

$$\phi(x, t) = v \tanh\left[\frac{m}{\sqrt{2}} \frac{(x - ut - x_0)}{\sqrt{1 - u^2}}\right], \tag{2.14}$$

which describes a Lorentz-contracted kink moving with velocity u. The energy of this solution,

$$E = \int dx \left[\frac{1}{2}\left(\frac{d\phi}{dt}\right)^2 + \frac{1}{2}\left(\frac{d\phi}{dx}\right)^2 + V(\phi)\right]$$
$$= \frac{M_{\text{cl}}}{\sqrt{1 - u^2}}, \tag{2.15}$$

is precisely what is required for a particle with mass M_{cl} moving with velocity u.

The key element for establishing the existence of the kink was that $V(\phi)$ had multiple degenerate vacua, and that the field approached different vacua at the two points of spatial infinity. Because of this intertwining of the topology of the vacua with the topology of spatial infinity, the resulting solitons are known as topological solitons. We can define a topological current

$$J^{\mu}_{\text{top}} = \frac{1}{2v} \epsilon^{\mu\nu} \partial_{\nu}\phi, \tag{2.16}$$

where $\epsilon^{\mu\nu} = -\epsilon^{\nu\mu}$ and $\epsilon^{01} = 1$. The antisymmetry of $\epsilon^{\mu\nu}$ guarantees that this current is divergenceless. Its normalization has been chosen so that the corresponding charge,

$$
Q_{\text{top}} = \int_{-\infty}^{\infty} dx\, J^0 = \frac{1}{2v} \int_{-\infty}^{\infty} dx\, \partial_1 \phi
$$
$$
= \frac{1}{2v} \left[\phi(\infty) - \phi(-\infty) \right], \tag{2.17}
$$

is equal to 1 and -1 for the kink and the antikink, respectively. This charge is manifestly conserved, because no finite energy process can change the asymptotic values of the field. In contrast with the conserved charges obtained via Noether's theorem, the topological charge is not directly associated with a symmetry of the Lagrangian.

These methods can be applied to the case of an arbitrary potential $V(\phi)$ with two or more degenerate global minima.[3] As before, let us assume that an additive constant has been used to set $V = 0$ at these minima. Using the fact that $\phi(\pm\infty)$ must each be at one of the minima of the potential, we again find that the static field equation

$$
0 = \frac{d^2\phi}{dx^2} - \frac{dV}{d\phi} \tag{2.18}
$$

implies that

$$
\frac{d\phi}{dx} = \pm\sqrt{2V(\phi)}. \tag{2.19}
$$

This yields the useful virial identity

$$
\int_{-\infty}^{\infty} dx \left[\frac{1}{2} \left(\frac{d\phi}{dx} \right)^2 \right] = \int_{-\infty}^{\infty} dx\, V(\phi) \tag{2.20}
$$

that relates the gradient and potential terms in the Lagrangian.

Integrating Eq. (2.19) yields

$$
x = \pm \int_{\phi_0}^{\phi} \frac{d\phi}{\sqrt{2V(\phi)}}, \tag{2.21}
$$

where $\phi(0) = \phi_0$ is arbitrary. For a smoothly varying potential V, the integral on the right-hand side diverges as $\phi(x)$ approaches any of the global minima. Hence, as x ranges from $-\infty$ to ∞, ϕ must vary monotonically from one global minimum of V to an adjacent one. In particular, it cannot approach the same minimum at both $x = -\infty$ and $x = \infty$; i.e., there are no static solutions with vanishing topological charge and a fortiori no static solutions for theories with a single vacuum.

---

[3] The potential may also have other, higher, local minima.

An alternative expression for the energy of a kink connecting vacua at ϕ_1 and ϕ_2 can be obtained with the aid of Eqs. (2.19) and (2.20). Assuming $\phi_2 > \phi_1$, we have

$$E = 2 \int_{-\infty}^{\infty} dx\, V(\phi) = 2 \int_{\phi_1}^{\phi_2} d\phi \left(\frac{d\phi}{dx}\right)^{-1} V(\phi) = \int_{\phi_1}^{\phi_2} d\phi\, \sqrt{2V(\phi)}. \qquad (2.22)$$

One particular example of note is the theory [3, 4] with

$$V(\phi) = -\frac{m^4}{\lambda} \left[\cos\left(\frac{\sqrt{\lambda}}{m}\phi\right) - 1 \right]. \qquad (2.23)$$

This has become known as the sine-Gordon theory, the play on the term Klein–Gordon being suggested by its Euler–Lagrange equation,

$$0 = \frac{d^2\phi}{dt^2} - \frac{d^2\phi}{dx^2} + \frac{m^3}{\sqrt{\lambda}} \sin\left(\frac{\sqrt{\lambda}}{m}\phi\right). \qquad (2.24)$$

The potential is periodic, with degenerate minima at $\phi = (2\pi m/\sqrt{\lambda})N \equiv Nv$, where N is any integer. Expanding about the minimum at $\phi = 0$, we have

$$V = \frac{1}{2}m^2\phi^2 - \frac{\lambda}{4!}\phi^4 + \frac{\lambda^2}{6!\,m^2}\phi^6 + \cdots. \qquad (2.25)$$

From this we see that the mass of the elementary particle is m, while λ determines the strength of the interactions between these particles.

Static kink and antikink solutions connecting adjacent minima can be obtained via Eq. (2.19). In particular, the kink solution interpolating between the vacua at Nv and $(N+1)v$ is [4]

$$\phi(x) = Nv + \frac{2v}{\pi} \tan^{-1}\left[e^{m(x-x_0)}\right] \qquad (2.26)$$

and has an energy

$$M_{\rm cl} = \frac{8m^3}{\lambda}. \qquad (2.27)$$

Although the functional form of the sine-Gordon kink, Eq. (2.26), seems rather different from that of the ϕ^4 kink, Eq. (2.11), the two are actually remarkably similar. This is shown in Fig. 2.4a, where the two types of kinks are plotted together. The couplings and masses have been chosen so that $\tilde{\lambda}/\lambda = \pi^4/8$ and $\tilde{m}/m = \pi/\sqrt{8}$, with tildes indicating the sine-Gordon parameters. The similarity between the kinks can be understood by noting that the two potentials, shown in Fig. 2.4b, are quite close to each other (after a shift in ϕ) in the region between two adjacent minima.

Because the $\lambda\phi^4$ model has only two vacua, its topological charge can only take on the values -1, 0, or 1. Although multisoliton configurations are possible, they must be composed of alternating kinks and antikinks. By contrast, the

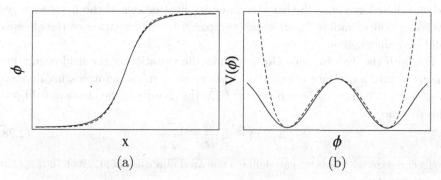

Fig. 2.4. A comparison of the kink solutions and potentials for the ϕ^4 (solid lines) and sine-Gordon (dashed lines) theories. The parameters are related as in the text, and the zeros of ϕ have been shifted so that the vacua of the two theories coincide.

sine-Gordon model admits configurations with more than one unit of topological charge. However, the discussion below Eq. (2.21) shows that these cannot include any static solutions. The physical explanation for this is that there is a repulsive force between kinks. Although this falls exponentially with the distance between the kinks, it never quite vanishes, so the potential energy of a two-kink configuration decreases monotonically as the kinks are drawn apart but never reaches a minimum at any finite separation.

2.2 Quantizing about the kink

In the context of quantum field theory, two questions naturally arise. First, to what extent is the explicit form of the classical solution meaningful in the quantum theory? Second, how is the existence of the classical soliton reflected in the states of the quantum theory?

In single-particle quantum mechanics it is meaningful to think of a particle as having a classical position if there is a state in which the uncertainties in the position and the velocity are both small compared to the relevant scales in the problem; this can be achieved if the particle mass is sufficiently large. The corresponding statement in quantum field theory is that we can meaningfully discuss a classical field profile if there is a state with sufficiently small uncertainties in both the field and its time derivative. This needs some explanation. If a quantum field is examined at very short distance scales the quantum fluctuations are enormous, and in fact diverge as the distance scale goes to zero. Nevertheless, this does not prevent us from treating electromagnetic fields as classical on macroscopic distances. This is possible because we can choose a distance scale L that simultaneously meets two criteria. First, it is small compared to the characteristic length scale of the classical field configuration, so that the latter is still evident even if it is only measured at a sequence of points separated by a distance L.

Second, it is large enough that the quantum fluctuations of the field, averaged over a region of radius L, are small compared to the variation of the classical field over that region.

For both the $\lambda\phi^4$ and sine-Gordon kinks the variation of the field occurs primarily within a distance of order m^{-1}, so we want to use an averaging distance $L \ll m^{-1}$. Because $d\phi/dx \sim mv \sim m^2/\sqrt{\lambda}$, the change in the classical field over this distance is

$$(\Delta\phi_L)_{\text{cl}} \sim L\frac{d\phi}{dx} \sim \frac{m^2 L}{\sqrt{\lambda}}. \qquad (2.28)$$

To compare with this we can define a smeared quantum field, given in n spatial dimensions by

$$\phi_L(\mathbf{x}) = \frac{1}{(2\pi L^2)^{n/2}} \int d^n y \, e^{-(\mathbf{y}-\mathbf{x})^2/2L^2} \phi(\mathbf{x}). \qquad (2.29)$$

An estimate of the quantum fluctuations of ϕ_L can be obtained by calculating them in the vacuum state of a free field with mass m. Thus, let us define

$$\begin{aligned}
(\Delta\phi_L)_{\text{qu}}^2 &= \langle 0|\phi_L(0)^2|0\rangle \\
&= \frac{1}{(2\pi L^2)^n} \int d^n y \, d^n z \, e^{-\mathbf{y}^2/2L^2} e^{-\mathbf{z}^2/2L^2} \langle 0|\phi(\mathbf{y})\phi(\mathbf{z})|0\rangle.
\end{aligned}$$
$$(2.30)$$

Using the fact that

$$\langle 0|\phi(\mathbf{y})\phi(\mathbf{z})|0\rangle = \int \frac{d^n k}{(2\pi)^n} \frac{e^{i\mathbf{k}\cdot(\mathbf{y}-\mathbf{z})}}{2\sqrt{\mathbf{k}^2 + m^2}}, \qquad (2.31)$$

we find that

$$(\Delta\phi_L)_{\text{qu}}^2 = \int \frac{d^n k}{(2\pi)^n} \frac{e^{-\mathbf{k}^2 L^2}}{2\sqrt{\mathbf{k}^2 + m^2}}. \qquad (2.32)$$

We are interested in the case where $L \ll m^{-1}$. From the behavior of the integral in this regime, we obtain

$$(\Delta\phi_L)_{\text{qu}} \sim \begin{cases} \sqrt{\ln(1/mL)}, & n = 1, \\ \\ L^{-(n-1)/2}, & n \geq 2. \end{cases} \qquad (2.33)$$

For our one-dimensional kinks we set $n = 1$. Our requirement that $(\Delta\phi_L)_{\text{qu}} \ll (\Delta\phi_L)_{\text{cl}}$ then becomes

$$\sqrt{\ln(1/mL)} \ll \frac{m^2 L}{\sqrt{\lambda}}. \qquad (2.34)$$

We can find an $L \ll m^{-1}$ that satisfies this condition as long as the dimensionless coupling λ/m^2 is small. This weak-coupling condition also ensures that the Compton wavelength $1/M_{\text{cl}}$ of the kink is much smaller than its classical width $1/m$.

This weak-coupling criterion extends to other one-dimensional solitons as well as to solitons in more spatial dimensions. If the coupling is strong, the classical soliton in general continues to be a solution of the field equations. However, its implications for the quantum theory are more difficult to untangle.

Let us therefore, at least for the time being, assume that we are in the weak-coupling regime. In this limit the classical soliton solution leads in a natural way to a family of quantum states with energies roughly equal to the classical soliton energy. These can be investigated using perturbative methods rather analogous to those that are used in studying the vacuum sector of the theory. Let us recall the standard procedure in that case.

We begin by finding the classical vacuum solution corresponding to a minimum of the potential. If there are several degenerate minima, we pick one, say $\phi = v$, as the starting point. If we define $\phi_{\text{vac}} \equiv v$, we can write

$$\phi(x,t) = \phi_{\text{vac}} + \eta(x,t) \tag{2.35}$$

and re-express the Lagrangian in terms of η. We then write

$$L = L_{\text{cl}} + L_{\text{quad}} + L_{\text{int}}, \tag{2.36}$$

where

$$L_{\text{cl}} = -\int dx\, V(\phi_{\text{vac}}) \tag{2.37}$$

is the classical vacuum energy,

$$L_{\text{quad}} = \int dx \left[\frac{1}{2}(\partial_\mu \eta)^2 - \frac{1}{2} V''(\phi_{\text{vac}}) \eta^2 \right] \tag{2.38}$$

contains the terms quadratic in η, and the interaction term L_{int}, which contains terms that are cubic and higher order in η, can be treated as a perturbation. There are no terms linear in η because we have expanded about a minimum of V.

We then change variables to diagonalize L_{quad}. To do this, we find a set of normal modes $f_j(x)$ that obey

$$\left[-\frac{d^2}{dx^2} + V''(\phi_{\text{vac}}) \right] f_j(x) = \omega_j^2 f_j(x) \tag{2.39}$$

and correspond to periodic solutions

$$\eta(x,t) = f_j(x) e^{-i\omega_j t} \tag{2.40}$$

of the linearized field equations. Choosing the f_j to be orthonormal, we expand the field as

$$\eta(x,t) = \sum_j \left[c_j(t) f_j(x) + \text{h.c.} \right], \tag{2.41}$$

where the $c_j(t)$ are operators and "h.c." represents the Hermitian conjugate term that must be added because η is a real field.[4] Substituting this expansion into Eq. (2.38) gives a sum of simple harmonic oscillator Lagrangians. After quantization, the energy eigenstates are labeled by the occupation numbers n_j of the various oscillators, with their energies being

$$E_{\text{quad}} = \sum_j \left(n_j + \frac{1}{2} \right) \omega_j. \tag{2.42}$$

There is an immediate problem, because the sum of the zero-point energies is infinite. Putting aside for the moment the details of how to regulate this infinity, we subtract from the Lagrangian an appropriate constant, δE, whose value is chosen to make the energy of the vacuum finite (and perhaps zero). We then go on to include the effects of the interaction terms, working order by order in perturbation theory. These not only give new contributions to the energy of the vacuum—which, if gravity is neglected, is unobservable in any case—but also modify the splitting between the vacuum and the lowest one-particle state; i.e., the mass of the elementary scalar. In general, it turns out that this correction is also infinite, and must be compensated by the addition of divergent counterterms to the Lagrangian. Thus, we now have

$$L = L_{\text{cl}} + L_{\text{quad}} + L_{\text{int}} - \delta E + L_{\text{ct}}, \tag{2.43}$$

where L_{ct} contains the field-dependent counterterms, with coefficients chosen to satisfy appropriate renormalization conditions.

The energy of the vacuum is then

$$E_{\text{vac}} = E_{\text{cl}}(\phi_{\text{vac}}) + \frac{1}{2} \sum_j \omega_j + \delta E + E_{\text{ct}}(\phi_{\text{vac}}) + \cdots, \tag{2.44}$$

where the ellipsis denotes terms that are at least first order in the coupling. Note that there can be an order unity contribution from L_{ct}, even though the counterterm coefficients arise from perturbative diagrams, if ϕ_{vac} is inversely proportional to the coupling.

We can now follow a similar procedure to obtain both the quantum corrections to the energy of a static kink and the spectrum of nearby states. Thus, we now write

$$\phi(x,t) = \phi_{\text{kink}}(x) + \tilde{\eta}(x,t), \tag{2.45}$$

expand L in powers of $\tilde{\eta}$, and find the normal modes that diagonalize the quadratic Lagrangian. All the above equations apply, provided that we replace ϕ_{vac} by ϕ_{kink}, and remember that the latter has a nontrivial x-dependence.

---

[4] In actuality, of course, the spectrum of normal frequencies includes a continuum portion, and the sums in this and subsequent formulas should be understood to include integrals where appropriate.

It is important to recognize that the counterterms, including δE, have already been fixed by the calculations in the vacuum sector. We are considering a single theory, with a single Lagrangian, and so must use the same values for the parameters, either finite or infinite, in both the vacuum and the kink sectors. Thus, we have

$$
\begin{aligned}
M_{\text{kink}} &= E_{\text{kink}} - E_{\text{vac}} \\
&= \Delta E_{\text{cl}} + \Delta E_{\text{zp}} + \Delta E_{\text{ct}} + O(\lambda),
\end{aligned}
\tag{2.46}
$$

where

$$
\Delta E_{\text{cl}} = E_{\text{cl}}(\phi_{\text{kink}}) - E_{\text{cl}}(\phi_{\text{vac}}) = M_{\text{cl}}
\tag{2.47}
$$

is the classical kink mass,

$$
\Delta E_{\text{zp}} = \frac{1}{2} \sum_j \omega_j^{\text{kink}} - \frac{1}{2} \sum_j \omega_j^{\text{vac}}
\tag{2.48}
$$

is the sum of the shifts of the oscillator zero-point energies, and

$$
\Delta E_{\text{ct}} = E_{\text{ct}}(\phi_{\text{kink}}) - E_{\text{ct}}(\phi_{\text{vac}})
\tag{2.49}
$$

is the contribution from the counterterms. Given the various divergent quantities that appear here, it is not at all obvious that the final result will be finite. Showing that it is indeed finite is a nontrivial test of our approach.

In order to see how this works, let us determine the leading quantum correction to the mass of the $\lambda\phi^4$ kink, following the approach of [1]. In the vacuum sector the normal modes are the familiar plane waves $f_k(x) = e^{ikx}$. The corresponding occupation numbers specify the number of particles with momentum k, where k ranges from $-\infty$ to ∞. Because the elementary particle mass is $\sqrt{2}\,m$, there is a continuous spectrum of frequencies

$$
\omega(k) = \sqrt{k^2 + 2m^2}
\tag{2.50}
$$

beginning at $\sqrt{2}\,m$. Substituting the mode functions into Eq. (2.41), and following the standard procedure, one arrives at the usual perturbation theory in terms of Feynman graphs. There is only a single divergent integral, which arises from the graph in Fig. 2.5. A convenient choice of renormalization scheme is to cancel this exactly by taking

$$
\mathcal{L}_{\text{ct}} = \frac{1}{2}\delta m^2\,\phi^2,
\tag{2.51}
$$

with

$$
\begin{aligned}
\delta m^2 &= -3i\lambda \int \frac{d^2 k}{(2\pi)^2} \frac{1}{k_\mu k^\mu - 2m^2 + i\epsilon} \\
&= \frac{3\lambda}{4\pi} \int_{-\infty}^{\infty} \frac{dk}{\sqrt{k^2 + 2m^2}} \\
&= \frac{3\lambda}{2\pi} \int_0^{\infty} \frac{dk}{\sqrt{k^2 + 2m^2}}.
\end{aligned}
\tag{2.52}
$$

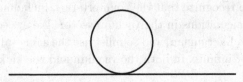

Fig. 2.5. The Feynman diagram that produces the only divergent integral in the $\lambda\phi^4$ theory.

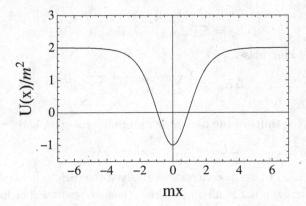

Fig. 2.6. The potential that enters the eigenvalue equation for the normal modes about the kink with $x_0 = 0$.

Turning now to the kink sector, we must first find the normal modes about the kink solution of Eq. (2.11). These modes are solutions of the eigenvalue equation

$$\left[-\frac{d^2}{dx^2} + U(x)\right] f_j(x) = \omega_j^2 f(x), \tag{2.53}$$

where

$$\begin{aligned}
U(x) &= V''(\phi_{\text{cl}}(x)) \\
&= -m^2 + 3\lambda\phi_{\text{cl}}^2 \\
&= m^2 \left\{2 - \frac{3}{\cosh^2[m(x - x_0)/\sqrt{2}]}\right\}
\end{aligned} \tag{2.54}$$

is shown in Fig. 2.6. We see immediately that the spectrum of eigenvalues includes a continuum beginning at $U(\pm\infty) = 2m^2$. There may also be discrete eigenvalues.

Without even using the explicit form of the kink solution, we can show that there must be a "zero mode", with $\omega = 0$. To do this, we recall that $\phi_{\text{kink}}(x)$ obeys the static field equation (2.18). Differentiating that equation with respect to x gives

$$0 = -\frac{d^2}{dx^2}\left(\frac{d\phi_{\text{kink}}}{dx}\right) + \frac{d}{dx}V'(\phi_{\text{kink}}(x))$$

$$= -\frac{d^2}{dx^2}\left(\frac{d\phi_{\text{kink}}}{dx}\right) + V''(\phi_{\text{cl}}(x))\left(\frac{d\phi_{\text{kink}}}{dx}\right), \tag{2.55}$$

which shows that

$$f_0(x) = \frac{d\phi_{\text{kink}}}{dx} \tag{2.56}$$

is a solution of Eq. (2.53) with eigenvalue $\omega^2 = 0$.

We also learn a bit more. Because $d\phi_{\text{kink}}/dx$ never vanishes, f_0 has no nodes. It is a familiar result from quantum mechanics that in a one-dimensional potential only the lowest eigenfunction has no nodes. Hence, in our problem there are no modes with $\omega^2 < 0$. This is fortunate, since an imaginary ω would correspond to an instability of the kink.

To proceed further, we have to actually solve the eigenvalue equation. Fortunately, Eq. (2.53) is equivalent to a Schrödinger equation with a potential $W(x) = \frac{1}{2}U(x)$, which turns out to be exactly soluble [5]. (This was one of the reasons for choosing the $\lambda\phi^4$ model as an example.) There is one more discrete eigenvalue, $\omega_1^2 = 3m^2/2$, with the (unnormalized) eigenfunction

$$f_1(x) = \frac{\sinh[m(x - x_0)/\sqrt{2}]}{\cosh^2[m(x - x_0)/\sqrt{2}]}. \tag{2.57}$$

The continuum modes are again labeled by a momentum k that runs over all positive and negative real values. The mode with momentum k has a frequency $\omega = \sqrt{k^2 + 2m^2}$ and corresponds to the eigenfunction

$$f_k(x) = e^{ikx}\left\{3m^2\tanh^2[m(x - x_0)/\sqrt{2}] - m^2 - 2k^2\right.$$
$$\left. -3\sqrt{2}\,imk\tanh[m(x - x_0)/\sqrt{2}]\right\}. \tag{2.58}$$

As $x \to \pm\infty$, this tends to

$$f_k(x) = \sqrt{4(m^2 - k^2)^2 + 18m^2k^2}\; e^{i[kx \pm \frac{1}{2}\delta(k)]}, \tag{2.59}$$

where the phase shift $\delta(k)$ is

$$\delta(k) = -2\arctan\left(\frac{3\sqrt{2}}{2}\frac{mk}{m^2 - k^2}\right). \tag{2.60}$$

We need to specify the branch of the arctangent. It is most convenient to choose this so that $\delta(k)$ tends to zero as $|k| \to \infty$. For positive k, the phase shift is monotonically decreasing, with $\delta(0^+) = 2\pi$ and $\delta(\infty) = 0$. For negative k, we set $\delta(-k) = -\delta(k)$. Note that this gives a discontinuity at $k = 0$.

Our expression for the kink energy contains divergent integrals of the zero-point energies of the continuum modes about the kink and the vacuum. These can be regulated by taking the one-dimensional space to have length a, and then imposing periodic boundary conditions on $\eta(x)$, with the limit $a \to \infty$ to be taken at the end of the calculation. This makes the spectra discrete and converts the integrals to sums, but the sums are still divergent. To make them finite, let us imagine that space consists of $2N$ points, so that there are only $2N$ normal modes, with N also to be eventually taken to infinity.

The modes about the vacuum are now labeled by discrete wavenumbers obeying

$$k_n^{\text{vac}} a = 2\pi n, \qquad n = \pm 1, \pm 2, \ldots, \pm N. \tag{2.61}$$

The "continuum" modes about the kink satisfy

$$k_n^{\text{kink}} a + \delta(k_n^{\text{kink}}) = 2\pi n, \qquad n = \pm 2, \pm 3, \ldots, \pm N. \tag{2.62}$$

The labeling of these follows from our conventions for the phase shift. These imply that $k_n^{\text{kink}} = -k_{-n}^{\text{kink}}$ and that $k_{n+1}^{\text{kink}} \approx k_n^{\text{vac}}$ for $0 < n \ll ma$, while for $n \gg ma$

$$(k_n^{\text{kink}} - k_n^{\text{vac}})a = -\delta(k_n^{\text{vac}}) + O(a^{-1}) \tag{2.63}$$

tends to zero. The absence of kink continuum modes with $n = \pm 1$ compensates for the two localized modes, with $\omega_0^{\text{kink}} = 0$ and $\omega_1^{\text{kink}} = \sqrt{3/2}\, m$, that have no counterparts in the vacuum sector.

The difference in zero-point energies can be written as

$$\Delta E_{\text{zp}} = \frac{1}{2}\left(\omega_0^{\text{kink}} - \omega_{-1}^{\text{vac}}\right) + \frac{1}{2}\left(\omega_1^{\text{kink}} - \omega_1^{\text{vac}}\right) + \sum_{n=2}^{N}\left(\omega_n^{\text{kink}} - \omega_n^{\text{vac}}\right). \tag{2.64}$$

The first two terms on the right-hand side are the contributions from the two localized modes about the kink and the two lowest vacuum modes. Up to terms of order $1/a$, they give

$$\frac{1}{2}\left(0 - \sqrt{2}\, m\right) + \frac{1}{2}\left(\frac{\sqrt{6}}{2}\, m - \sqrt{2}\, m\right) = \left(\frac{\sqrt{6}}{4} - \sqrt{2}\right) m. \tag{2.65}$$

The last term in Eq. (2.64) is the contribution from the kink continuum modes and the remainder of the vacuum modes, with the prefactor of $1/2$ canceled by a factor of 2 because the sum is only over positive values of n. To evaluate this continuum contribution, we use Eq. (2.63) and write

$$\sum_{n=2}^{N}\left(\omega_n^{\text{kink}} - \omega_n^{\text{vac}}\right) = \sum_{n=2}^{N}\left[\sqrt{(k_n^{\text{kink}})^2 + 2m^2} - \sqrt{(k_n^{\text{vac}})^2 + 2m^2}\right]$$
$$= -\sum_{n=2}^{N}\sqrt{(k_n^{\text{vac}})^2 + 2m^2}\left\{\frac{k_n^{\text{vac}}\delta(k_n^{\text{vac}})}{[(k_n^{\text{vac}})^2 + 2m^2]\, a} + O(a^{-2})\right\}. \tag{2.66}$$

If we now take the limits $N \to \infty$ and $a \to \infty$, this sum is converted to an integral, with

$$\frac{1}{a} \sum_{n=2}^{N} \to \frac{1}{2\pi} \int_0^\infty dk. \tag{2.67}$$

The right-hand side of Eq. (2.66) then becomes

$$-\frac{1}{2\pi} \int_0^\infty dk \, \frac{k\delta(k)}{\sqrt{k^2 + 2m^2}}$$
$$= \frac{1}{2\pi} \int_0^\infty dk \, \sqrt{k^2 + 2m^2} \, \frac{d\delta}{dk} - \frac{1}{2\pi} \delta(k)\sqrt{k^2 + 2m^2} \bigg|_{k=0}^{k=\infty}. \tag{2.68}$$

The integral on the right-hand side of this equation is

$$\frac{1}{2\pi} \int_0^\infty dk \, \sqrt{k^2 + 2m^2} \, \frac{d\delta}{dk}$$
$$= -\frac{3\sqrt{2}}{2\pi} m \int_0^\infty \frac{dk}{\sqrt{k^2 + 2m^2}} - \frac{3\sqrt{2}}{4\pi} m \int_0^\infty \frac{m^2 \, dk}{\sqrt{k^2 + 2m^2}(k^2 + \frac{1}{2}m^2)}$$
$$= -\frac{3\sqrt{2}}{2\pi} m \int_0^\infty \frac{dk}{\sqrt{k^2 + 2m^2}} - \frac{\sqrt{6}}{6} m, \tag{2.69}$$

while the surface term gives

$$-\frac{1}{2\pi} \left[\lim_{k \to \infty} \delta(k)\sqrt{k^2 + 2m^2} - \lim_{k \to 0^+} \delta(k)\sqrt{k^2 + 2m^2} \right] = -\frac{3\sqrt{2}}{2\pi} m + \sqrt{2}\, m. \tag{2.70}$$

Adding the continuum contributions from Eqs. (2.69) and (2.70) to the contribution from the discrete modes in Eq. (2.65), we obtain

$$\Delta E_{\text{zp}} = \left(\frac{\sqrt{6}}{12} - \frac{3\sqrt{2}}{2\pi} \right) m - \frac{3\sqrt{2}\, m}{2\pi} \int_0^\infty \frac{dk}{\sqrt{k^2 + 2m^2}}. \tag{2.71}$$

This is still divergent. However, we have not yet added the contributions from the counterterm Lagrangian, which are given by

$$\Delta E_{\text{ct}} = -\frac{1}{2}\delta m^2 \int dx \left[\phi_{\text{kink}}(x)^2 - \phi_{\text{vac}}(x)^2 \right]$$
$$= \left(-\frac{1}{2} \right) \left[\frac{3\lambda}{2\pi} \int_0^\infty \frac{dk}{\sqrt{k^2 + 2m^2}} \right] \left(-\frac{2\sqrt{2}\, m}{\lambda} \right). \tag{2.72}$$

This is of order unity, because the factor of λ in δm^2 is canceled by the $1/\lambda$ from ϕ_{kink}^2, and exactly cancels the divergence from the zero-point energies. Putting our results together, we have

$$M_{\text{kink}} = \frac{2\sqrt{2}}{3}\frac{m^3}{\lambda} + \left(\frac{\sqrt{6}}{12} - \frac{3\sqrt{2}}{2\pi}\right)m + O\left(\frac{\lambda}{m}\right) \tag{2.73}$$

for the energy of the ground state in the kink sector.

This is the state with all of the oscillator modes in their ground state. There are also higher energy states in which some of the occupation numbers n_j are nonzero. For the continuum modes, these are scattering states containing elementary bosons moving in the presence of the kink. Quanta of the discrete mode with $\omega = \sqrt{3/2}\,m$ can be interpreted as elementary bosons bound to the kink. The remaining mode, that with $\omega = 0$, is a different matter. A naive use of Eq. (2.42) would imply that the energy was independent of the occupation number. This would give an infinite set of degenerate states, a puzzling result.

This mode also leads to pathologies when one tries to use perturbation theory to calculate the next-order contribution to the kink mass. Proceeding straightforwardly leads to a Feynman diagram expansion similar to the usual one, except that the propagators and the vertices are position-dependent. The problem is that there are infrared divergences that arise from the contribution of the $\omega = 0$ mode to the propagator.

The explanation for these difficulties is that an oscillator with zero frequency is not an oscillator at all, so the line of reasoning that leads to states labeled by occupation number breaks down. The resolution of these problems is the subject of the next section.

2.3 Zero modes and collective coordinates

The anomalies associated with the $\omega = 0$ mode are not peculiar to the $\lambda\phi^4$ kink. Equation (2.55) does not depend on the form of the potential, and thus shows that a kink solution in any potential has a zero mode. The physical significance of this mode can be seen by noting that

$$\frac{d\phi_{\text{kink}}}{dx} = -\frac{d\phi_{\text{kink}}}{dx_0}, \tag{2.74}$$

so that excitation of the zero mode corresponds to a shift of the kink position. Because the underlying theory is translationally invariant, this shift leaves the potential energy unchanged.

This mode can be handled by recognizing that it is not an oscillator mode, and replacing its oscillator coefficient $c_0(t)$ by a collective coordinate $z(t)$ that represents the position of the kink [6–12]. Instead of our previous mode expansion, we write

$$\begin{aligned}
\phi(x,t) &= \phi_{\text{kink}}(x - z(t)) + \eta(x - z(t)) \\
&= \phi_{\text{kink}}(x - z(t)) + \sideset{}{'}\sum_j c_j(t) f_j(x - z(t)),
\end{aligned} \tag{2.75}$$

where the prime indicates that the summation is only over modes with nonzero ω. Here both the $c_j(t)$ and $z(t)$ are to be understood as quantum mechanical operators. We will also need the time derivative of this expression,

$$\dot{\phi}(x,t) = \dot{z}(t)\frac{d\phi_{\text{kink}}(x - z(t))}{dz}$$
$$+ \sum_j{}' \left[\dot{c}_j(t)f_j(x - z(t)) + \dot{z}(t)c_j(t)\frac{\partial f_j(x - z(t))}{\partial z}\right]. \quad (2.76)$$

The next step is to substitute these expansions into the Lagrangian and to pick out the terms that are quadratic in c_j, $\dot{c}_j$, or $\dot{z}$. (The translation invariance guarantees that the Lagrangian does not depend on z.) After the spatial integration, the terms quadratic in the c_j or the $\dot{c}_j$ (and independent of $\dot{z}$) are just as before, except for the omission of the zero-mode contributions. The $\dot{\phi}^2$ term in the Lagrangian appears to give cross-terms

$$\dot{z}\dot{c}_j \int dx \frac{d\phi_{\text{kink}}(x - z)}{dz} f_j(x - z) = -\dot{z}\dot{c}_j \int dx \frac{d\phi_{\text{kink}}(x - z)}{dx} f_j(x - z) \quad (2.77)$$

involving $\dot{z}\dot{c}_j$, but these vanish because the nonzero modes that we have retained are orthogonal to the zero mode, $d\phi_{\text{kink}}/dx$. Finally, the term quadratic in $\dot{z}$ is

$$\frac{1}{2}\dot{z}^2 \int dx \left(\frac{d\phi_{\text{kink}}}{dz}\right)^2 = \frac{1}{2}\dot{z}^2 \int dx \left(\frac{d\phi_{\text{kink}}}{dx}\right)^2 \equiv \frac{1}{2}A\dot{z}^2. \quad (2.78)$$

By making use of the virial identity of Eq. (2.20), A can be rewritten as

$$A = \frac{1}{2}\int dx \left(\frac{d\phi_{\text{kink}}}{dx}\right)^2 + \int dx\, V(\phi_{\text{kink}}) = M_{\text{cl}}. \quad (2.79)$$

Putting all this together, we have

$$L_{\text{quad}} = \frac{1}{2}M_{\text{cl}}\,\dot{z}^2 + \sum_j{}' \left(\frac{1}{2}\dot{c}_j^2 - \frac{1}{2}\omega_j^2 c_j^2\right). \quad (2.80)$$

The collective coordinate appears only in a single term, which is just the standard kinetic energy of a nonrelativistic particle with mass M_{cl}. Its conjugate momentum,

$$P = M_{\text{cl}}\,\dot{z}, \quad (2.81)$$

is a conserved quantity, commuting with the Hamiltonian. Because z can range from $-\infty$ to ∞, P has a continuous spectrum of eigenvalues. The energy eigenstates $|P, \{n_j\}\rangle$ are labeled by a momentum and a set of occupation numbers. The states with all of the n_j vanishing correspond to a kink with momentum P. States with nonzero occupation numbers have elementary particles of the theory in addition to the kink.

The order-unity corrections to the energy of the kink states arise from the zero-point energies of the oscillating modes. These are just as calculated in the previous section, so the energy of this state is

$$E = M_{\text{kink}} + \frac{P^2}{2M_{\text{cl}}} + O(P^4) + O(\lambda), \qquad (2.82)$$

where M_{kink} is given by Eq. (2.73).

To the order that we have calculated, this is just the low-momentum expansion of the relativistic energy of a particle with momentum P and mass M_{kink}. It is notable that this result depends crucially on the virial identity, which was needed to relate the coefficient of P^2 to the kink mass; we will encounter other examples of such apparently fortuitous identities in higher dimensions.

The interaction terms in the Lagrangian lead to several types of corrections to Eq. (2.82). First, there are $O(\lambda)$ and higher-order corrections to M_{kink}. There are also corrections to the coefficient of P^2, so that M_{cl} is replaced by M_{kink}. Finally, there are corrections proportional to P^4 and higher powers of P, just as required from the expansion of the relativistic energy $\sqrt{M_{\text{kink}}^2 + P^2}$.

The zero mode we have studied here resulted from the fact that the Lagrangian had a translational symmetry that was not a symmetry of the kink. In general, we will find zero modes arising whenever a soliton breaks one or more symmetries of the Lagrangian. In all such cases, one introduces a corresponding collective coordinate whose conjugate momentum becomes nonzero when the zero mode is excited in a time-dependent manner [13]. The symmetry in question may be a spacetime symmetry, as in the present case, or it may be an internal symmetry, with the corresponding momentum being the appropriate conserved charge.

Although most zero modes are directly related to symmetries in this manner, this is not always the case. In particular, there are some theories, described in more detail in Chap. 8, in which solitons at rest do not interact with each other. These theories have multisoliton solutions that have a set of zero modes, with corresponding collective coordinates, for each individual soliton. Because the solitons do interact when in motion, the conjugate variables, corresponding to the momenta and charges of the component solitons, are not necessarily conserved.

2.4 Fermions and fermion zero modes

The analysis of the preceding two sections can be generalized in a straightforward manner to accommodate additional fields, assuming that the kink remains a solution of the classical equations. For the case of an added bosonic field, preserving the kink solution requires an appropriate choice of its interaction terms in the Lagrangian. Because fermionic fields always vanish classically, they never interfere with the existence of a classical soliton solution.

One does a normal mode analysis of the additional fields in the background defined by the kink solution. Continuum modes correspond to scattering states of particles moving in the presence of the kink, and discrete modes with nonzero frequency correspond to elementary particles bound to the kink. Any bosonic zero modes are usually related to internal symmetries, and are handled by the introduction of collective coordinates, as described above. However, one often finds fermionic zero modes, which have somewhat different consequences.

As a concrete example [14], consider adding a single fermion field to the ϕ^4 theory that gave rise to our kink solution, with the Lagrangian density being

$$\mathcal{L} = \frac{1}{2}(\partial_\mu \phi)^2 - \frac{\lambda}{4}(\phi^2 - v^2)^2 + i\bar{\psi}\gamma^\mu \partial_\mu \psi - G\bar{\psi}\phi\psi. \tag{2.83}$$

The discrete $\phi \to -\phi$ symmetry of the purely bosonic theory is preserved if the fermion field is simultaneously transformed according to $\psi \to \gamma^5 \psi$, where $\gamma^5 \equiv -\gamma^0 \gamma^1$.

In the vacuum sector with $\phi = v$ the normal modes of ψ correspond to periodic solutions of the Dirac equation

$$i\gamma^\mu \partial_\mu \psi - M\psi = 0, \tag{2.84}$$

where $M = Gv$ is the fermion mass. If we write

$$\psi(x,t) = \chi_j(x)e^{-i\omega_j t}, \tag{2.85}$$

then $\chi_j(x)$ must satisfy

$$\left[-i\gamma^0\gamma^1\partial_x + M\gamma^0\right]\chi_j = \omega_j \chi_j. \tag{2.86}$$

This has a continuous spectrum of plane wave solutions χ_q with $-\infty < q < \infty$ and $\omega_q = \sqrt{q^2 + M^2}$. In addition, there is a second set of eigenfunctions $\chi_q^c \equiv -i\gamma^1 \chi_q$ with negative frequency $-\omega_q$. The field and its adjoint can be expanded in terms of these modes as

$$\psi(x,t) = \sum_q \left[b_q(t)\chi_q(x) + d_q^\dagger(t)\chi_q^c(x)\right],$$

$$\bar{\psi}(x,t) = \sum_q \left[b_q^\dagger(t)\bar{\chi}_q(x) + d_q(t)\bar{\chi}_q^c(x)\right]. \tag{2.87}$$

Requiring that ψ and $\bar{\psi}$ obey the canonical anticommutation relations leads to the usual algebra for the b_q, the d_q, and their adjoints. States are labeled by sets of occupation numbers that can each be either zero or one. The creation operators $b_q^\dagger$ and $d_q^\dagger$ have been introduced in such a way that they both increase the energy by an amount ω_q, with the former creating positive-energy fermions and the latter positive-energy antifermions.

In the presence of a kink, the normal modes correspond to solutions of

$$\left[-i\gamma^0\gamma^1\partial_x + G\gamma^0\phi_{\text{kink}}(x)\right]\chi_j = \omega_j \chi_j. \tag{2.88}$$

As in the vacuum sector, if χ_j is a solution with positive eigenvalue ω_j, then $\chi_j^c = -i\gamma^1\chi_j$ is a solution with eigenvalue $-\omega_j$. Thus, the modes with nonzero eigenvalues all come in pairs, corresponding to a fermion state and a charge conjugate antifermion state. However, if there are any modes with $\omega_j = 0$, it is possible that χ_j and $-i\gamma^1\chi_j$ might be linearly dependent, in which case there would not be a doubling of the mode. In fact, it is easy to see that the zero modes can be chosen to be eigenstates of $-i\gamma^1$. In two spacetime dimensions the spinor representation is two-dimensional, so $-i\gamma^1$ has two eigenvalues, 1 and -1, with two-component eigenvectors $s_\pm$. With $\omega_j = 0$, Eq. (2.88) is solved by [14]

$$\chi_0 = e^{\mp G \int_0^x dx'\, \phi_{\text{kink}}(x')}\, s_\pm. \tag{2.89}$$

For the kink solution, which is positive at large positive x and negative at large negative x, this is a normalizable mode with the upper choice of signs; for an antikink, with the opposite behavior, the lower choice of signs gives a normalizable solution. Finally, a vacuum solution, with ϕ having the same sign at both $x = \infty$ and $x = -\infty$, has no normalizable zero modes, in agreement with the previous discussion. In all of these cases, the presence or absence of the zero mode does not depend on the detailed structure of the background ϕ field, but only its values at spatial infinity. Like the existence of the kink itself, these zero modes are closely tied to the topology.

With a zero mode present, the mode expansions of Eq. (2.88) are replaced by

$$\psi(x,t) = a(t)\chi_0(x) + \sum_q \left[b_q(t)\chi_q(x) + d_q^\dagger(t)\chi_q^c \right],$$
$$\bar\psi(x,t) = a^\dagger(t)\bar\chi_0(x) + \sum_q \left[b_q^\dagger(t)\bar\chi_q(x) + d_q(t)\bar\chi_q^c \right], \tag{2.90}$$

where the summations are only over modes with nonzero eigenvalue. The commutation relations among the b_q, the d_q, and their adjoints is now supplemented by

$$\{a, a^\dagger\} = 1, \qquad \{a, a\} = \{a^\dagger, a^\dagger\} = 0, \tag{2.91}$$

with a and $a^\dagger$ anticommuting with all of the b_q and d_q. As in the vacuum sector, states can be labeled by occupation numbers that are either one or zero. For the modes with nonzero frequencies the interpretation of these occupation numbers as indicating the presence or absence of particles and antiparticles is just as before.

The situation is somewhat different for the zero mode. As with the other modes, the anticommutation relations imply the existence of two states that are connected by a and $a^\dagger$. However, because the two states are now degenerate in energy (and momentum), it is not really meaningful to say that one corresponds to the presence of a particle while the other does not. Instead, it seems more reasonable to simply say that all of the kink states have become doubly degenerate. If we indicate this extra degree of freedom by a $+$ or $-$, the states with momentum P and vanishing occupation numbers for all bosonic modes and all nonzero fermionic modes are related by

$$a^\dagger|P-\rangle = |P+\rangle,$$
$$a|P+\rangle = |P-\rangle. \tag{2.92}$$

One way to express this is to assign a fractional fermion number, $\pm 1/2$, to these two basic states, with integer increments then arising from the action of the $b_q^\dagger$ and $d_q^\dagger$ [14].

One can also envision situations with more than one fermionic zero mode; this is often the case for solitons in higher dimensions. If there are N fermionic zero modes, each soliton state acquires a degeneracy of 2^N. A similar phenomenon arises in string theory, where the zero-frequency Ramond modes of the superstring produce the degeneracy needed to obtain a multiplet of fermionic states.

It is interesting to note that a concrete realization of fermion fractionalization is found in polyacetylene, a conducting material consisting of parallel chains of atoms. Because the motion of the electrons is primarily along these chains, it can be treated as a quasi-one-dimensional material. It has kink-like defects and fermion zero modes very much analogous to those of the relativistic field theories, although differing in that the polyacetylene electrons, since they are actually living in three dimensions, have two spin states. The fermion fractionalization is manifested through an unusual relation between the charge and spin at a kink [15].

2.5 Kinks in more spacetime dimensions

The kink solutions in two spacetime dimensions are, trivially, also solutions of the field equations in $D > 2$ spacetime dimensions, with the field being independent of all but one of the coordinates. The energy density is localized in one spatial direction, but invariant under translations in the orthogonal directions, so the kink can no longer be interpreted as a particle. Instead, it represents a boundary, or domain wall, separating two regions in which the field approaches different vacuum values.

The expression for the kink energy now finds an interpretation as the energy density on the domain wall. Although this has dimensions of $(\text{mass})^{D-1}$, the same expression suffices because the dimensionality of the coupling constant is also dimension-dependent.

The domain wall clearly breaks translation invariance in the direction orthogonal to the wall. In the directions parallel to the wall, not only translation invariance, but also Lorentz invariance, is preserved. This can be seen, for example, in the energy–momentum tensor, which for our scalar field theories takes the form

$$T_{\mu\nu} = \partial_\mu\phi\,\partial_\nu\phi - \eta_{\mu\nu}\mathcal{L}, \tag{2.93}$$

with $\eta_{\mu\nu} = \text{diag}\,(1, -1, -1, \ldots, -1)$ being the Minkowskian metric in D space-time dimensions. Let us take x^{D-1} to be the direction orthogonal to the wall. Using Eqs. (2.1) and (2.19), we find that

$$T_{\mu\nu} = \hat{\eta}_{\mu\nu} \left(\frac{\partial \phi}{\partial x^{D-1}} \right)^2, \qquad (2.94)$$

where

$$\hat{\eta}_{\mu\nu} = \text{diag}\,(1, -1, -1, \ldots, -1, 0) \qquad (2.95)$$

can be viewed as the Minkowskian metric restricted to the first $D-1$ spacetime dimensions.

From this expression for $T_{\mu\nu}$ we see that there is an anisotropic pressure that vanishes in the x^{D-1} direction but that is negative, and equal in magnitude to the energy density, in all other directions. Thus, the wall preserves an SO($D-2$,1) subgroup of the original SO($D-1$,1) Lorentz symmetry. The fact that the pressure is negative has interesting gravitational consequences, which will be discussed in Chap. 7.

The normal mode analysis of the fluctuations of the fields about the kink is easily extended to fluctuations about the domain wall. Each two-dimensional kink mode with eigenvalue ω gives rise to a family of modes with frequencies $\sqrt{\omega^2 + \mathbf{k}_T^2}$, where $\mathbf{k}_T$ represents the wavenumbers in the spatial directions parallel to the wall. If the two-dimensional mode was a bound state mode localized about the kink, the D-dimensional mode can be interpreted as a particle of mass ω whose motion is confined to the domain wall. The translational zero mode goes over to a family of massless modes with frequency $|\mathbf{k}|$; these correspond to small deformations of the planar domain wall.

If there are fermions in the theory, a fermion zero mode about the kink is promoted to a massless particle localized on the domain wall. If the number of spacetime dimensions is odd, this particle has a definite chirality. To see this, note that in a spacetime with an even number $D = 2k$ dimensions, the natural generalization of the γ^5 of four-dimensional spacetime is

$$\Gamma = i^{k-1} \gamma^0 \gamma^1 \ldots \gamma^{D-1}, \qquad (2.96)$$

which anticommutes with the γ^A ($A = 0, 1, \ldots, D-1$) and obeys

$$\Gamma^2 = I. \qquad (2.97)$$

It thus has eigenvalues ± 1. Massless fermions can be described by spinors that are eigenvectors of Γ and have a chirality determined by the corresponding eigenvalue. By contrast, in an odd number of dimensions the last of the Dirac matrices is just i times the γ^5 analogue defined in one lower dimension. The product of all the γ^A is then proportional to the unit matrix, and there is no analogue of γ^5.

Now consider a domain wall extension of the kink solution, with the number D of spacetime dimensions being odd and the coordinate orthogonal to the wall being x^{D-1}. From the analysis of Sec. 2.4, we know that the fermion zero mode is an eigenvector of $i\gamma^{D-1}$, with the eigenvalue determined by whether the wall is built upon a kink or an antikink. But this is the γ^5 analogue in $D-1$ dimensions, so this mode, when viewed as a particle moving in the $(D-1)$-dimensional world volume of the domain wall, has a definite chirality determined by the kink or antikink nature of the wall.

This finds application in lattice gauge theory, where the domain wall fermion method yields four-dimensional chiral fermions by means of zero modes centered on a domain wall in five spacetime dimensions.

2.6 Multikink dynamics

The methods of Sec. 2.2 make it possible to study the interactions of a kink with the elementary particles of the theory, as long as the coupling constants are small enough for perturbative calculations to be valid. One would also like to be able to study the interactions between kinks, or between a kink and an antikink, including such processes as kink–kink and kink–antikink scattering, or the production of many elementary scalars via kink–antikink annihilation.

There are some limiting cases where this can be approached analytically [16–18]. For example, if two solitons are far apart, with not too large a relative velocity, the tail of one can be treated as a small perturbation on the other, making it possible to calculate the force between them and determine whether it is attractive or repulsive. However, the more general case of two kinks colliding at finite velocity is much more challenging

A case that is particularly amenable to analysis is the sine-Gordon theory. If two kinks collide, conservation of topological charge requires that the final state also contain two kinks. One might also expect some of the kink kinetic energy to be converted into radiation, which at the classical level would be manifested as small fluctuations of the ϕ field. However, this is not what happens. There is an exact classical solution [19] of the field equation,

$$\phi(x,t) = \frac{2v}{\pi} \tan^{-1} \left\{ u \sinh \left(\frac{mx}{\sqrt{1-u^2}} \right) \left[\cosh \left(\frac{mut}{\sqrt{1-u^2}} \right) \right]^{-1} \right\}, \qquad (2.98)$$

that describes two kinks with velocity u that are well separated at large negative t, collide at $t = 0$, and then re-emerge and move away from each other with the same velocity and asymptotically unchanged field profiles, as illustrated in Fig. 2.7. Thus, these are actually solitons in the more restricted sense of the word. The total energy of this solution is[5]

$$E = \frac{16m^3}{\lambda} \frac{1}{\sqrt{1-u^2}} = \frac{2M_{\text{cl}}}{\sqrt{1-u^2}}, \qquad (2.99)$$

---

[5] Verifying this result at arbitrary t is quite tedious. However, since energy is conserved it is sufficient to do the calculation for $t = 0$, where it is quite straightforward.

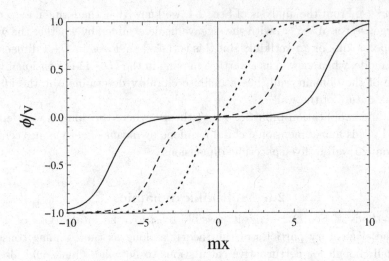

Fig. 2.7. The sine-Gordon two-kink solution for velocity $u = 0.3$. The profiles are for times $t = \pm 20m^{-1}$ (solid curve), $t = \pm 10m^{-1}$ (dashed curve), and $t = 0$ (dotted curve).

just as expected for two noninteracting solitons moving with velocity u. Because the solitons are identical, there is an ambiguity as to whether the kinks pass through each or instead collide and recoil backwards. Examination of the large-time behavior of Eq. (2.98) shows that there is a net time advance

$$\Delta t = -\frac{2\sqrt{1 - u^2}}{mu} \ln |u|. \tag{2.100}$$

Since one can show that the long-range force between kinks is repulsive, Eq. (2.100) seems inconsistent with the former interpretation, but it is quite reasonable if the solution is interpreted as two solitons that approach each other but reverse direction before they meet.

In a kink–antikink collision there is no topological impediment to annihilation of the two, but here again there is an analytic solution [19],

$$\phi(x,t) = \frac{2v}{\pi} \tan^{-1}\left\{ \frac{1}{u} \sinh\left(\frac{mut}{\sqrt{1 - u^2}}\right) \left[\cosh\left(\frac{mx}{\sqrt{1 - u^2}}\right)\right]^{-1} \right\}, \tag{2.101}$$

that describes two solitons that collide, pass through each other, and then emerge undeformed, as shown in Fig. 2.8. There is again a net time advance given by Eq. (2.100). Since one can show that the long-range interaction between a kink and an antikink is attractive, the appearance of an advance makes sense. The energy for this case is also given by Eq. (2.99).

The fact that the kink–antikink interaction is attractive suggests that bound states might exist. Indeed, continuing Eq. (2.101) to imaginary u, with $u = is$, gives

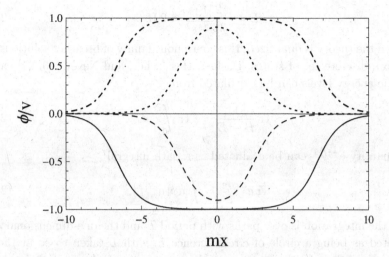

Fig. 2.8. The sine-Gordon solution for a left-moving kink and a right-moving antikink, with velocity $u = 0.3$. The profiles are for times $t = -20m^{-1}$ (solid curve), $t = -5m^{-1}$ (dashed curve), $t = 5m^{-1}$ (dotted curve), and $t = 20m^{-1}$ (dashed-dotted curve). At $t = 0$, $\phi(x) = 0$ for all x.

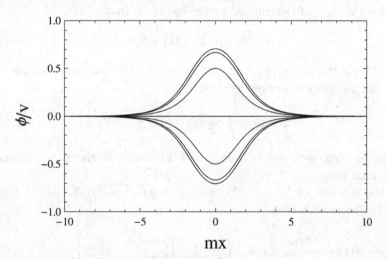

Fig. 2.9. The sine-Gordon breather solution for $s = 0.5$. The curves show the field profile at equally spaced times, one-twelfth of a period apart.

$$\phi(x,t) = \frac{2v}{\pi}\tan^{-1}\left\{\frac{1}{s}\sin\left(\frac{mst}{\sqrt{1+s^2}}\right)\left[\cosh\left(\frac{mx}{\sqrt{1+s^2}}\right)\right]^{-1}\right\}. \qquad (2.102)$$

This solution, known as the breather or doublet, can be interpreted as a kink and an antikink bound together and oscillating as illustrated in Fig. 2.9. The energy of the breather is

$$E = \frac{2M_{\text{cl}}}{\sqrt{1+s^2}} < 2M_{\text{cl}}. \tag{2.103}$$

When the theory is quantized, this continuous family of breather solutions goes over to a discrete set of states. Dashen, Hasslacher, and Neveu [20, 21] showed that the energy levels can be calculated from

$$G(E) = \text{Tr}\,\frac{1}{H-E} = i\,\text{Tr}\int_0^\infty dT\,e^{i(E-H)T}. \tag{2.104}$$

The quantity e^{-iHT} can be evaluated as a path integral

$$\text{Tr}\,e^{-iHT} = \int [d\phi]\,e^{iS(T)}, \tag{2.105}$$

where the integration is over paths with period T and the one-dimensional space is treated as being a circle of circumference L, with L taken to ∞ at the end of the calculation. In the stationary phase approximation the path integral is dominated by the classical periodic solutions; i.e., solutions in which the breather goes through an integral number of complete oscillations and traverses space an integral number of times in time T. In the leading approximation, the breather states have masses M determined by the condition that

$$S(\tau(M)) + M\tau(M) = 2\pi n, \tag{2.106}$$

where $\tau(M)$ is the period of a breather with energy $E = M$. Since for one degree of freedom an integral over one cycle gives

$$\oint p\,dq = \oint p\dot{q}\,dt = \oint (H+L)\,dt, \tag{2.107}$$

this can be seen as a generalization to field theory of the Bohr–Sommerfeld quantization condition.

For the solution of Eq. (2.102), with $\tau = 2\pi\sqrt{1+s^2}/ms$ and M given by Eq. (2.103), one finds that

$$S(\tau) = \frac{32\pi m^2}{\lambda}\left\{\cos^{-1}\left(\frac{2\pi}{m\tau}\right) - \left[\left(\frac{m\tau}{2\pi}\right)^2 - 1\right]^{-1}\right\}. \tag{2.108}$$

The second quantity in the curly brackets is equal to $-M\tau(M)$. Applying Eq. (2.106) then gives

$$M_n^{(0)} = \frac{16m^3}{\lambda}\sin\left(\frac{n\lambda}{16m^2}\right), \tag{2.109}$$

with the proviso that n be such that the argument of the sine is less than $\pi/2$.

A better approximation is obtained by including the effects of the fluctuations about the classical path. These are analogous to the order-unity zero-point corrections to the static ϕ^4 kink that were calculated in Sec. 2.2. It turns out that these can be calculated in closed form for the sine-Gordon breather [21]. Remarkably, their only effect is to replace $\gamma \equiv \lambda/m^2$ by

$$\gamma' = \frac{\lambda}{m^2}\left(1 - \frac{\lambda}{8\pi m^2}\right)^{-1}. \tag{2.110}$$

In fact, the first quantum corrections to the sine-Gordon kink mass can be obtained by the same substitution. Thus, the kink has a mass

$$M_{\text{kink}} = \frac{8m^3}{\lambda} - \frac{m}{\pi} + O(\lambda/m) = \frac{8m}{\gamma'} + O(\lambda/m), \tag{2.111}$$

while to the same order the breather states have masses

$$M_n = \frac{16m}{\gamma'}\sin\left(\frac{n\gamma'}{16}\right), \qquad n = 1, 2, \ldots, < \frac{8\pi}{\gamma'}. \tag{2.112}$$

For weak coupling, $\lambda/m^2 \ll 1$, we have

$$M_1 = m\left[1 - \frac{1}{6}\left(\frac{\lambda}{16m^2}\right)^2 + O(\lambda^3/m^6)\right], \tag{2.113}$$

so that in the limit $\lambda \to 0$ the lowest breather state becomes degenerate with the elementary boson of the theory. In fact, it *is* the elementary boson, just seen in a different guise. The higher excited states, with masses

$$M_n = M_1\left[n - \frac{1}{6}(n^3 - n)\left(\frac{\lambda}{16m^2}\right)^2 + O(\lambda^3/m^6)\right], \tag{2.114}$$

can be understood as bound states of n elementary bosons. As a check of this statement, Dashen, Hasslacher, and Neveu did a perturbative calculation of the binding energy of two elementary sine-Gordon bosons. Working to $O(\lambda^4)$, they obtained

$$\frac{2M_1 - M_2}{M_1} = \left(\frac{\lambda}{16m^2}\right)^2 + \frac{4}{\pi}\left(\frac{\lambda}{16m^2}\right)^3 + \left(\frac{12}{\pi^2} - \frac{1}{12}\right)\left(\frac{\lambda}{16m^2}\right)^4$$
$$+ O(\lambda^5/m^{10}), \tag{2.115}$$

in perfect agreement with Eq. (2.112).

The special features of the sine-Gordon dynamics—the absence of energy loss or distortion in kink–kink and kink–antikink scattering, and the existence of dissipationless time-dependent breather solutions (as well as the ability to obtain such explicit results in closed form)—are due to the fact that the theory is a completely integrable system, with an infinite number of nontrivial conserved quantities. To explore other theories, such as the $\lambda\phi^4$ model, one must generally resort to numerical methods, even at the classical level.

In the $\lambda\phi^4$ theory any two-soliton configuration must consist of a kink and an antikink, with their relative position determined by the fields at $x = \pm\infty$. For example, if $\phi(\pm\infty) = -v$, the kink (going from $-v$ to v) must always be to the left of the antikink (going from v to $-v$). If the two collide, they will recoil and reverse direction. The collision will not be completely elastic; some energy will be always be lost to radiation (i.e., small-amplitude oscillations of the scalar field). Throughout this process, the region between the solitons remains (approximately) in the vacuum at $\phi = v$. This should be contrasted with the sine-Gordon case, where the region between the solitons is in a different vacuum before and after the collision.

Behavior somewhat like the sine-Gordon case is possible in theories where $V(\phi)$ has more than two degenerate minima, not necessarily related by any symmetry. For example, consider a theory with degenerate vacua at $\phi = v_A$, v_B, and v_C, with $v_A < v_B < v_C$. There will be two types of kinks (A-to-B and B-to-C), and two types of antikinks. If a right-moving B-to-C kink collides with a left-moving C-to-B antikink, the emerging particles may be a right-moving A-to-B kink and a left-moving B-to-A antikink. The analogous behavior in the collision of domain walls separating different (not necessarily degenerate) vacua in four spacetime dimensions leads, in certain theories, to an interesting new possibility, termed classical transitions [22, 23], that are an alternative to the transitions via quantum mechanical bubble nucleation that will be discussed in Chap. 12.

There are no known examples of stable analogues of the sine-Gordon breather in other theories. However, it has been found that some theories support oscillating solutions that have a surprisingly long lifetime before their energy is eventually radiated away. These solutions, known as oscillons, can arise even in theories with a single vacuum, and hence no kink solutions [24–26]. For some recent work on these, see [27–29].

2.7 The sine-Gordon–massive Thirring model equivalence

In the previous section we saw that a weak-coupling analysis of the sine-Gordon theory reveals that the lowest breather state is in fact identical to the elementary boson of the theory. Even more remarkable results are suggested by an analysis of the spectrum in the strong-coupling limit.

A word of caution is in order. We might well expect that in the strong-coupling regime higher-order quantum corrections would be important, thus rendering both our expression for the kink mass and the WKB result for the breather masses meaningless. However, the results described below suggest that this may not be the case, and that the sine-Gordon theory is sufficiently special that the WKB formula for the mass might be exact, as indeed turns out to be the case [30, 31]. At the same time, though, it should be kept in mind that in the strong-coupling regime the kink Compton wavelength is greater than the classical soliton width and the direct significance of the classical field profile in the quantum state is uncertain.

Equation (2.112) for the breather state masses gives

$$M_n = 2M_{\text{kink}} \sin\left[\frac{n\pi}{2} \frac{\lambda/(8\pi m^2)}{1 - \lambda/(8\pi m^2)}\right]. \tag{2.116}$$

The requirement that

$$n < \frac{8\pi m^2}{\lambda} - 1 \tag{2.117}$$

implies that there are no breather states if $\lambda \geq 4\pi m^2$. Hence, the "elementary particle" of the theory, which we previously identified with the $n = 1$ breather state, is absent in this case. Indeed, already for $\lambda > 2\pi m^2$, the kink, and not the elementary ϕ, is the lightest particle in the theory.

To examine the spectrum for λ just below $4\pi m^2$, let us write

$$\frac{\lambda}{4\pi m^2} = \frac{1}{1 + \delta/\pi}. \tag{2.118}$$

The energy of the lowest breather mode can then be written as

$$M_1 = M_{\text{kink}}\left[2 - \delta^2 + \frac{4\delta^3}{\pi} + O(\delta^4)\right]. \tag{2.119}$$

This suggests that for λ near the limiting value, the lowest (and only) breather state should perhaps be interpreted as a weakly bound state of a kink and an antikink.

Putting this result aside for a moment, let us consider the massive Thirring model [32]. This is a theory of a fermion field ψ in $(1+1)$-dimensional spacetime, with a Lagrangian density

$$\mathcal{L} = \bar{\psi}(i\gamma^\mu \partial_\mu - M)\psi - \frac{g}{2}(\bar{\psi}\gamma^\mu \psi)^2. \tag{2.120}$$

The particle spectrum includes an elementary fermion and its antiparticle, each with mass M. In addition, if $g > 0$, so that the interaction is attractive, there are also fermion–antifermion bound states. The lowest of these (and the only one if g is sufficiently small), has a mass

$$M_{\text{bound}} = M\left[2 - g^2 + \frac{4g^3}{\pi} + O(g^4)\right]. \tag{2.121}$$

Note the resemblance between Eqs. (2.119) and (2.121). If we take this to be more than coincidence, we could identify δ and g, and write

$$\frac{\lambda}{4\pi m^2} = \frac{1}{1 + g/\pi}. \tag{2.122}$$

The weak-coupling $(\lambda/m^2 \to 0)$ limit of the sine-Gordon theory would then correspond to the strong-coupling $(g \to \infty)$ limit of the massive Thirring model, while the $g \rightarrow 0$ weak-coupling limit of the latter would correspond to the limit $\lambda/m^2 \rightarrow 4\pi$. This would then suggest the correspondences

$$\text{kink} \longleftrightarrow \text{elementary } \psi$$
$$\text{antikink} \longleftrightarrow \text{elementary } \bar{\psi}$$
$$\text{elementary } \phi \longleftrightarrow \bar{\psi}\text{-}\psi \text{ bound state}$$
$$\text{topological charge} \longleftrightarrow \text{fermion number}$$

between the two theories.[6]

The above remarks are suggestive, but hardly conclusive. They are made rigorous by an analysis of Coleman [33]. Adopting Eq. (2.122) and making the identifications

$$\frac{\sqrt{\lambda}}{2\pi m} \, \epsilon^{\mu\nu} \, \partial_\nu \phi = -\bar{\psi}\gamma^\mu\psi,$$
$$\frac{m^4}{\lambda} \cos\left(\frac{\sqrt{\lambda}}{m}\phi\right) = -M\bar{\psi}\psi \qquad (2.123)$$

(with appropriate renormalization and normal ordering of the composite operators), he showed that the matrix elements of the two theories agree to all orders of perturbation theory.

These identifications relate expressions involving the bosonic field ϕ with fermion bilinears. Mandelstam [34] showed that one could go further and construct the fermionic field from the bosonic field. In a basis where the gamma matrices are $\gamma^0 = \sigma_1$ and $\gamma^1 = i\sigma_2$, the upper and lower components of $\psi(x)$ are given by

$$\psi_1(x) = C : \exp\left[-\frac{i\beta}{2}\phi(x) - \frac{2\pi i}{\beta}\int_{-\infty}^x dz\, \dot{\phi}(z)\right] :,$$
$$\psi_2(x) = -iC : \exp\left[\frac{i\beta}{2}\phi(x) - \frac{2\pi i}{\beta}\int_{-\infty}^x dz\, \dot{\phi}(z)\right] :, \qquad (2.124)$$

where colons indicate normal ordering and C is a constant whose value need not concern us here. By using the formula

$$e^A e^B = e^{[A,B]} e^B e^A, \qquad (2.125)$$

which holds when $[A, B]$ is a c-number, one can easily verify that $\psi(x)$ and $\psi(y)$ anticommute at equal times for $x \neq y$. Verifying that the correct anticommutator is obtained for $x = y$ is more subtle, but this can be done by an appropriate choice of the constant C and careful handling of the singularities.

---

[6] The correspondence between the bosonic kink and the fermionic ψ may seem particularly strange. Recall, however, that there is no spin in two-dimensional spacetime, so the distinction between fermions and bosons is based only on statistics. A change in statistics can be compensated by an interaction; e.g., the Pauli exclusion principle for fermions can be replaced by an appropriate repulsive interaction between bosons.

Thus, the sine-Gordon and massive Thirring model are not two different theories, but rather two "dual" formulations that describe precisely the same physics. To state this differently, an experimenter in two-dimensional spacetime would not be able to distinguish between the two. The fact that the duality relates the strong-coupling regime of one formulation to the weak-coupling regime of the other provides a useful calculational tool, since quantities that are not perturbatively accessible in one formulation may be calculable in the other.

The duality also teaches us important conceptual lessons. Elementary introductions to quantum field theory typically begin with free field theory, where there is a one-to-one correspondence between quantum fields and "elementary" particles. Even after going on to treat the interacting theory, there is often a bias towards viewing these particles as the fundamental ones, with other objects, whether found as bound states or from semiclassical quantization of a soliton, seen as rather different entities. The sine-Gordon–massive Thirring duality shows that this is misguided, since a particle that is solitonic in one formulation is elementary in another. A solitonic particle may be no more and no less fundamental than one corresponding to a quantum of a weakly coupled field. The various descriptions—elementary particle, soliton, bound state—should be viewed as characterizations that are useful within the context of a particular formulation in certain regions of parameter space. In other regions of parameter space (e.g., when no formulation is in a weak-coupling regime) it may be that none of these characterizations are either useful or appropriate.

A final lesson concerns the existence of degenerate vacua. The sine-Gordon theory appears to have an infinite number of vacuum states, whereas the massive Thirring model has a unique vacuum. It might therefore seem strange that the two theories could be equivalent. However, there is no difficulty here, because the sine-Gordon vacua are all physically equivalent and cannot be distinguished by any local measurement. Indeed, if one views $\phi(x)$ as a periodic variable, with ϕ and $\phi + v$ identified, there is just a single sine-Gordon vacuum.

3
Solitons in more dimensions—Vortices and strings

The key to the existence of the kink solutions was that $\phi(x)$ approached different vacua at $x = -\infty$ and $x = \infty$, making it impossible to smoothly deform the kink into a vacuum solution. This strategy can be generalized to obtain solitons in higher dimensions. In this chapter I consider the simplest example, a theory with U(1) symmetry in two spatial dimensions.

3.1 First attempt—global vortices

With one spatial dimension, spatial infinity consists of two disconnected points. In two dimensions spatial infinity is connected and can be described in polar coordinates as the circle at $r = \infty$. If we want a nonsingular configuration in which ϕ takes on different vacuum values at different points along this circle, we will need a theory with a continuous family of vacuum states. A simple example is the theory of a complex scalar field,

$$\phi(x) = \phi_1(x) + i\phi_2(x) = \rho(x)e^{i\alpha(x)}, \tag{3.1}$$

governed by the Lagrangian density

$$\mathcal{L} = \frac{1}{2}(\partial_\mu \phi)^\dagger (\partial^\mu \phi) - \frac{\lambda}{4}\left(|\phi|^2 - \frac{\mu^2}{\lambda}\right)^2. \tag{3.2}$$

This theory has a global U(1) symmetry. If μ^2 and λ are both positive, the scalar field potential is minimized when

$$|\phi| = v \equiv \sqrt{\frac{\mu^2}{\lambda}} \tag{3.3}$$

and the U(1) symmetry is spontaneously broken. There is a continuous set of vacuum states, given by $\rho = v$ and an arbitrary uniform value for α. The elementary excitations are a scalar with mass $m_\phi = \sqrt{2}\,\mu$ and a massless Goldstone boson.

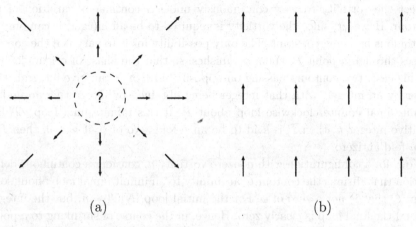

(a)   (b)

Fig. 3.1. Two configurations of the complex scalar field. The lengths of the arrows indicate the magnitude $|\phi|$, while their directions reflect the orientation in the ϕ_1-ϕ_2 plane. No matter what form the fields take in the central region, the configuration on the left cannot be smoothly deformed to a vacuum solution, such as that on the right.

Generalizing from the one-dimensional configuration of Fig. 2.1, let us consider the two-dimensional configuration shown in Fig. 3.1a. The lengths of the arrows represent the magnitude of ϕ, while their directions indicate the orientation in the ϕ_1-ϕ_2 plane. Regardless of the behavior of the fields in the interior region, it is clear that the fields cannot be continuously deformed to a vacuum solution, such as that shown in Fig. 3.1b. Hence, we may expect to find a static solution that is different from the vacuum by starting with this configuration and then continuously deforming it until the energy is minimized. Such a solution is known as a vortex. More specifically, the vortices in this theory with a global U(1) symmetry are termed global vortices, in contrast with the vortices, to be encountered in Sec. 3.3, that arise when the U(1) symmetry is gauged.

There is a nice topological formulation of this situation. For any smooth configuration of ϕ, we can define the line integral

$$N[C] = \frac{1}{2\pi} \oint_C d\boldsymbol{\ell} \cdot \boldsymbol{\nabla}(\arg \phi) = \frac{1}{2\pi} \oint_C d\boldsymbol{\ell} \cdot \boldsymbol{\nabla}\alpha, \qquad (3.4)$$

where the contour C is the circle at spatial infinity. If ϕ is nonzero everywhere on this circle, so that its phase is well defined at each point of the curve, $N[C]$ counts the number of full rotations that this phase makes during one clockwise circuit around C. This must be an integer n, which may be termed the vorticity or winding number. For the example of Fig. 3.1a, $n = 1$, while for any vacuum solution $n = 0$. Because n is quantized, smooth variations of the configuration cannot take an $n = 1$ configuration to the vacuum.

We can also consider contours other than the circle at infinity. As long as ϕ is nonzero everywhere on the contour, $N[C]$ must still be an integer. One would expect the vorticity to vary continuously under a continuous variation of the contour. However, since the vorticity is required to be an integer, it can only be continuous by being constant. The only possibility for it to vary is if the contour passes through a point P where ϕ vanishes, so that the phase of ϕ is undefined. In this case, two contours passing on opposite sides of P can give integrals that differ by an integer, with that integer being obtained by integrating around an infinitesimal counterclockwise loop about P. If this infinitesimal loop yields a positive integer n, then P is said to be an n-fold zero of ϕ; if $n < 0$, then P is an n-fold antizero.

Now, for a configuration with nonzero vorticity n, consider a continuous deformation that shrinks the contour C at infinity to an infinitesimal loop about some point Q that is not a zero of ϕ. For the initial loop, $N[C] = n$, but the integral around the final loop is clearly zero. Hence, in the course of shrinking to a point, the contour must have passed through a number of zeros of ϕ. If we count zeros (antizeros) with a positive (negative) sign and appropriate multiplicity, the net number of zeros is equal to the vorticity.

Going beyond the arguments for the existence of a vortex to actually finding a static solution requires solving the complex static field equation

$$0 = \boldsymbol{\nabla}^2\phi - \lambda(|\phi|^2 - v^2)\phi. \tag{3.5}$$

This is equivalent to a pair of real partial differential equations in two variables. In contrast with the case of the kink, finding the general solution is a daunting task.

Matters can be simplified by using an ansatz that exploits the symmetries of the theory. The simplest guess would be to look for a rotationally invariant solution, with ϕ a function only of r. However, such a configuration has zero vorticity and so is not what we want. Instead, let us adopt the ansatz

$$\phi(\mathbf{x}) = f(r)e^{i\theta}, \tag{3.6}$$

which is invariant under the combination of a spatial rotation by an angle α and a U(1) phase rotation by $-\alpha$. At this point $f(r)$ can be a complex function. However, if we also impose the requirement that the field be invariant under a reflection about the x-axis combined with complex conjugation of the field, so that $\phi(r, \theta) = \phi^*(r, -\theta)$, then $f(r)$ must be real. Because we now have only a single unknown function, we anticipate replacing the two partial differential equations with a single ordinary differential equation.

An essential element here is the symmetry of the problem. Given a set of field equations, one can always impose some simplifying ansatz and seek a solution. In general, however, substitution of the ansatz into the field equations will yield more equations than there are unknown functions, and there will be no consistent nontrivial solution. This difficulty is avoided if the ansatz is the most general one consistent with some symmetry of the Lagrangian.

To understand this, consider a theory involving one or more fields, assembled into a column vector $\chi(x)$, that is invariant under a symmetry group G. Now let $\bar{\chi}$ be a field configuration that is itself invariant under G. An arbitrary perturbation about $\bar{\chi}$ can be decomposed into a part $\delta_1\chi$ that is invariant under G and a part $\delta_2\chi$ that is not. Expanding the action about $\bar{\chi}$ gives

$$S[\chi] = \int dt\, dx \left[\mathcal{L}[\bar{\chi}] + \left.\frac{\delta\mathcal{L}}{\delta\chi}\right|_{\bar{\chi}} (\delta_1\chi + \delta_2\chi) + O(\delta\chi^2) \right]. \tag{3.7}$$

Because $\mathcal{L}$ and $\bar{\chi}$ are both G-invariant, the term linear in $\delta_2\chi$ vanishes after the integration. Hence, to show that $\bar{\chi}$ is a stationary point of the action it is sufficient to show that it is stationary under G-invariant variations.

Although arguments such as these can show that a symmetric configuration is a solution, determining whether the solution is stable requires going outside the symmetric ansatz. It is quite possible, as we will see in Sec. 3.4, for a solution to be a local minimum of the energy among rotationally invariant configurations, but only a saddle point among all solutions.

There are two routes to obtaining the field equation for our ansatz. One is to simply substitute the ansatz into Eq. (3.5). A second approach, which in many cases is algebraically simpler, exploits the facts that there are no terms linear in $\delta_2\chi$ in the action and that the ansatz includes all possible symmetric configurations. Hence, we can find the equations for a symmetric stationary point by substituting the ansatz directly into the action and then varying with respect to the various unknown functions in the ansatz.

Either approach yields the single equation

$$0 = \frac{d^2f}{dr^2} + \frac{1}{r}\frac{df}{dr} - \frac{f}{r^2} + \lambda(v^2 - f^2)f. \tag{3.8}$$

The boundary conditions are that $f(0) = 0$, so that ϕ is nonsingular at the origin, and $f(\infty) = v$, so that the field approximates a vacuum solution at large distance. It is a straightforward matter to integrate this equation numerically.

However, there is a problem. Written in terms of ρ and α, the static energy functional is

$$E = \int d^2x \left[\frac{1}{2}(\boldsymbol{\nabla}\rho)^2 + \frac{1}{2}\rho^2(\boldsymbol{\nabla}\alpha)^2 + \frac{\lambda}{4}(\rho^2 - v^2)^2 \right]. \tag{3.9}$$

For our ansatz, the second term in the integrand is

$$\frac{1}{2}\rho^2(\boldsymbol{\nabla}\alpha)^2 = \frac{1}{2}\frac{f^2}{r^2}. \tag{3.10}$$

At large r, this is approximately $v^2/(2r^2)$, causing the energy integral to diverge logarithmically. Furthermore, this divergence is not an artifact of the particular ansatz that we chose; a similar divergence occurs for any configuration with nonzero vorticity.

Despite this divergence, solutions similar to this may be physically relevant. For example, one can consider a system with a finite density of global vortices, and thus a finite energy per vortex. Such configurations, but with the two-dimensional vortices extended into three-dimensional strings, could arise in a cosmological context via the Kibble mechanism, which will be discussed in Chap. 7. The logarithmic divergence of the single-vortex energy would then lead to long-range interactions between the strings that could be viewed as being mediated by the massless Goldstone boson of the theory. However, let us now turn instead to the problem of finding two-dimensional solitons with finite energy.

3.2 Derrick's theorem

The failure to find a finite energy soliton in the previous section is a special case of a general result due to Derrick [35]. Consider a theory in D spatial dimensions involving one or more scalar fields ϕ_a, with a Lagrangian density of the form

$$\mathcal{L} = \frac{1}{2} G_{ab}(\phi) \, \partial_\mu \phi_a \, \partial^\mu \phi_b - V(\phi), \tag{3.11}$$

where the eigenvalues of G are all positive definite for any value of ϕ, and $V = 0$ at its minima.

Any finite energy static solution of the field equations is a stationary point of the potential energy

$$E = I_K + I_V, \tag{3.12}$$

where

$$I_K[\phi] = \frac{1}{2} \int d^D x \, G_{ab}(\phi) \, \partial_j \phi_a \, \partial_j \phi_b \tag{3.13}$$

and

$$I_V[\phi] = \int d^D x \, V(\phi) \tag{3.14}$$

are both positive. Since the solution is a stationary point among all configurations, it must, *a fortiori*, also be a stationary point among any subset of these configurations to which it belongs. Therefore, given a solution $\bar{\phi}(\mathbf{x})$, consider the one-parameter family of configurations

$$f_\lambda(\mathbf{x}) = \bar{\phi}(\lambda \mathbf{x}) \tag{3.15}$$

that are obtained from the solution by rescaling lengths. The potential energy of these configurations is given by

$$E(\lambda) = I_K[f_\lambda] + I_V[f_\lambda]$$

$$= \lambda^{2-D} I_K[\bar{\phi}] + \lambda^{-D} I_V[\bar{\phi}]. \tag{3.16}$$

Since $f_1 = \bar{\phi}(\mathbf{x})$ is a solution, $\lambda = 1$ must be a stationary point of $E(\lambda)$, which implies that

$$0 = (D - 2) I_K[\bar{\phi}] + D I_V[\bar{\phi}]. \tag{3.17}$$

For $D = 1$, this gives a more general form of the virial identity of Eq. (2.20). For $D = 2$, it tells us that $I_V = 0$, which implies that ϕ must be at a vacuum value everywhere. This rules out a vortex solution in the model of Sec. 3.1 since, as was already demonstrated, a nonsingular vortex solution must have at least one point where ϕ vanishes.[1] For $D \geq 3$, Eq. (3.17) can only be satisfied if both I_K and I_V vanish, which rules out anything but a constant vacuum solution.

Thus, in order to be able to have solitons in two or more dimensions we need to add more structure to the theory. One successful approach is to consider a gauge theory (either Abelian or non-Abelian) with a Lagrangian density of the form

$$\mathcal{L} = -\frac{1}{2}\text{tr}\, F_{\mu\nu}F^{\mu\nu} + \frac{1}{2}(D_\mu\phi)^\dagger(D^\mu\phi) - V(\phi). \tag{3.18}$$

For a static solution with vanishing electric fields F_{0j}, and hence $A_0 = 0$, the energy is

$$E = I_F + I_K + I_V, \tag{3.19}$$

where now

$$I_F[A] = \frac{1}{2}\int d^D x \,\text{tr}\, F_{ij}^2,$$

$$I_K[\phi, A] = \frac{1}{2}\int d^D x \,(D_j\phi)^\dagger(D_j\phi),$$

$$I_V[\phi] = \int d^D x\, V(\phi). \tag{3.20}$$

Given a solution $\bar{\phi}(x)$, $\bar{A}_j(x)$, we define scaled fields

$$f_\lambda(x) = \bar{\phi}(\lambda x),$$
$$g_{j\lambda}(x) = \lambda\bar{A}_j(\lambda x). \tag{3.21}$$

Proceeding as in the pure scalar case, we find that

$$E(\lambda) = I_F[g_\lambda] + I_K[f_\lambda, g_\lambda] + I_V[f_\lambda]$$

$$= \lambda^{4-D}I_F[\bar{A}] + \lambda^{2-D}I_K[\bar{\phi}, \bar{A}] + \lambda^{-D}I_V[\bar{\phi}], \tag{3.22}$$

which is stationary at $\lambda = 1$ if

$$0 = (D-4)I_F[\bar{A}] + (D-2)I_K[\bar{\phi}, \bar{A}] + DI_V[\bar{\phi}]. \tag{3.23}$$

This identity allows the possibility of static solutions with gauge and scalar fields for $D = 2$ or 3. In four spatial dimensions, the only possibility—and a physically very important one—is a pure Yang–Mills theory, with no scalar fields, in which case E is scale invariant and independent of λ. In five or more dimensions the theory of Eq. (3.18) has no nontrivial solutions of this type.

---

[1] There are, however, two-dimensional theories, such as the O(3) sigma model, that do have nontrivial solutions in which the field is everywhere at a vacuum value [36].

3.3 Gauged vortices

Let us return to the U(1) theory of Sec. 3.1, but with the U(1) now being a gauge symmetry which, for convenience, I will describe with the language of electromagnetism. The derivatives in Eq. (3.2) are replaced by covariant derivatives, so that

$$\mathcal{L} = -\frac{1}{4}F_{\mu\nu}F^{\mu\nu} + \frac{1}{2}(D_\mu\phi)^\dagger(D^\mu\phi) - \frac{\lambda}{4}\left(|\phi|^2 - \frac{\mu^2}{\lambda}\right)^2, \qquad (3.24)$$

with

$$D_\mu\phi = \partial_\mu\phi + ieA_\mu\phi. \qquad (3.25)$$

As before, the scalar field potential leads to a nonzero vacuum expectation value $\langle\phi\rangle = v = \sqrt{\mu^2/\lambda}$ and spontaneous breaking of the U(1) symmetry. The elementary particles are a scalar with mass $m_\phi = \sqrt{2}\,\mu = \sqrt{2\lambda}\,v$ and a vector with mass $m_A = ev$.

The Lagrangian is invariant under gauge transformations of the form

$$\begin{aligned}\phi(x) &\to e^{ie\Lambda(x)}\phi(x),\\ A_\mu(x) &\to A_\mu(x) - \partial_\mu\Lambda(x).\end{aligned} \qquad (3.26)$$

Under such a transformation, the line integral $N[C]$ of Eq. (3.4) transforms as

$$N[C] \to N[C] + \frac{e}{2\pi}\oint_C d\boldsymbol{\ell}\cdot\boldsymbol{\nabla}\Lambda. \qquad (3.27)$$

For a nonsingular gauge transformation, with $\Lambda(x)$ single-valued everywhere, the second term vanishes, so the vorticity is a gauge-invariant quantity.

The difficulty with the global vortex was that the energy density from the $(\boldsymbol{\nabla}\alpha)^2$ term only fell as $1/r^2$ at large distance. This problem can be avoided in the gauge theory, where the ordinary derivative is replaced by the covariant derivative,

$$\mathbf{D}\phi = e^{i\alpha}\left[\boldsymbol{\nabla}\rho + i\rho(\boldsymbol{\nabla}\alpha - e\mathbf{A})\right]. \qquad (3.28)$$

If we set

$$\mathbf{A} = \frac{1}{e}\boldsymbol{\nabla}\alpha \qquad (3.29)$$

(up to exponentially small corrections) at large distances, and have ρ rapidly approaches v, then $\mathbf{D}\phi$ vanishes exponentially with r, and the divergence in the energy is eliminated. Because this asymptotic vector potential is pure gauge (i.e., gauge equivalent to $\mathbf{A} = 0$), it does not generate a magnetic field at large r.

Equation (3.29) implies that

$$N[C] = \frac{1}{2\pi}\oint_C d\boldsymbol{\ell}\cdot\boldsymbol{\nabla}\alpha = \frac{e}{2\pi}\oint_C d\boldsymbol{\ell}\cdot\mathbf{A} = \frac{e}{2\pi}\int d^2x\,B, \qquad (3.30)$$

where the last equality follows from Stokes's theorem, with the integral being over the area enclosed by the curve C, and $B = -F_{12}$ denoting the magnetic field. With C taken to be the circle at $r = \infty$, this tells us that the total magnetic flux Φ is related to the vorticity n by

$$\Phi = \frac{2\pi n}{e}. \tag{3.31}$$

The quantization of the vorticity, an essentially topological result, thus results in quantization of magnetic flux.[2]

Let us now look for an $n = 1$ vortex solution by trying a simplifying ansatz [37]. Examination of the field equations shows that when looking for a static solution one can consistently set $A_0 = 0$.[3] Let us assume this to be the case and generalize the global vortex ansatz of Eq. (3.6) to

$$\phi(\mathbf{x}) = v e^{i\theta} f(evr),$$
$$A_j(\mathbf{x}) = \epsilon_{jk} \hat{x}^k \frac{a(evr)}{er}, \tag{3.32}$$

with the coefficient functions f and a both real. This is the most general $n = 1$ ansatz that is both (1) rotationally symmetric in the sense that the effects of a spatial rotation can be completely compensated by a spatially uniform gauge transformation and (2) invariant under a reflection symmetry about the x-axis accompanied by complex conjugation of the scalar field. The latter symmetry also requires that $A_1(r, \theta) = -A_1(r, -\theta)$ and $A_2(r, \theta) = A_2(r, -\theta)$. For later reference, note that the magnetic field is

$$B = \frac{1}{er} \frac{da}{dr}. \tag{3.33}$$

Because we are only interested in static solutions, it suffices to substitute our ansatz into the static energy functional, obtaining

$$\begin{aligned}
E &= \int d^2 x \left[\frac{1}{2e^2 r^2} \left(\frac{da}{dr} \right)^2 + \frac{v^2}{2} \left(\frac{df}{dr} \right)^2 + \frac{v^2}{2} \frac{(1-a)^2}{r^2} f^2 + \frac{\lambda v^4}{4} (1 - f^2)^2 \right] \\
&= \pi v^2 \int du\, u \left[\frac{(a')^2}{u^2} + (f')^2 + \frac{(1-a)^2}{u^2} f^2 + \frac{\lambda}{2e^2} (1 - f^2)^2 \right], \tag{3.34}
\end{aligned}$$

with primes denoting derivatives with respect to $u = evr$. The integral in the second line involves the ratio λ/e^2, but is otherwise independent of the parameters of the theory. We can immediately conclude that

$$E = v^2 F(\lambda/e^2) = \frac{m_\phi^2}{2\lambda} F(\lambda/e^2), \tag{3.35}$$

---

[2] It may seem strange to have obtained flux quantization from a classical theory without making any reference to quantum mechanics. Actually, there is a hidden factor of $\hbar$ here, because the charge e_{field} that appears in the classical Lagrangian is not the same as the charge e_{particle} of the quanta of the field. The two have different dimensions, and are related by $e_{\text{particle}} = \hbar e_{\text{field}}$.

[3] This can be viewed as requiring the solution to be invariant under a combination of time translation and the reflection $A_0(\mathbf{x}) \to -A_0(\mathbf{x})$.

with the function $F(s)$ expected to be of order unity. As with the kink, the ratio of the soliton mass to that of the elementary particle is inversely proportional to the coupling.

Variation of the energy functional with respect to f and a yields

$$0 = \frac{d^2 a}{du^2} - \frac{1}{u}\frac{da}{du} + (1-a)f^2, \tag{3.36}$$

$$0 = \frac{d^2 f}{du^2} + \frac{1}{u}\frac{df}{du} - \frac{(1-a)^2}{u^2}f + \frac{\lambda}{e^2}(1-f^2)f. \tag{3.37}$$

Requiring the fields to be nonsingular at the origin gives the boundary conditions

$$f(0) = 0, \qquad a(0) = 0. \tag{3.38}$$

To have finite total energy, we must require that $|\phi|$ approach its vacuum value and $\mathbf{D}\phi$ tend to zero as $r \to \infty$, giving two more boundary conditions,

$$f(\infty) = 1, \qquad a(\infty) = 1. \tag{3.39}$$

[As a check on our calculations, note that these boundary conditions on $a(u)$, together with Eq. (3.33), give a total magnetic flux of $2\pi/e$, which is the correct value for an $n = 1$ vortex.]

Examining these equations near the origin, we see that $f(u) \sim u$ and $a(u) \sim u^2$ for small u. The behavior at large distance is a bit more subtle. Linearizing Eq. (3.36) about the asymptotic values of a and f leads to a Bessel equation, from which one finds that $1 - a \sim \sqrt{u}\,e^{-u} \sim \sqrt{r}\,e^{-m_A r}$ and hence that

$$B \sim \frac{1}{\sqrt{r}}e^{-m_A r}. \tag{3.40}$$

If $\lambda/e^2 < 2$ (i.e., if $m_\phi < 2m_A$), a similar linearization of Eq. (3.36) gives $1 - f \sim e^{-m_\phi r}/\sqrt{r}$. However, if $\lambda/e^2 > 2$ this would give a falloff faster than that required by the naïvely higher-order term containing $(1 - a)^2$, which can no longer be neglected. The net result is that [38, 39]

$$1 - f \sim \begin{cases} \dfrac{1}{\sqrt{r}}e^{-m_\phi r}, & m_\phi < 2m_A, \\[2ex] \dfrac{1}{r}e^{-2m_A r}, & m_\phi > 2m_A. \end{cases} \tag{3.41}$$

The field equations cannot be solved analytically. A numerical solutions for $\lambda/e^2 = 1/2$ is shown in Fig. 3.2. The form of this solution can be readily understood. The relation between vorticity and magnetic flux requires that there be a region of nonzero B. Because the gauge field acquires a mass when $|\phi|$ is nonzero, it is energetically best for the magnetic flux to be concentrated in a vortex core

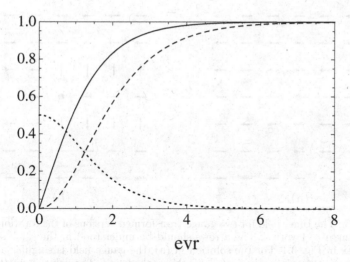

Fig. 3.2. The coefficient functions f (solid line) and a (dashed line) and e times the magnetic field B (dotted line) for the vortex solution with $\lambda = e^2/2$.

region where $|\phi|$ is small. Since the magnetic energy depends on B^2, the magnetic energy is reduced by making this region as large as possible. Competing with this, and favoring a smaller vortex core, is the fact that it costs energy for $V(\phi)$ to be away from its minimum. The relative strength of these two effects is determined by λ/e^2.

Although this solution was obtained by making use of the manifestly rotationally invariant ansatz of Eq. (3.32), it should be kept in mind that the form of the solution can be changed by a gauge transformation. In an extreme example, taking $\Lambda(r, \theta) = \theta/e$, with $-\pi < \theta < \pi$, in Eq. (3.26) makes ϕ real throughout space, but at the cost of introducing a singularity in A_θ along the line $\theta = \pi$, as illustrated in Fig. 3.3a. This singularity can be avoided by using a gauge transformation that only makes ϕ real in a region that excludes the origin and the negative x-axis, as illustrated in Fig. 3.3b.

3.4 Multivortex solutions

We can also look for static solutions with $n > 1$. For $n = 1$ we could argue that the minimum energy configuration with unit vorticity would necessarily give a static solution. This argument fails for $n > 1$ because it might turn out that the lower bound on the energy is only achieved when n unit vortices are infinitely separated from each other. This possibility is excluded if we consider only rotationally invariant configurations, which necessarily have the n zeros of ϕ all coincident at the origin. Let us therefore look for a configuration that is a minimum of the energy among all rotationally invariant configurations with

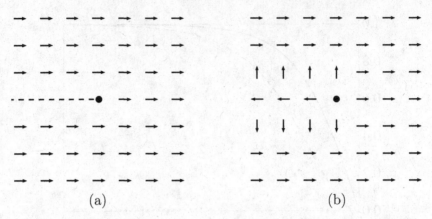

$$(a) \qquad\qquad\qquad (b)$$

Fig. 3.3. The Higgs field for two gauge-transformed versions of the rotationally invariant $n = 1$ vortex. The arrows should be understood in the same sense as those in Fig. 3.1. For the solution in (a), the gauge field is singular along the negative x-axis, as indicated by the dashed line. The solution in (b) is everywhere nonsingular.

vorticity n. By the arguments given in the previous section, this will necessarily be a stationary point—although not necessarily a minimum—among the full set of configurations.

We proceed by replacing the ansatz of Eq. (3.32) with

$$\phi(\mathbf{x}) = e^{in\theta} v f(evr),$$
$$A_j(\mathbf{x}) = \epsilon_{jk}\, \hat{x}^k\, \frac{a(evr)}{er}. \qquad (3.42)$$

Equations (3.36) and (3.37) are replaced by

$$0 = \frac{d^2 a}{du^2} - \frac{1}{u}\frac{da'}{du} + (n - a)f^2, \qquad (3.43)$$

$$(3.44)$$

$$0 = \frac{d^2 f}{du^2} + \frac{1}{u}\frac{df}{du} - \frac{(n - a)^2}{u^2}f + \frac{\lambda}{e^2}(1 - f^2)f, \qquad (3.45)$$

with the boundary conditions unchanged except that $a(\infty) = n$. The large-distance behavior of a and f is similar to that for the unit vortex. Near the origin we still have $a(r) \sim r^2$, but now $f(r) \sim r^n$. In fact, near $r = 0$ we have

$$\phi \sim (x + iy)^n. \qquad (3.46)$$

As noted above, these solutions are not necessarily stable against dissociation into n unit vortices. Some initial insight into this possibility is gained by considering the forces between two well-separated vortices. There are two such forces: an electromagnetic one mediated by the long-range tail of B, and a scalar force mediated by the long-range tail of $v - |\phi|$. The sign of the force between like-charged static objects depends on whether the boson mediating the force has even or odd spin. Hence, the electromagnetic field gives a repulsive force between the vortices, while the scalar force is attractive. Because the fields fall exponentially with distance, both forces are weak, with the dominant force being the one with the slower falloff. From Eqs. (3.40) and (3.41), we see that this will give an attraction if $m_\phi < m_A$, and a repulsion if $m_\phi > m_A$.

If the two vortices are close together, or even overlapping, the problem is nonlinear, so the fact that the vortices repel at large distance does not rule out the possibility they might attract at short distances. However, numerical studies [40] show that this is not the case. For any value of their separation, two vortices repel or attract according to whether m_ϕ/m_A is greater than or less than unity. The rotationally invariant multivortex solution is unstable against dissociation in the former case, but stable in the latter.

The borderline case, with $m_\phi = m_A$ (i.e., $\lambda/e^2 = 1/2$), is particularly interesting. Not only does the force between two vortices vanish identically for any value of the separation, but one can show that there are static n-vortex solutions for any choice of the n vortex positions, with the energy in every case being exactly n times the mass of a unit vortex [41, 42]. This case, which can be understood more fully in terms of supersymmetry, is discussed further in Chap. 8.

These results illuminate a phenomenon in condensed matter physics. The Ginzburg–Landau model gives a phenomenological description of superconductors near the transition temperature. This model has a complex order parameter $\phi(\mathbf{x})$, with $|\phi|^2$ being proportional to the density of Cooper pairs, so that $\phi = 0$ in the normal, nonsuperconducting, state. For static systems the free energy, as a functional of ϕ and the electromagnetic potential A_j, takes the same form as the static energy that follows from the Lagrangian of the three-dimensional extension of our U(1) gauge theory. The various parameters in our Lagrangian can be translated into quantities that depend on the microscopic properties of the superconducting material. The superconductor is type I or type II according to whether λ/e^2 is less than or greater than 1/2. These two types are distinguished by their behavior in a magnetic field. Type I reverts to the normal state in a magnetic field above a relatively low critical value, while a type II superconductor can persist up to a much higher magnetic field in a mixed state where the magnetic field is confined to vortex lines that are the three-dimensional extensions of our two-dimensional vortices. This behavior can be understood in terms of the intervortex forces. In a type I superconductor the vortices attract, and so combine into extended regions in which the superconductor breaks down. If the superconductor is type II, the vortices repel and tend to arrange themselves in a lattice that places them as far apart as possible consistent with the net magnetic flux incident on the superconductor.

3.5 Quantization and zero modes

The mode analysis and quantization of the gauge vortex proceeds much as in the case of the kink, but with a few new twists associated with the zero modes.

There are two zero modes corresponding to translation of the vortex in the two spatial directions. As before, these are treated by introducing collective coordinates z^j corresponding to the location of the vortex. However, in contrast to the previous case, the solution describing a moving vortex cannot be obtained simply by replacing $\mathbf{r}$ in the static solution with $\mathbf{r} - \mathbf{z}(t)$. Because this is a gauge theory, the fields must satisfy the Gauss's law constraint,

$$0 = \partial_\mu F^\mu{}_0 - \frac{1}{2}ie(\phi^* D_0\phi - \phi D_0\phi^*). \tag{3.47}$$

In most gauges this requires a nonzero A_0. For the present discussion we will not need to actually solve for A_0, but only to be aware that it might be present.

The kinetic energy associated with the excitation of these zero modes is

$$\Delta E = \int d^2x \left(\frac{1}{2}F_{0j}^2 + \frac{1}{2}|D_0\phi|^2\right). \tag{3.48}$$

If the only time dependence of $A_j(\mathbf{r} - \mathbf{z})$ and $\phi(\mathbf{r} - \mathbf{z})$ is through $\mathbf{v} = d\mathbf{z}/dt$,

$$\begin{aligned}
F_{0j} &= \partial_0 A_j - \partial_j A_0 \\
&= -v^k \partial_k A_j - \partial_j A_0 \\
&= -v^k F_{kj} - \partial_j(v^k A_k + A_0),
\end{aligned}$$

$$\begin{aligned}
D_0\phi &= \partial_0\phi + ieA_0\phi \\
&= -v^k \partial_k\phi + ieA_0\phi \\
&= -v^k D_k\phi + ie(v^k A_k + A_0)\phi.
\end{aligned} \tag{3.49}$$

Substituting these expressions into Eq. (3.48) and integrating by parts, we obtain

$$\begin{aligned}
\Delta E = &\int d^2x \left[\frac{1}{2}(v^k F_{kj})^2 + \frac{1}{2}|v^k D_k\phi|^2\right] \\
&+ \frac{1}{2}\int d^2x \, (v^k A_k + A_0)\left[\partial_j(F_{j0} + v^l F_{lj}) + \frac{1}{2}ie\phi^*(D_0\phi - v^l D_l\phi)\right. \\
&\left. - \frac{1}{2}ie\phi(D_0\phi^* - v^l D_l\phi^*)\right].
\end{aligned} \tag{3.50}$$

The quantity in square brackets in the second integral vanishes as a consequence of Gauss's law and the fact that F_{jk} and $D_j\phi$ satisfy the static field equations. Next, note that the rotational invariance of the vortex solution implies that

$$\int d^2x \, |D_1\phi|^2 = \int d^2x \, |D_2\phi|^2 \tag{3.51}$$

and

$$\int d^2x \, [(D_1\phi^*)(D_2\phi) + (D_2\phi^*)(D_1\phi)] = 0, \tag{3.52}$$

and hence that

$$\int d^2x \, |v^k D_k \phi|^2 = \frac{1}{2} \mathbf{v}^2 \int d^2x \, |D_k \phi|^2. \tag{3.53}$$

Even without this rotational symmetry, the fact that $F_{ij} = -\epsilon_{ij} B$ is sufficient to show that

$$\int d^2x \, (v^k F_{kj})^2 = \frac{1}{2} \mathbf{v}^2 \int d^2x \, F_{jk}^2. \tag{3.54}$$

Finally, the virial identity of Eq. (3.23) implies that

$$\int d^2x \, \frac{1}{4} F_{jk}^2 = \int d^2x \, V(\phi). \tag{3.55}$$

Using these three integral identities, we obtain

$$\Delta E = \frac{1}{2} \mathbf{v}^2 \int d^2x \left[\frac{1}{4} F_{jk}^2 + \frac{1}{2} |D_k \phi|^2 + V(\phi) \right]$$
$$= \frac{1}{2} M_{\text{cl}} \, \mathbf{v}^2, \tag{3.56}$$

where M_{cl} is the classical approximation to the mass of the vortex. Just as with the kink, a virial identity is essential for ensuring that the soliton mass enters the kinetic energy in the proper manner.

The translation modes are not the only zero modes. Given any function $\Lambda(\mathbf{x})$, a small fluctuation of the form

$$\delta_\Lambda A_j = -\partial_j \Lambda, \qquad \delta_\Lambda \phi = ie\Lambda \phi \tag{3.57}$$

is just a gauge transformation. This leaves the potential energy unchanged, and so is a zero mode. Because the new configuration resulting from such a fluctuation is physically equivalent to the original one, these modes are less relevant. They are eliminated if a gauge condition is imposed on the theory at the outset.

However, it is sometimes more convenient to focus on the full set of fluctuations about a given classical solution and to divide these into a set of gauge modes (to be ignored) and a set of nongauge zero modes that are orthogonal to all gauge modes. If we define the inner product between two modes $\chi_r \equiv \{\delta_r A_j, \delta_r \phi\}$ and $\chi_s \equiv \{\delta_s A_j, \delta_s \phi\}$ to be

$$\langle \chi_r | \chi_s \rangle = \int d^2x \left[\delta_r A_j \, \delta_s A_j + \frac{1}{2} (\delta_r \phi^* \, \delta_s \phi + \delta_r \phi \, \delta_s \phi^*) \right], \tag{3.58}$$

then a mode $\{\delta A_j, \delta \phi\}$ is orthogonal to the gauge mode generated by Λ if

$$0 = \int d^2x \left[-\partial_j \Lambda \, \delta A_j - \frac{1}{2} ie\Lambda(\phi^* \, \delta \phi - \phi \, \delta \phi^*) \right]$$
$$= \int d^2x \, \Lambda \left[\partial_j \delta A_j - \frac{1}{2} ie(\phi^* \, \delta \phi - \phi \, \delta \phi^*) \right], \tag{3.59}$$

where the second line follows by an integration by parts that is valid if $\Lambda(\mathbf{r})$ vanishes at spatial infinity.[4] This condition will be met for any choice of $\Lambda(\mathbf{r})$ if the mode satisfies

$$0 = \partial_j \delta A_j - \frac{1}{2} ie(\phi^* \delta\phi - \phi\,\delta\phi^*), \tag{3.60}$$

which is known as the background gauge condition.

A nongauge zero mode that does not satisfy this condition can be brought into background gauge by the addition of a suitable gauge mode. Thus, the naïve expression for the zero mode corresponding to spatial translation in the kth direction would be $\delta_k A_j = \partial_k A_j$, $\delta_k \phi = \partial_k \phi$, which is not in background gauge. Instead, we want a mode of the form

$$\delta_k A_j = \partial_k A_j - \partial_j \Lambda_{(k)},$$
$$\delta_k \phi = \partial_k \phi + ie\Lambda_{(k)}\phi, \tag{3.61}$$

with a suitably chosen gauge function $\Lambda_{(k)}$. The desired gauge function turns out to be $\Lambda_{(k)} = A_k$, with the result that

$$\delta_k A_j = F_{kj}, \qquad \delta_k \phi = D_k \phi. \tag{3.62}$$

The static field equations then insure that the background gauge condition is satisfied.

3.6 Adding fermions

Just as we saw with the kink, adding fermion fields to the theory can have interesting consequences associated with the presence of zero modes. As an example, let us suppose that we have a spontaneously broken $U(1)_W$ gauge theory with a gauge field W_μ and a complex scalar field ϕ with $U(1)_W$ charge $2q$. [The notation is intended to emphasize that the gauge field and the $U(1)$ symmetry are not those of electromagnetism; the true electromagnetic field will be introduced later.] This theory has a vortex solution of the form

$$\phi = v e^{i\theta} f(r),$$
$$W_i = \epsilon_{ij}\, \hat{x}^k \frac{a(r)}{2qr},$$
$$W_0 = 0, \tag{3.63}$$

with $f(0) = a(0) = 0$ and $f(\infty) = a(\infty) = 1$. Following Jackiw and Rossi [43], let us introduce a fermion field ψ that couples to ϕ and W_μ via

$$\mathcal{L}_{\text{fermion}} = i\bar{\psi}\gamma^\mu D_\mu \psi - \frac{i}{2} g\phi\bar{\psi}\psi^c + \frac{i}{2} g\phi^*\bar{\psi}^c\psi, \tag{3.64}$$

---

[4] If Λ does not vanish at spatial infinity, the gauge mode is non-normalizable because of the $\delta_\Lambda \phi$ term. In theories where the gauge symmetry is unbroken or only partially broken there can be normalizable modes associated with gauge functions that are nonzero at spatial infinity. These global gauge modes turn out to be physically significant, as will be explained in Chap. 5.

where g is real and

$$D_\mu\psi = (\partial_\mu + iqW_\mu)\psi. \tag{3.65}$$

Note that $U(1)_W$ invariance requires that the charge of ψ be half that of ϕ.

Because we are in three spacetime dimensions, ψ has two complex components, and the Dirac matrices are 2×2. The charge conjugate field ψ^c is given by

$$\psi^c = C\bar\psi^T. \tag{3.66}$$

If we take γ^0 and γ^2 to be the symmetric matrices σ_z and $-i\sigma_x$ and γ^1 to be the antisymmetric $i\sigma_y$, we can choose $C = i\gamma^0\gamma^2 = i\sigma_y$.

The Dirac equation that follows from this Lagrangian is

$$0 = i\gamma^\mu D_\mu\psi - ig\phi\psi^c. \tag{3.67}$$

In a vacuum background, with $\phi = v$ and $W_\mu = 0$, this gives

$$i\gamma^\mu\partial_\mu\psi = gv\gamma^2\psi^*,$$
$$i\gamma^\mu\partial_\mu(\gamma^2\psi^*) = gv\psi, \tag{3.68}$$

and hence

$$(\gamma^\mu\partial_\mu)^2\psi = \partial^\mu\partial_\mu\psi = -g^2v^2\psi, \tag{3.69}$$

showing that the fermion mass is $\mu = gv$. Solving for the normal modes, one finds that the positive and negative frequencies are paired, as expected, with the eigenfunctions interchanged by the transformation $\psi \to \gamma^0\psi$.

Now consider the spectrum in the presence of a vortex. The nonzero positive and negative energy eigenvalues are still paired, with their eigenfunctions related by multiplication by γ^0. Any zero-energy modes can be chosen to be eigenfunctions of $\gamma^0 = i\gamma^1\gamma^2$, with eigenvalue ± 1. The Dirac equation for a zero mode $\psi^{(0)}$ can be written as

$$0 = \left[D_1 - \gamma^1\gamma^2 D_2\right]\psi^{(0)} - ig\phi\gamma^1\gamma^2(\psi^{(0)})^*. \tag{3.70}$$

Now note that

$$D_1 \pm iD_2 = e^{\pm i\theta}\left[\partial_r \pm \frac{i}{r}\partial_\theta \pm \frac{a(r)}{2r}\right]. \tag{3.71}$$

Hence, we will get a common factor of $e^{i\theta}$ in all the terms of Eq. (3.70) if we require $\psi^{(0)}$ to be of the form $\psi^{(0)} = u(r)\chi$, with the constant spinor χ satisfying

$$i\gamma^1\gamma^2\chi = \chi = \chi^*. \tag{3.72}$$

We then have

$$0 = \left[\partial_r + \frac{a(r)}{2r}\right]u - gvf(r)u^*. \tag{3.73}$$

This is solved by

$$u(r) = C_1 \exp\left\{ \int_0^r dr' \left[gvf(r') - \frac{a(r')}{2r'} \right] \right\}$$
$$+ iC_2 \exp\left\{ -\int_0^r dr' \left[gvf(r') + \frac{a(r')}{2r'} \right] \right\}, \qquad (3.74)$$

where C_1 and C_2 are real constants. To have a normalizable solution we must set $C_1 = 0$, so that u is pure imaginary. Because a and f approach their asymptotic values exponentially fast, $u(r)$ is localized about the vortex, falling as $e^{-\mu r}/\sqrt{r}$.

One might wonder whether there are other zero modes. The analysis of [43] shows that there are not, and also shows that in an axially symmetric background with n vortices superimposed at the origin there are precisely n zero modes. In fact, one can derive an index theorem [44] that extends this result to a background with n vortices at arbitrary positions.

Let us now go to four spacetime dimensions, with the vortex solution interpreted as a z-independent vortex line situated along the z-axis. As we will see, the fermionic zero mode in $(2+1)$ dimensions leads to a massless particle bound to the vortex line in $(3+1)$ dimensions.

In four dimensions we have a four-component fermion field ψ that can be written as a sum of two spinors of definite chirality that obey

$$\gamma^5 \psi_R = \psi_R,$$
$$\gamma^5 \psi_L = -\psi_L. \qquad (3.75)$$

The fermionic part of the Lagrangian becomes

$$\mathcal{L}_{\text{fermion}} = i\bar{\psi}_R \gamma^\mu D_\mu \psi_R + i\bar{\psi}_L \gamma^\mu D_\mu \psi_L - g\phi\bar{\psi}_L \psi_R - g\phi^* \bar{\psi}_R \psi_L. \qquad (3.76)$$

This has an invariance under the transformation $\psi \to e^{i\beta\gamma^5}\psi$, $\phi \to e^{-2i\beta}\phi$ that forbids any bare (i.e., independent of ϕ) fermion mass term. The Dirac equation is now

$$0 = i\gamma^\mu D_\mu \psi_L - g\phi\psi_R,$$
$$0 = i\gamma^\mu D_\mu \psi_R - g\phi^* \psi_L. \qquad (3.77)$$

Let us first look for a solution $\psi^{(0)}$ that is independent of t and z. Guided by the $(2+1)$-dimensional example, we take $\psi_R^{(0)}$ and $\psi_L^{(0)}$ to be functions only of r and to be eigenvectors of $i\gamma^1\gamma^2$, with

$$i\gamma^1\gamma^2 \psi_L^{(0)} = \psi_L^{(0)},$$
$$i\gamma^1\gamma^2 \psi_R^{(0)} = -\psi_R^{(0)}. \qquad (3.78)$$

This leads to[5]

$$0 = \left[\partial_r + \frac{a(r)}{2r}\right]\psi_L^{(0)} - igvf(r)\gamma^1\psi_R^{(0)},$$

$$0 = \left[\partial_r + \frac{a(r)}{2r}\right]\psi_R^{(0)} - igvf(r)\gamma^1\psi_L^{(0)}. \tag{3.79}$$

Comparison with Eq. (3.73) for the $(2+1)$-dimensional zero mode suggests that we take

$$\psi_R^{(0)} = -i\gamma^1\psi_L^{(0)}. \tag{3.80}$$

Doing so, we obtain the normalizable solution

$$\psi_L^{(0)} = C\exp\left\{-\int_0^r dr'\left[gvf(r') + \frac{a(r')}{2r'}\right]\right\}\chi, \tag{3.81}$$

where χ is a constant left-handed spinor obeying

$$i\gamma^1\gamma^2\chi = \chi. \tag{3.82}$$

Let us now introduce dependence on t and z and look for a solution of the form

$$\psi_L = b(t,z)\psi_L^{(0)}(r),$$

$$\psi_R = b(t,z)\psi_R^{(0)}(r). \tag{3.83}$$

Substituting this into Eq. (3.77) gives

$$0 = (\gamma^0\partial_0 + \gamma^3\partial_3)b(t,z)\chi. \tag{3.84}$$

We now note that a left-handed spinor satisfying Eq. (3.82) also obeys

$$\gamma^0\chi = -\gamma^3\chi. \tag{3.85}$$

It follows that Eq. (3.84) is solved by taking $b(t,z)$ to be an arbitrary function of $t-z$. In particular, the modes of definite energy have $b = e^{-i\omega(t-z)}$. Those with $\omega > 0$ correspond to "right-moving" particles with energy $E = \omega$ that move along the vortex string at the speed of light, and must therefore be massless.[6] The $\omega < 0$ modes correspond, after the standard reinterpretation, to antiparticles with energy $E = |\omega|$, also right-moving. In order to obtain left-movers, a different particle species, transforming oppositely under $U(1)_W$, is required.

At the beginning of this section, it was emphasized that the $U(1)_W$ whose breaking allowed the existence of the vortex solution was not electromagnetism.

---

[5] Because ψ_L and ψ_R have opposite $U(1)_W$ charges, W_μ enters their covariant derivatives with opposite signs. When combined with the opposite signs in Eq. (3.78), the net effect is that $a(r)$ enters with the same sign in both of these equations.

[6] The notion of right- and left-moving along the string should not be confused with the right and left chirality denoted by the subscripts on ψ_R and ψ_L.

Now suppose that the fermion field, but not ϕ or W_μ, also carries a true electric charge e. The modes built upon the (2+1)-dimensional zero mode then correspond to either a particle with electric charge e or an antiparticle with charge $-e$. Let us also suppose that there is a second fermion field η, also with electric charge e, that has opposite $U(1)_W$ coupling and so produces left-moving particles along the string.

In the ground state in the vortex background, all of the modes of ψ and η are unoccupied. In the Dirac sea picture, this can be viewed as having all negative-energy states filled and all positive-energy states empty. If an external electric field $\mathbf{E}$ along the positive z-direction is imposed, it will increase the momentum, and thus the energy, of the right-moving ψ states, and decrease the magnitude of the momentum, and thus the energy, of the left-moving η states. This will promote some of the occupied negative-energy ψ states to occupied positive-energy states, thus creating real right-moving particles with charge e, while some of the empty positive-energy states of η will become empty negative-energy states, corresponding to the creation of left-moving "holes" or antiparticles, of charge $-e$. This process conserves electric charge but, because the particles of opposite charge move in opposite directions, creates a net electric current. Now suppose that the electric field is turned off while the ψ particle energies are still all less than μ. The positive energy ψ particles cannot leave the string, because that would require an energy at least equal to the free ψ mass, and they cannot move down to lower energy states on the string, because these are all occupied. A similar argument applies to the η antiparticles. Hence, these particles remain in the states in which they were, and the current persists even in the absence of the applied electric field. Thus, the string has become a superconductor [45].

What would have happened if ψ were the only fermion field in the theory? The process described above would have produced ψ particles with charge e but no compensating charge $-e$ antiparticles, and so would have violated electric charge quantization. However, consistency of the four-dimensional theory rules out this possibility. Because the $U(1)_W$ gauge symmetry is chiral, renormalizability requires that all triangle anomalies be absent, and so necessitates the inclusion of additional fermion fields. One can easily check that the conditions for anomaly cancellation are just what is needed to ensure electric charge conservation along the string [45].

Of course, the existence of the fermion zero mode did not really depend on the fact that the $U(1)_W$ was gauged; examination of Eq. (3.74) shows that $u(r)$ would still be normalizable even if the background were that of a global string with $a(r) = 0$. In that case additional fermion fields are not needed for the four-dimensional theory to be consistent. However, without such fields electric charge conservation requires that charge flow on to the string from the exterior region [46, 47].

It is also possible to produce superconducting strings by adding additional bosonic fields to the theory [45]. For a detailed discussion of this, see [48].

4

Some topology

In the previous two chapters I argued for the existence of solitons by starting with a configuration in which ϕ approached different vacuum values at different parts of spatial infinity, with these values being chosen in such a way that the configuration could not be continuously deformed into a uniform vacuum solution over all of space. For both the kink and the U(1) vortex it was fairly easy to visualize the situation and convince oneself of the topological stability of the soliton. However, we want to be able to consider theories in which the space of vacuum solutions is more complicated, as well as theories in three (or more) spatial dimensions. In such cases the topological constraints can be harder to visualize. It is therefore useful to formulate matters more precisely. As I will explain in this chapter, the mathematical language of homotopy provides an ideal framework for this.

4.1 Vacuum manifolds

The general situation that we want to consider is a field theory with n scalar fields that can be assembled into an n-component column vector ϕ and a scalar field potential $V(\phi)$ that has a family of degenerate minima that form a manifold $\mathcal{M}$. Let us assume that this degeneracy is a consequence of a symmetry group G that is spontaneously broken to a subgroup H by the vacuum expectation value of ϕ. (One can certainly construct potentials with degenerate minima that are not related by a symmetry, but in such cases the degeneracy is usually broken by higher-order quantum corrections.) Given a value of ϕ, the action of an element g of G transforms ϕ to $D(g)\phi$, where $D(g)$ is the appropriate n-dimensional representation of G. In particular, if ϕ_0 minimizes V, then so does $D(g)\phi_0$ for any choice of g. Furthermore, our assumption that the degeneracy is entirely due to the symmetry implies that all minima of V are of this form.

One might then guess that there is a one-to-one correspondence between elements of G and minima of V. This is only true if G is completely broken. If G

is only partially broken, there is an unbroken subgroup H that can be defined by the requirement that it leaves ϕ_0 invariant; i.e., that $D(h)\phi_0 = \phi_0$ for any element h of H. Consequently, for any given $g \in G$ and $h \in H$, gh and g have the same effect; i.e., $D(gh)\phi_0 = D(g)\phi_0$. We can therefore define equivalence classes of elements of G by defining two elements g_1 and g_2 to be equivalent if $g_2 = g_1 h$ for some $h \in H$. The set of such equivalence classes is the coset space G/H. There is a one-to-one correspondence between these equivalence classes and the minima of V, so

$$\mathcal{M} = G/H. \tag{4.1}$$

A word of caution is appropriate here. In perturbative treatments of field theories the distinction between different Lie groups that share the same Lie algebra [e.g., SU(2) and SO(3)] is often ignored. It will be essential in our topological considerations to keep track of these distinctions and to be clear as to precisely which groups G and H are. This will be discussed in more detail in Sec. 4.3. However, it will helpful to first introduce the concepts of homotopy and the fundamental group.

4.2 Homotopy and the fundamental group $\pi_1(\mathcal{M})$

Let us start by considering closed paths, or loops, on a manifold $\mathcal{M}$. In particular, let us pick a point x_0 on $\mathcal{M}$ and restrict our attention to paths that begin and end at x_0. Any such path can be specified by a continuous function $f(t)$ taking values in $\mathcal{M}$, with $0 \leq t \leq 1$ and $f(0) = f(1) = x_0$. Figure 4.1 shows three such paths. In this figure, $\mathcal{M}$ is the two-dimensional Euclidean plane, but with the shaded region deleted. It is evident that path b can be smoothly deformed so that it coincides with path a. However, path c cannot be smoothly deformed into a, since that would require passing part of the path through the forbidden shaded region, which is not part of $\mathcal{M}$. The concept of "smoothly deformable" can be made more precise as follows. Let $f(t)$ and $g(t)$ be continuous paths beginning and ending at x_0. They can be smoothly deformed into one

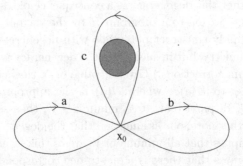

Fig. 4.1. Loops a and b are homotopic to each other, but not to loop c.

another if and only if there is a continuous function $k(s,t)$ with $0 \le s, t \le 1$ such that

$$k(0,t) = f(t),$$
$$k(1,t) = g(t),$$
$$k(s,0) = k(s,1) = x_0. \qquad (4.2)$$

Thus, $k(s,t)$ can viewed as a sequence of loops, labeled by s, that begin and end at x_0, with f being the first in the sequence and g the last. Paths f and g are said to be *homotopic at* x_0, and the family of paths that define the function k is a *homotopy*.

One can define a product on the space of paths. Given paths f and g, their product is defined as

$$(f \circ g)(t) = \begin{cases} f(2t), & 0 \le t \le 1/2, \\ g(2t-1), & 1/2 \le t \le 1. \end{cases} \qquad (4.3)$$

In other words, $f \circ g$ is the path obtained by going around f and then going around g. An inverse path f^{-1} can be defined as going around f in the reverse direction; i.e., $f^{-1}(t) = f(1-t)$.

The next step is to divide the paths on $\mathcal{M}$ into *homotopy classes*, with the homotopy class $[f]$ denoting the set of paths that are homotopic to f. The definition of the product of paths can be carried over to the product of homotopy classes. Because the product of any path homotopic to f with any path homotopic to g is homotopic to $f \circ g$, we have

$$[f] \circ [g] = [f \circ g]. \qquad (4.4)$$

Note that not every path in $[f \circ g]$ is a product path. Equation (4.3) shows that the product of two paths must pass through x_0 three times (at $x = 0, 1/2$, and 1). However, there are certainly paths homotopic to this product that only go through x_0 at their beginning and end, and not at any intermediate points.

There is a trivial "path", $f(t) = x_0$, that just stays at x_0 for its entire length; let us denote the homotopy class of this path by $[I]$. It is easy to see that $[g] \circ [I] = [I] \circ [g] = [g]$ and $[g] \circ [g^{-1}] = [I]$ for any $[g]$. It is also clear that the multiplication of homotopy classes is associative.

Thus, multiplication of homotopy classes has a group structure, with $[I]$ being the identity. This group is denoted $\pi_1(\mathcal{M}, x_0)$, and called the *fundamental group of* $\mathcal{M}$ *at* x_0. Because there are generalizations $\pi_n(\mathcal{M}, x_0)$, which we will encounter later on, the fundamental group is also known as the *first homotopy group*.

If $\mathcal{M}$ is a connected manifold, the fundamental group does not depend on the choice of the base point x_0. To see this, note that any loop beginning and ending at x_0 can be mapped onto a loop beginning and ending at y_0 by adding a path from y_0 to x_0 at the beginning of the loop and then traversing this path backward from x_0 to y_0 at the end of the loop. This mapping preserves all of the homotopy relations between loops, and thus defines an isomorphism between the fundamental groups at x_0 and y_0. If these groups are Abelian, this isomorphism does not depend on which path from y_0 to x_0 is chosen, but the detailed correspondence between homotopy classes of the two groups can be path-dependent if the groups are non-Abelian.

Because the fundamental group does not depend on the choice of the base point, one usual writes simply $\pi_1(\mathcal{M})$, and refers to the fundamental group of $\mathcal{M}$. If all loops on a manifold can be deformed to the trivial loop, then there is only a single homotopy class. We denote this by writing $\pi_1(\mathcal{M}) = 0$, and the manifold is said to be simply connected.

Let us consider some examples:

(i)   Any loop on the Euclidean plane can be continuously shrunk to a point, so R^2 is simply connected. The same is clearly true for all higher dimensional Euclidean spaces and, although a loop on a line may be a bit harder to visualize, for the real line R^1. Thus,

$$\pi_1(R^n) = 0. \tag{4.5}$$

(ii)   Now consider the space shown in Fig. 4.1, the Euclidean plane with a disk removed. Loops can be characterized by the number of times they wind around the hole left by the disk, with counterclockwise (clockwise) windings counted positively (negatively). Thus, in the figure loops a and b have winding number 0, while loop c has winding number 1. Two loops with the same number of windings are homotopic, and under multiplication of loops the winding numbers add. Thus, the fundamental group of this manifold is just Z, the additive group of the integers. The same would be true if we put an outer boundary on the manifold and reduced it to a ring $r_1 < r < r_2$ enclosing the hole. Indeed, we could just shrink the ring to a circle, S^1, without changing the homotopy group. Thus,

$$\pi_1(S^1) = Z. \tag{4.6}$$

(iii)   Consider next a two-sphere,[1] S^2. It may be obvious that any loop on this sphere can be shrunk to a point. If it isn't, imagine deleting from the sphere some point through which the loop does not pass. The sphere with a point deleted is topologically equivalent to the plane, so the result follows from Example (i). Similar arguments applied in higher dimensions show that

$$\pi_1(S^n) = 0, \qquad n \geq 2. \tag{4.7}$$

(iv)   Figure 4.2 shows the "figure-eight space", a plane with two disks removed. Two loops, a encircling the left hole, and b encircling the right one, are shown.

---

[1] Recall that an n-sphere, S^n, is an n-dimensional manifold that can be viewed as a spherical hypersurface in $n + 1$ Euclidean dimensions. In particular, a one-sphere is a circle and a two-sphere is the surface of a solid ball in ordinary three-dimensional space.

Fig. 4.2. The "figure-eight space". Loops a and b do not commute, and the fundamental group is non-Abelian.

Now consider the product loop $a \circ b \circ a^{-1} \circ b^{-1}$ that goes once clockwise and once counterclockwise around each hole. After some attempts at deforming the loops, you should be able to convince yourself that this product loop cannot be deformed to a point. On the other hand, the loop $a \circ a^{-1} \circ b \circ b^{-1}$ is clearly homotopic to the trivial loop. Thus, the fundamental group of this manifold contains elements $[a]$ and $[b]$ that do not commute, and so is non-Abelian.

4.3 Fundamental groups of Lie groups

Lie groups can be viewed as manifolds, and their fundamental groups are of particular interest. In examining these, the distinction between groups that share the same Lie algebra is crucial.[2] Let us start with the most familiar example of two groups with the same Lie algebra, SU(2) and SO(3).

SU(2) is the group of 2×2 unitary matrices with unit determinant. Any such matrix can be written in the form

$$U = b_0 + i\mathbf{b} \cdot \boldsymbol{\sigma}, \tag{4.8}$$

where the σ_j $(j = 1, 2, 3)$ are the Pauli matrices and

$$b_0^2 + b_1^2 + b_2^2 + b_3^2 = 1. \tag{4.9}$$

This last equation is just that for the unit three-sphere, so we see that as a manifold $SU(2) = S^3$.

SO(3) is the group of rotations in three dimensions. Any element of the group can be identified by giving a unit vector $\hat{\mathbf{n}}$ that specifies the rotation axis, and a (counterclockwise) rotation angle about that axis that lies in the range $0 \leq \psi \leq \pi$. Note that rotations by π about $\hat{\mathbf{n}}$ and $-\hat{\mathbf{n}}$ have the same effect, and correspond to the same group element. As a manifold, SO(3) can be mapped onto a three-dimensional ball of radius π. The center corresponds to the identity element (rotation by $\psi = 0$ about any axis). All other elements lie on a radial line along $\hat{\mathbf{n}}$, with the distance from the origin being equal to ψ. Because $(\hat{\mathbf{n}}, \psi = \pi)$

---

[2] A more detailed discussion of Lie groups and Lie algebras is given in Appendix A.

and $(-\hat{\mathbf{n}}, \psi = \pi)$ are the same group element, antipodal points on the surface of the ball must be identified.

The relation between the two groups is seen by writing Eq. (4.8) as

$$U = \cos(\psi/2) + i\hat{\mathbf{n}} \cdot \boldsymbol{\sigma} \sin(\psi/2). \tag{4.10}$$

While the range $0 \leq \psi \leq \pi$ covers all of SO(3), twice that range, $0 \leq \psi \leq 2\pi$, is needed to obtain all of SU(2). With this enlarged range, the two SU(2) matrices U, given by $(\hat{\mathbf{n}}, \psi)$, and $-U$, corresponding to $(-\hat{\mathbf{n}}, 2\pi - \psi)$, map to the same element of SO(3). Thus, SU(2) is a double cover of SO(3).

The relation between the topologies of the two groups can be understood in terms of this mapping. The two elements of SU(2) that are mapped to the same element of SO(3) lie on antipodal points of the three-sphere defined by Eq. (4.9). By taking one element from each such pair, we see that SO(3) corresponds to the upper half of the three-sphere, including the "equator" (which is actually a two-sphere), but with the caveat that antipodal points on the equator must be identified. This yields the previous construction of SO(3) in terms of a three-dimensional ball.[3]

The relation between the two groups can also be understood from a more algebraic point of view. The center of a group is defined as the set of group elements that commute with all elements of the group; this is in fact a subgroup. The center of SU(2) consists of two elements, the identity matrix I, and the matrix $z = -I$, with the latter corresponding to a rotation by 2π about any axis. These form the cyclic group with two elements, Z_2. Now suppose that we define an equivalence relation under which every SU(2) matrix U is equivalent to $zU = -U$. Because z commutes with every element of the group, this equivalence is compatible with the group multiplication, and the equivalence classes themselves form a group, $SU(2)/Z_2$, which is just SO(3) itself.

Because SU(2) is topologically a three-sphere, we know from Example (iii) of the previous section that it is simply connected; every closed loop can be continuously contracted to a point. What does this tell us about SO(3)? By using the mapping of elements from SU(2) to SO(3), any path on SU(2) can be mapped to a path on SO(3). If the path is a closed loop on SU(2), it is obviously a closed loop on SO(3); since it is contractible on SU(2), it must also be contractible, and homotopic to the trivial loop, on SO(3). But consider a path on SU(2) that starts at some U_0 and ends at the antipodal point, zU_0. This is not a loop in SU(2), but because U_0 and zU_0 are mapped to the same element of SO(3), it is mapped to a closed loop in SO(3). However, this cannot be a contractible loop in SO(3), because that would imply that it could be smoothly deformed to a trivial SO(3) loop, which must correspond to a trivial SU(2) loop.

---

[3] It may be easier to visualize this by going to one fewer dimension. The upper half of a two-sphere (the upper half of the surface of a globe) is topologically the same as a disk enclosed by a circle, which is the two-dimensional "ball".

Hence, SO(3) must have at least two homotopy classes. In fact, that is all that it has. Going around this loop twice in SO(3) corresponds to a path in SU(2) that runs from U_0 to zU_0 and then back to U_0. This is a closed loop in SU(2), and hence must be contractible. Thus, we have

$$\pi_1(\mathrm{SU}(2)) = 0, \tag{4.11}$$

$$\pi_1(\mathrm{SO}(3)) = \pi_1(\mathrm{SU}(2)/Z_2) = Z_2. \tag{4.12}$$

These ideas can be extended to other Lie groups. For every Lie algebra, there is a unique simply connected group, known as the universal covering group. Let G be this group, and let K be either its center or a subgroup of its center. If G is semisimple, K is a finite group. In this case, by defining the elements g and kg to be equivalent, where g and k are arbitrary elements of G and K, we obtain a group G/K that is not simply connected. In fact, by an extension of the arguments given for SU(2) and SO(3),

$$\pi_1(G/K) = K. \tag{4.13}$$

If K is the full center of the covering group, then G/K is known as the adjoint group.

Turning now to the representations of these groups, recall that the irreducible representations of SU(2) can be labeled by a "spin" s that can be either an integer or a half-integer, but that the half-integer spin representations, for which a rotation by 2π is represented by the matrix $-I$, are not true (i.e., single-valued) representations of SO(3). In the general case, the true representations of G/K are those representations of G for which every element of K is represented by a unit matrix.

The groups most often encountered in high energy physics applications are the unitary and orthogonal groups. The group SU(N) is simply connected for any $N \geq 2$. Its center consists of the matrices $e^{2\pi ik/N}I_N$ ($k = 0, 1, \ldots, N-1$), which form the cyclic group Z_N. We have already discussed the representations for $N = 2$. For $N = 3$, recall that any irreducible representation can be constructed from the direct product of p fundamental **3** and q antifundamental **$\bar{3}$** representations by suitable symmetrization or antisymmetrization and extraction of traces. The triality of such a representation is defined to be $p - q$ (mod 3). Only the representations with zero triality are representations of the adjoint group SU(3)/Z_3. These ideas can be extended to larger N in an obvious manner, although it should be noted that if N is not prime there are groups intermediate in size between the adjoint group and the covering group.

The group U(1), the multiplicative group with elements $e^{i\alpha}$, is topologically a circle. Hence, we see from Eq. (4.6) that

$$\pi_1(\mathrm{U}(1)) = Z, \tag{4.14}$$

so that U(1) is not simply connected. Its simply connected covering group is R, the additive group of real numbers, with $U(1) = R/Z$.

None of the orthogonal groups are simply connected. SO(2) is identical to U(1), and so its fundamental group is Z. For $N \geq 3$, we have

$$\pi_1(SO(N)) = Z_2, \qquad N \geq 3. \tag{4.15}$$

The covering group of SO(N) is known as Spin(N). The first few of these are more often recognized in other forms. We have already seen that Spin(3) = SU(2). The next, Spin(4), is identical to SU(2)×SU(2), while Spin(5) = Sp(4), and Spin(6) = SU(4). For further discussion of these and other compact Lie groups, see Appendix A.

As mentioned previously, in perturbative treatments of gauge field theories one is actually only concerned with the Lie algebra, and the distinction between the Lie groups that share that algebra is unimportant. This distinction often matters when using topology to study solitons. There is often some freedom in specifying the symmetry group. For the full symmetry group G of the theory, the only requirement is that all fields in the theory must transform under true representations of G. Hence, G can always be chosen to be the universal covering group, but it can also be taken to be a quotient group if some classes of representations are absent. The choice that is made for G will, however, determine the choice for the unbroken group H.

For future reference, it should be noted that the results in Eqs. (4.12) and (4.13) are instances of a more general result. Let G be a connected and simply connected Lie group, and let H be a subgroup of G. If H is not a connected group, it has a connected subgroup H_0 that contains the identity; if H is connected, $H_0 = H$. The cosets H/H_0 form a group that is often termed the *zeroth homotopy group*, $\pi_0(H)$. Then[4]

$$\pi_1(G/H) = \pi_0(H). \tag{4.16}$$

If H is connected, $\pi_0(H) = 0$. If H is a discrete group, $\pi_0(H) = H$.

4.4 Vortices and homotopy

In Chap. 3, I argued that a configuration with the asymptotic behavior shown in Fig. 3.1a could not be continuously deformed to a uniform vacuum solution, and hence that varying such a configuration until a local minimum of the energy was reached would produce a nontrivial solution. Let us now rephrase this argument in the language of homotopy.

In two dimensions, spatial infinity can be described as a circle at $r = \infty$. As θ varies from 0 to 2π, the values of the field $\phi(r = \infty, \theta)$ on this circle trace out a loop in the vacuum manifold $\mathcal{M}$. Roughly speaking, the argument is that if this loop is in a different homotopy class than the vacuum, then there must be a soliton solution. However, this is not quite right.

---

[4] For a proof of this theorem, see [49].

Homotopy, as defined by Eq. (4.2), is an equivalence relation between two loops that have a common end point. In the field theory context, this would relate two field configurations such that $\phi_1(\infty,0) = \phi_2(\infty,0) = \phi_0$ for some fixed ϕ_0. However, our arguments for the existence of a vortex involved continuous deformations of the initial configuration, but without any requirement that there be a point where the value of the field was held fixed. Thus, from a physical point of view our primary interest is in *free homotopy*, where the base point condition is omitted; two loops f and g are said to be freely homotopic if there is a continuous function that satisfies the first two, but not necessarily the third, lines of Eq. (4.2).

If two loops are homotopic with a fixed point, they are obviously also freely homotopic. The converse need not be true, even if the loops share a common base point. This is illustrated in Fig. 4.3. As long as it is attached to the base point x_0, loop g cannot be deformed into loop f. However, it certainly can be deformed into f if it is released from the base point. Hence, the two loops are freely homotopic, but not homotopic with a base point.

These two loops can be related with the aid of loop a; thus, g is homotopic to $a \circ f \circ a^{-1}$. More generally, two loops h and k with a common base point are said to be conjugate if there is a loop c such that h is homotopic to $c \circ k \circ c^{-1}$. This relationship defines a set of *conjugacy classes*. If the fundamental group is Abelian, each conjugacy class contains a single homotopy class, but a conjugacy class can contain several homotopy classes if $\pi_1(\mathcal{M})$ is non-Abelian.

I remarked previously that there is an isomorphism between fundamental groups with different base points, but that this isomorphism is not unique if the fundamental group is non-Abelian; i.e., the mapping of homotopy classes in $\pi_1(\mathcal{M},x_0)$ to homotopy classes in $\pi_1(\mathcal{M},y_0)$ may depend on the choice of the path connecting x_0 and y_0. However, this ambiguity is entirely within a conjugacy class; elements of a given conjugacy class at x_0 are always mapped to the same conjugacy class at y_0, regardless of the choice of path.

Another issue to be addressed is gauge invariance. We have seen that to obtain a finite energy vortex we need to work in a gauge theory. This means that many

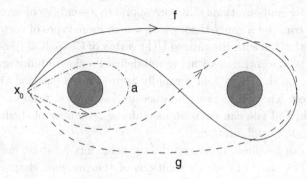

Fig. 4.3. Loops f and g are not homotopic. However, they are freely homotopic, because g is homotopic to afa^{-1}.

different field configurations can represent the same physical vortex. How do we know that all of these correspond to the same conjugacy class?

Let $\phi_1(r,\theta)$ and $\phi_2(r,\theta)$ be two such configurations, so that

$$\phi_2(r,\theta) = g(r,\theta)\phi_1(r,\theta), \tag{4.17}$$

with the gauge transformation $g(r,\theta)$ being a smooth function. Now define a gauge transformation $k(s)$ that is a continuous function of s such that $k(0) = g^{-1}(r=0)$ and $k(1) = I$; the connectedness of the gauge group guarantees that such a function exists. Then [50]

$$F(s,\theta) = g\left(\frac{s}{1-s},\theta\right) k(s)\phi_1(\infty,\theta) \tag{4.18}$$

is a smooth function of s and θ with the property that

$$\begin{aligned} F(0,\theta) &= \phi_1(\infty,\theta), \\ F(1,\theta) &= \phi_2(\infty,\theta). \end{aligned} \tag{4.19}$$

Hence, $\phi_1(\infty,\theta)$ and $\phi_2(\infty,\theta)$ are freely homotopic, and our two gauge-equivalent configurations correspond to the same conjugacy class.

Let us briefly summarize where we are at this point. Every finite energy configuration can be assigned to a conjugacy class according to the behavior of the scalar field at spatial infinity, with this assignment being invariant under nonsingular gauge transformations. Configurations within the same conjugacy class can be smoothly deformed into each other, but those in different classes cannot be. If there is only a single conjugacy class, there are no topologically stable vortices. If there is more than one conjugacy class, then a configuration that minimizes the energy within a nontrivial class (i.e., a class that does not contain the trivial constant configuration) gives a topologically stable vortex solution.

This connects the existence of topological vortices to the existence of a fundamental group with more than one element. However, we have not yet connected the group structure of $\pi_1(\mathcal{M})$ to the physical properties of vortices. To do this, we must consider configurations that correspond to assemblies of several vortices.

To this end, consider a gauge theory with one or more types of vortex solutions. Let us assume that, as with the gauged $U(1)$ vortex of Chap. 3, the energy density of each vortex is concentrated within a well-defined region of finite area, and that outside this region the Higgs field ϕ rapidly approaches a (spatially nonuniform) vacuum solution. The interactions between well-separated vortices will then be relatively weak, and one can envision assembling a number of them together to form a multivortex configuration.

Before this can be done, their asymptotic behaviors must be made compatible. For example, the $U(1)$ vortex solutions of the previous chapter cannot be smoothly joined when each is written in a gauge where they take the form of Eq. (3.32). However, they can be combined if they are first gauge-transformed into a form, such as that in Fig. 3.3b, where ϕ has a fixed phase θ_0 outside a wedge containing the vortex core.

More generally, consider a theory with vortices corresponding to nontrivial elements of $\pi_1(\mathcal{M})$. Let us assume that as one traverses a large circle enclosing vortex 1 the field ϕ traces out a loop in the vacuum manifold $\mathcal{M}$ that begins and ends at ϕ_1, that going around a circle enclosing vortex 2 the field traces out a loop beginning and ending at ϕ_2, and so on. In the U(1) theory, where $\mathcal{M}$ is a circle, one can choose the starting points of these loops so that the ϕ_j are all the same. This is not true in general, since for a larger $\mathcal{M}$ the various loops need not have any values of ϕ in common. However, as long as $\mathcal{M}$ is connected, the various ϕ_j can each be smoothly connected by a path f_j in $\mathcal{M}$ to a common value ϕ_0.

We can then use the following prescription to construct a multivortex configuration. To begin, deform each vortex so that ϕ lies on the vacuum manifold outside a circle of finite radius surrounding the vortex core. Next, gauge-transform the field of vortex j so that $\phi = \phi_j$ everywhere outside a wedge-shaped region that contains the vortex core. Next, surround this wedge with a larger wedge and gauge transform ϕ so that it smoothly varies along the path f_j from ϕ_j to ϕ_0 as one goes from the inner wedge to the outer wedge. Outside the outer wedge for each vortex $\phi = \phi_0$, so there is no problem in assembling the wedges into a smooth configuration

Figure 4.4 illustrates this for the case of three wedges. As one moves from the point A_j to B_j, the field traces out a loop in $\mathcal{M}$ corresponding to an element $[h_j]$ of $\pi_1(\mathcal{M}, \phi_j)$. Going along the path $C_j A_j B_j D_j$ then gives the corresponding element $[h_j]$ of $\pi_1(\mathcal{M}, \phi_0)$. Traversing the full loop indicated in the figure gives the product

$$[h_{\text{tot}}] = [h_1] \circ [h_2] \circ [h_3]. \tag{4.20}$$

For the U(1) theory of the previous chapter, with $\pi(\mathcal{M}) = Z$, this product is just the addition of the vorticities.

There were two points in this construction where somewhat arbitrary choices were made. One of these was the choice of the path f_j in $\mathcal{M}$ linking ϕ_j and ϕ_0, thus defining a map between the homotopy groups with different base points. As noted previously, such mappings may not be unique, and another choice for this path could have yielded a different element of $\pi_1(\mathcal{M}, \phi_0)$, although one that was in the same conjugacy class.

The second arbitrary choice was in the orientation of the wedges for the vortices. For example, if the wedge for vortex 2 in Fig. 4.4 had been oriented downward, Eq. (4.20) would have been replaced by

$$[h_{\text{tot}}] = [h_1] \circ [h_3] \circ [h_2]. \tag{4.21}$$

Neither of these ambiguities is of any consequence if the homotopy group is Abelian. The mapping between $\pi_1(\mathcal{M}, \phi_1)$ and $\pi_1(\mathcal{M}, \phi_0)$ is then path-independent and unique, and the products in Eqs. (4.20) and (4.21) are equal.

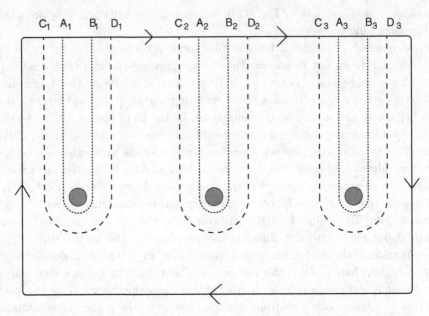

Fig. 4.4. The patching together of three vortex solutions, as described in the text. The shaded circles, running left to right, represent the cores of vortices 1, 2, and 3. Along the dotted lines the scalar field is equal to ϕ_1, ϕ_2, and ϕ_3, respectively. Along the dashed lines, and in the region outside these dashed lines, the field is equal to ϕ_0.

The elements of $\pi_1(\mathcal{M})$ define topological charges for the vortices that combine unambiguously in multivortex configurations. The possibilities when $\pi_1(\mathcal{M})$ is non-Abelian are more complex and will not be described in detail here. A detailed discussion of these can be found in [49].

4.5 Some illustrative vortex examples

It may be useful to illustrate the results of the previous sections with some examples.

(i) *U(1) = SO(2) symmetry broken by a complex scalar field*: This is the case studied in detail in Chap. 3. The vacuum expectation value of the Higgs field is of the form $\langle\phi\rangle = ve^{i\alpha}$, with $0 \leq \alpha < 2\pi$, so the vacuum manifold $\mathcal{M} = S^1$, and $\pi_1(\mathcal{M}) = \pi_1(S^1) = Z$. [Alternatively, we could make use of Eq. (4.16) by taking the gauge group G to be the covering group of U(1), which is the additive group of the real numbers, and the unbroken subgroup H to be the additive group of the integers.] The elements of $\pi_1(\mathcal{M})$ can be identified with the vorticity defined by Eq. (3.4). This was normalized so that it takes on integer values, and is additive when vortices are combined.

(ii) *SO(N) symmetry broken to SO(N − 1)*: This can be achieved with a scalar field ϕ that transforms under the vector representation of SO(N) and has a vacuum expectation value that satisfies an equation of the form $\sum_{a=1}^{N} \phi_a^2 = v^2$. This defines an $(N − 1)$-sphere, and so

$$\mathcal{M} = SO(N)/SO(N − 1) = S^{N-1}. \tag{4.22}$$

For $N = 2$ we have the previous example, but for $N \geq 3$ we have $\pi_1(\mathcal{M}) = 0$, the trivial group, and there are no topologically stable vortices. This can also be seen by examining Fig. 3.1a. For $N = 2$, the arrows are confined to the plane of the paper and cannot be smoothly rotated to be all parallel. For $N = 3$, these arrows live in three dimensions and can be rotated to be perpendicular to the plane of the paper, giving a topologically trivial configuration that can be smoothly deformed into a pure vacuum.

(iii) Z_2 *vortices*: Consider an SO(3) gauge theory with two scalar fields, ϕ and χ, that transform as SO(3) vectors governed by the scalar field potential

$$V(\phi, \chi) = \frac{\lambda_\phi}{4}(\phi^2 − v_\phi^2)^2 + \frac{\lambda_\chi}{4}(\chi^2 − v_\chi^2)^2 + g(\phi \cdot \chi)^2. \tag{4.23}$$

The sign of g determines the relative orientation of ϕ and χ at the minima of V. If it is negative, these must be parallel, and the symmetry is broken to U(1). If instead $g > 0$, ϕ and χ must be orthogonal vectors of lengths v_ϕ and v_χ, respectively, and the SO(3) symmetry is completely broken. Hence, $\mathcal{M} = G/H = SO(3)$, and $\pi_1(\mathcal{M}) = Z_2$.[5] We thus have Z_2 vortices, with the property that a combination of two vortices is topologically trivial and can be smoothly deformed to the vacuum.

In this latter case we can write down a rotationally invariant ansatz,

$$\phi = v_\phi (0, 0, 1),$$
$$\chi = v_\chi f(r) (\cos \theta, \sin \theta, 0),$$
$$\mathbf{A}_j = \epsilon_{jk} \hat{x}^k \frac{a(r)}{r} (0, 0, 1), \tag{4.24}$$

in which ϕ is constant in space while χ and $\mathbf{A}_j$ are essentially embeddings of the U(1) Higgs and gauge fields into the larger group. We could just as easily use an ansatz in which χ was constant and ϕ was the field whose orientation was twisted from point to point. It is easy to see that there are solutions of both types (and possibly others, as well). Stability is a more difficult issue, and a detailed study would be needed to see whether the heavier of these two solutions is stable under small perturbations against decay to a lighter vortex.

---

[5] Alternatively, we could take G to be the covering group SU(2), in which case H would be the Z_2 subgroup composed of the two elements I and −I. The result for $\pi_1(\mathcal{M})$ would then follow from Eq. (4.16).

Now consider the ansatz

$$\phi = v_\phi\,(0,0,1),$$
$$\chi = v_\chi f(r)\,(\cos n\theta, \sin n\theta, 0),$$
$$\mathbf{A}_j = \epsilon_{jk}\hat{x}^k \frac{a(r)}{r}\,(0,0,1), \tag{4.25}$$

that corresponds to an embedding of a vorticity n solution of the U(1) theory. Because we have a Z_2 topological charge, this solution has vanishing topological charge if n is even, and unit charge if n is odd. In particular, the $n = -1$ antivortex solution is topologically equivalent to the vortex solution of Eq. (4.24). But let us suppose that the two Higgs fields have very different energy scales, with $v_\phi \gg v_\chi$, while $\lambda_\phi \sim \lambda_\chi$. Because $\langle\phi\rangle$ breaks SO(3) to U(1), at energies well below v_ϕ one would appear to have a U(1) gauge theory with a single complex Higgs field formed from the components of χ orthogonal to $\langle\phi\rangle$. That theory would have vortices with ordinary integer topological charge, and solutions with different values of n could not be deformed into one another. The vortex and the antivortex would not be equivalent.

These two points of view are reconciled by considering the homotopy that takes an $n = 2$ SO(3) configuration to the vacuum. In the course of this homotopy, the untwisting of the χ field must be accompanied, at intermediate stages, by a twisting of the ϕ field. At large distances, where the covariant derivatives vanish, this can be accomplished by a gauge transformation and is unproblematic. However, within the vortex core there is gradient energy associated with the twisting of ϕ. Because the scale of this energy is set by v_ϕ, there is a large potential energy barrier that must be traversed. As a result, the $n = 2$ configuration, although not topologically stable, is dynamically protected from decay into the vacuum sector.[6]

This illustrates that while topology is an important guide, it is not the whole story.

(iv) *Weinberg–Salam theory and semilocal vortices*: The electroweak interactions of the Standard Model have a gauged SU(2) symmetry, with coupling constant g, and a gauged U(1) symmetry, with coupling constant g'. The Higgs field is a complex doublet ϕ. The minima of $V(\phi)$ occur when $\phi^\dagger\phi = v^2/2$, where $v \neq 0$ is determined by the Higgs field mass and self-coupling. Writing $\phi = (\phi_1, \phi_2)^t$, we have

$$(\mathrm{Re}\,\phi_1)^2 + (\mathrm{Im}\,\phi_1)^2 + (\mathrm{Re}\,\phi_2)^2 + (\mathrm{Im}\,\phi_2)^2 = \frac{v^2}{2}. \tag{4.26}$$

This is the equation for a three-sphere. Since $\pi_1(S^3) = 0$, there are no topological vortices in this theory.

---

[6] The protection against decay is only absolute at the classical level. The decay could proceed quantum mechanically via tunneling through the potential energy barrier, although the rate for this would be exponentially small.

However, there are some interesting related possibilities. First, consider a similar theory, but with vanishing SU(2) coupling g. There would then be a local U(1) symmetry, but only a global SU(2) symmetry. As we have seen, finiteness of the energy requires that the covariant derivative of the scalar field falls sufficiently rapidly at large distance. This could be achieved if the long-distance twisting of ϕ were entirely within the gauged U(1) group, but not if ϕ were rotated by elements of the SU(2), because the effects of the gradients could not be compensated by a coupling to an SU(2) gauge field. Consequently, a vortex configuration with

$$\phi = \begin{pmatrix} 0 \\ f(r)e^{i\theta} \end{pmatrix} \tag{4.27}$$

and an appropriate U(1) gauge field might actually be stable, because the homotopy connecting the asymptotic field to a uniform vacuum configuration would pass through configurations with SU(2) twisting of ϕ, and thus infinite energy. Configurations such as these have been termed *semilocal vortices* [51].

However, depending on the parameters of the theory, a configuration such as that in Eq. (4.27) could be unstable even without unwinding. The usual topological argument for the stability of such a configuration uses the fact that while avoiding a singularity at the origin forces $f(0)$ to vanish, the need to minimize the energy from the potential forces f to rapidly approach v as r increases. Now suppose that instead of setting $\phi_1 = 0$ we consider configurations of the form

$$\phi = \begin{pmatrix} g(r) \\ f(r)e^{i\theta} \end{pmatrix} \tag{4.28}$$

with g not necessarily vanishing. The potential is minimized if $f^2 + g^2 = v^2$, so there is not necessarily any energetic penalty in letting f be small, or even vanishing, as long as there is a compensating increase in g. Hence, a configuration with nonzero vorticity in f might be unstable against the transformation of the point zero at the origin into an ever increasing region of vanishing f. At spatial infinity the measured vorticity would remain nonzero, but it would eventually become unobservable at any finite r. A detailed numerical analysis shows that such an instability is in fact present if the scalar self-coupling is large compared to the U(1) gauge coupling. On the other hand, for small scalar self-coupling this instability is absent and semilocal vortices can exist [52–54].

Still focusing on this latter case, let us restore the SU(2) gauge coupling, but with $g \ll g'$. Although there is no longer an infinite energy barrier preventing the unwinding of the field of Eq. (4.27), one might well expect there to be a finite barrier that would preserve the semilocal vortex. This has been confirmed by numerical analysis. However, this analysis also shows that this barrier disappears before g reaches the observed electroweak value. Hence, the SU(2)×U(1) electroweak theory does not have stable vortex solutions [55–57].

(v) *Alice strings*: Consider an SO(3) gauge theory with a Higgs field transforming according to the five-dimensional irreducible representation [58]. This field can be represented as a traceless symmetric 3×3 matrix that transforms as

$$\phi \rightarrow R\phi R^t, \qquad (4.29)$$

where R is an SO(3) rotation matrix. Now let $V(\phi)$ be such that it is minimized when two of the eigenvalues of ϕ are equal, so that ϕ is of the form

$$\phi = R \begin{pmatrix} a & 0 & 0 \\ 0 & a & 0 \\ 0 & 0 & -2a \end{pmatrix} R^t, \qquad (4.30)$$

or, equivalently,

$$\phi = a(I - 3\mathbf{e}\mathbf{e}^t), \qquad (4.31)$$

with $\mathbf{e}$ a real unit three-vector. Thus each possible vacuum is specified by a unit vector $\mathbf{e}$, but with the caveat that $-\mathbf{e}$ and $\mathbf{e}$ specify the same vacuum. To identify the vacuum manifold $\mathcal{M}$, we first note that the space of unit three-vectors is the unit two-sphere. Making $\mathbf{e}$ and $-\mathbf{e}$ equivalent corresponds to identifying antipodal points on the two-sphere, giving a space that is the same as a disk (a circle and the area it encloses) with antipodal points on its boundary identified. A closed loop on this space can include a jump from one point on the boundary to the antipodal point, as shown in Fig. 4.5. By continuous deformations of the path (e.g., by bringing $B \sim B'$ and $C \sim C'$ together and then moving the combined point away from the disk boundary) one can subtract or add additional jumps, but only in pairs. Thus, there are two homotopy classes of loops: those which include an odd number of such jumps, and those with either an even number or no jump at all. The homotopy group $\pi_1(\mathcal{M})$ is therefore Z_2. [A similar argument gives an alternative demonstration that the fundamental group of SO(3) is Z_2.]

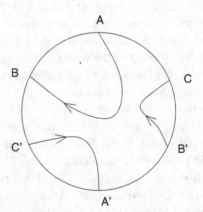

Fig. 4.5. A closed loop on a disk with antipodal points identified. As drawn, the loop has three "jumps". If B′ and C (and thus B and C′) are brought together, two of these can be eliminated. However, it is never possible to subtract or add an odd number of jumps.

We can identify the unbroken gauge group by considering the vacuum with

$$\phi = \phi_0 \equiv \begin{pmatrix} a & 0 & 0 \\ 0 & a & 0 \\ 0 & 0 & -2a \end{pmatrix} \equiv a(I - \mathbf{e}_0\mathbf{e}_0^t). \tag{4.32}$$

The unbroken subgroup H clearly includes the U(1) subgroup consisting of rotations about $\mathbf{e}_0$; i.e., about the 3-axis. However, it also includes another component, disconnected from the first, consisting of rotations by π about any axis perpendicular to the 3-axis. Because these rotations by π do not commute with the elements of the U(1), H is not the direct product U(1)$\times Z_2$, but rather the semidirect product group U(1)$\rtimes Z_2 = $ Pin(2).

Now consider a vortex solution in which ϕ at large distances is of the form

$$\begin{aligned} \phi(r,\theta) &\approx R(\theta)\phi_0 R(\theta)^{-1} \\ &= a[I - 3\mathbf{e}(\theta)\mathbf{e}(\theta)^t], \end{aligned} \tag{4.33}$$

where $\mathbf{e}(\theta) = R(\theta)\mathbf{e}_0$. This will be topologically stable if $R(\theta)$ traces out an uncontractible loop as θ ranges from 0 to 2π; i.e., if $\mathbf{e}(2\pi) = -\mathbf{e}(0)$. At each value of θ the unbroken U(1) subgroup is the set of rotations about $\mathbf{e}(\theta)$, generated by a charge operator $Q(\theta)$. If one makes a full circuit of the vortex, the result is a reversal of $\mathbf{e}$, and thus a change in the sign of Q. Hence, the U(1) charges are double-valued and the sign of a particle's charge is ambiguous. This can be represented by imagining a branch cut starting at the vortex and either ending at another vortex or running out to infinity, and saying that a particle's charge changes sign when the particle crosses the branch cut. The exact location of the branch cut is not physically meaningful and not gauge invariant, but the existence of a cut is. The vortices (or rather, the strings that are their extension into three dimensions) that give rise to this phenomenon have been termed "Alice strings", in an allusion to Lewis Carroll's novel [58].

Even though the sign of a charge is not gauge-invariant, it is physically meaningful to ask whether a pair of particles have the same or opposite charges (e.g., the two particles cannot annihilate if they have the same charge, but might be able to if they have opposite charges). However, suppose that we have an Alice string in three spatial dimensions that closes on itself to make a circular loop. The branch cut in two spatial dimensions becomes a surface in three dimensions that is bounded by the string loop, but whose location is otherwise arbitrary. Now start with two particles with equal charges and take one of them along a path that goes around the string loop and then returns to its original position. This reverses the sign of the particle's charge, so it would seem that the two particles have opposite sign, and so zero net charge. On the other hand, the electric flux through a Gaussian surface enclosing both the string loop and the two particles would be unchanged, as should be expected from charge conservation. It is tempting to try to reconcile these facts by saying that the U(1) charge of the particle was transferred to the string loop when the particle crossed the surface of branch cuts. However, the time of this transfer would depend on the arbitrary choice for locating the surface. Instead, we are led to the conclusion that we can assign a physically meaningful gauge-invariant charge to the entire system comprising the two particles and the loop, but that this charge cannot be localized within the system in a gauge-invariant manner. In another allusion to Carroll, such charges have been termed "Cheshire charges" [59].

For further discussion of Alice strings and Cheshire charges, see [60–62].

(vi)   *A non-Abelian fundamental group*: Consider the same SO(3) gauge theory as in the previous example, but with a potential that is minimized when the eigenvalues of ϕ are all distinct, so that

$$\phi = R \begin{pmatrix} a_1 & 0 & 0 \\ 0 & a_2 & 0 \\ 0 & 0 & a_3 \end{pmatrix} R^t, \qquad (4.34)$$

with $a_1 \neq a_2 \neq a_3$. This is invariant only under the identity or rotations by π about either the x-, the y-, or the z-axis, so the unbroken subgroup of SO(3) is now a discrete group with four elements.

For determining $\pi_1(\mathcal{M})$ it is actually more convenient to take the original gauge group G to be the covering group, SU(2), so that we can make use of Eq. (4.16). This doubles the size of the unbroken group H, which now has eight elements, corresponding to the SU(2) matrices $\pm I$, $\pm \sigma_x$, $\pm \sigma_y$, and $\pm \sigma_z$. These form the quaternion group, Q, which is clearly non-Abelian. Because $H = Q$ is discrete, Eq. (4.16) gives

$$\pi_1(G/H) = \pi_0(H) = Q \qquad (4.35)$$

and we have a non-Abelian fundamental group, as promised. Vortices in this theory are discussed in [63].

4.6 Higher homotopy groups

There is a natural generalization of the fundamental group. The latter classifies closed loops, which are maps from a circle, S^1, to a given manifold. The higher homotopy groups, $\pi_n(\mathcal{M})$, classify maps from an n-sphere, S^n, to the manifold.

Let us first consider the second homotopy group, $\pi_2(\mathcal{M}, x_0)$. Although the base point x_0 is explicitly indicated, we will see below that, as with π_1, the group is independent of the choice of base point. This is the natural object for classifying particle-like solitons in three spatial dimensions, where spatial infinity is a two-sphere. (The line-like strings that are the extensions of vortices to three dimensions are still classified by π_1.)

We start by mapping the two-sphere to a finite region of the plane. One way to visualize this is to imagine puncturing a hole in the surface of a balloon,

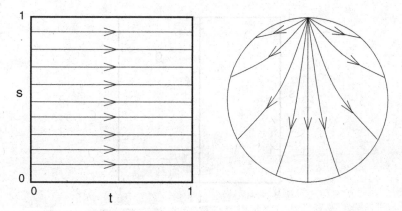

Fig. 4.6. The mapping from a square to a sphere. The entire perimeter of the square is mapped onto the north pole of the sphere, and the horizontal lines of constant s are mapped onto the curves shown on the sphere. As one goes along each of these curves, t runs from 0, at the north pole, to $1/2$, at the edge of the diagram, and then to 1 as the curve returns to the north pole on the (hidden) opposite side of the sphere.

and then stretching and flattening out this surface until the infinitesimal circle surrounding the puncture becomes the perimeter of a square. This square can be covered by coordinates s and t, with $0 \leq s, t \leq 1$, as illustrated in Fig. 4.6. A smooth function on the sphere becomes a smooth function of s and t with the additional constraint that it has the same value everywhere on the perimeter; i.e., $f(0,t) = f(1,t) = f(s,0) = f(s,1)$. A convenient mapping of these coordinates back to the sphere is shown in Fig. 4.6. Here s labels the loops, with $s = 0$ and $s = 1$ being degenerate loops that reduce to a point and $s = 1/2$ being the largest loop. The distance along each loop is parameterized by t.

Homotopy is defined by generalizing Eq. (4.2). Let $f(s,t)$ and $g(s,t)$ be two maps, both of which are equal to x_0 everywhere on the perimeter of the square. These are homotopic if there is a function $k(s,t,u)$ with $0 \leq s, t, u \leq 1$ such that

$$k(s,t,0) = f(s,t),$$
$$k(s,t,1) = g(s,t),$$
$$k(s,0,u) = k(s,1,u) = k(0,t,u) = k(1,t,u) = x_0. \tag{4.36}$$

Just as before, this relation defines a set of homotopy classes.

The product of two maps can be defined by

$$(f \circ g)(s,t) = \begin{cases} f(s,2t), & 0 \leq t \leq 1/2, \\ g(s,2t-1), & 1/2 \leq t \leq 1. \end{cases} \tag{4.37}$$

This is illustrated graphically in Fig. 4.7. As was true with loops, this product can be carried over to homotopy classes. We want to show that these

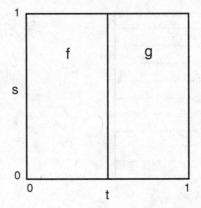

Fig. 4.7. Diagrammatic illustration of the product of two maps $f(s,t)$ and $g(s,t)$. The functions are equal to x_0 everywhere on the heavy solid lines.

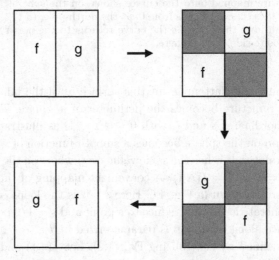

Fig. 4.8. A demonstration that π_2 is always Abelian. The functions are equal to x_0 everywhere on the heavy solid lines and in the dark shaded regions. The transformations indicated by the arrows are all smooth deformations of one map into the next.

homotopy classes with this product form a group. Associativity is straightforward to demonstrate. The identity element is the homotopy class containing the constant map $f_0(s,t) = x_0$. The existence of an inverse follows from the observation that $f(s,t) \circ f(s,-t)$ is homotopic to the identity map. Thus, we have a based homotopy group $\pi_2(\mathcal{M}, x_0)$.

However, note the series of maps shown in Fig. 4.8. Each is clearly homotopic to the previous one, with the net result being that $f \circ g$ and $g \circ f$ are homotopic. Thus, in contrast with the first homotopy group, π_2 is always Abelian.

As with the first homotopy group, $\pi_2(\mathcal{M}, x_0)$ does not depend on the choice of x_0, provided that $\mathcal{M}$ is a connected manifold. In the case of π_1, this could be seen by considering a path connecting two possible base points. For π_2, this path must be inflated to become a thin tube, but the demonstration is otherwise analogous. We can therefore drop the reference to the base point, and simply write $\pi_2(\mathcal{M})$.

Nevertheless, just as with π_1, the isomorphism between $\pi_2(\mathcal{M}, x_0)$ and $\pi_2(\mathcal{M}, y_0)$ can depend on the path of the tube from x_0 to y_0. If two such paths can be combined to form a noncontractible loop from x_0 to y_0 and back to x_0, then there will be maps that can be deformed into each other if the base point condition is removed, but not otherwise. Thus, these maps will be freely homotopic, but not homotopic. This situation can arise even if the first homotopy group of $\mathcal{M}$ is Abelian, although it is necessary that this group be nontrivial [48].

When ambiguities of this sort are absent, three-dimensional particle-like solitons can be associated with a unique homotopy class, with an associated additive topological charge. On the other hand, if there is an ambiguity, the soliton is associated with a set of homotopy classes, and there are subtleties in defining the topological charges of multisoliton configurations. This is seen, for example, in configurations containing both magnetic monopoles and Alice strings [61].

This discussion of π_2 can be carried over, with the obvious generalizations, to the π_n with $n \geq 3$. Like π_2, these higher homotopy groups are always Abelian. In particular, we will encounter π_3 in our study of Yang–Mills instantons.

4.7 Some results for higher homotopy groups

In our later discussions we will need a number of results concerning the homotopy groups of spheres. Let us begin with $\pi_n(S^n)$, which classifies maps from one n-sphere to another. For $n = 1$, we saw previously that $\pi_1(S^1) = Z$, with the integer associated with a map being the winding number N that counts the number of times that the first S^1 winds around the second. If the mapping at a point is counted with a positive or negative sign according to whether the two circles are being traversed in the same or opposite directions, respectively, then the total count at every point on the target circle is N. Explicitly, if θ denotes the angle on the first circle, and $\alpha(\theta)$ the angle to which this is mapped on the second circle, the winding number that was defined in Eq. (3.4) can be written as

$$N = \frac{1}{2\pi} \int_0^{2\pi} d\theta \, \frac{d\alpha}{d\theta}. \tag{4.38}$$

A winding number can be defined in an exactly analogous fashion for maps from S^2 to S^2. Using the standard spherical coordinates θ and ϕ for the first sphere and $\alpha(\theta, \phi)$ and $\beta(\theta, \phi)$ for the second, we can define the winding number

$$N = \frac{1}{4\pi} \int d^2\Omega \, \frac{\sin\alpha}{\sin\theta} \left(\frac{d\alpha}{d\theta} \frac{d\beta}{d\phi} - \frac{d\beta}{d\theta} \frac{d\alpha}{d\phi} \right)$$

$$= \frac{1}{4\pi} \int d\theta \, d\phi \, \sin\alpha \left(\frac{d\alpha}{d\theta} \frac{d\beta}{d\phi} - \frac{d\beta}{d\theta} \frac{d\alpha}{d\phi} \right). \tag{4.39}$$

It is useful to rewrite this in terms of Cartesian components. Let us define the unit vector

$$\hat{\mathbf{e}}(\mathbf{r}) = (\sin\alpha\cos\beta, \sin\alpha\sin\beta, \cos\alpha), \tag{4.40}$$

with $\alpha(\mathbf{r})$ and $\beta(\mathbf{r})$ functions of three-dimensional Euclidean space. Converting Eq. (4.39) to Cartesian coordinates, we find that the winding number of $\hat{\mathbf{e}}(\mathbf{r})$ on a sphere of fixed radius is

$$
\begin{aligned}
N &= \frac{1}{8\pi}\epsilon^{ijk}\epsilon^{abc}\int dS^i\,\hat{e}^a\,(\partial_j\hat{e})^b\,(\partial_k\hat{e})^c \\
&= \frac{1}{8\pi}\epsilon^{ijk}\int dS^i\,\hat{\mathbf{e}}\cdot\partial_j\hat{\mathbf{e}}\times\partial_k\hat{\mathbf{e}},
\end{aligned}
\tag{4.41}
$$

where dS^i is the surface element on the sphere.

This integral is invariant under smooth variation of $\hat{\mathbf{e}}$. To see this, consider the variation $\hat{\mathbf{e}}\to\hat{\mathbf{e}}' = \hat{\mathbf{e}}+\mathbf{v}$ where $\mathbf{v}$ is infinitesimal. Because $\hat{\mathbf{e}}$ and $\hat{\mathbf{e}}'$ are both unit vectors, we must require that $\mathbf{v}$, like $\partial_k\hat{\mathbf{e}}$, be orthogonal to $\hat{\mathbf{e}}$. To first order in $\mathbf{v}$, the change in N is

$$
\begin{aligned}
\delta N &= \frac{1}{8\pi}\epsilon^{ijk}\int dS^i\,(2\hat{\mathbf{e}}\cdot\partial_j\hat{\mathbf{e}}\times\partial_k\mathbf{v} + \mathbf{v}\cdot\partial_j\hat{\mathbf{e}}\times\partial_k\hat{\mathbf{e}}) \\
&= \frac{1}{4\pi}\epsilon^{ijk}\int dS^i\,\hat{\mathbf{e}}\cdot\partial_j\hat{\mathbf{e}}\times\partial_k\mathbf{v} \\
&= \frac{1}{4\pi}\epsilon^{ijk}\int dS^i\,\partial_j\,(\hat{\mathbf{e}}\cdot\mathbf{v}\times\partial_k\hat{\mathbf{e}}) \\
&= 0.
\end{aligned}
\tag{4.42}
$$

The last term on the first line vanishes because the three factors are all orthogonal to $\hat{\mathbf{e}}$, and the last line follows from the fact that the integration is over a surface without boundary.

Furthermore, N is invariant under perturbations of the integration surface, as long as $\hat{\mathbf{e}}$ remains well defined. Hence, if we have a field $\phi(\mathbf{r})$ that transforms as an SO(3) vector and define $\hat{\mathbf{e}} = \phi/|\phi|$, the winding number of $\hat{\mathbf{e}}$ over a surface is invariant under deformations of the surface that do not take it through a zero of ϕ. Arguments analogous to those for the vortex case then show that N is equal to the total number of zeros of ϕ in the region enclosed by the surface of integration, with each zero being counted with a plus or minus sign according to the sign of the winding on an infinitesimal sphere enclosing the zero.

These results are readily generalized to arbitrary n, so we have

$$\pi_n(S^n) = Z. \tag{4.43}$$

There is a second result that generalizes from the $n = 1$ case. By removing a point from S^n to obtain R^n, and noting that R^n is simply connected, we showed that all maps from S^1 into S^n with $n > 1$ are contractible. Essentially the same argument shows that maps from any sphere to a higher-dimensional sphere are contractible; i.e., that

$$\pi_k(S^n) = 0, \qquad n > k. \tag{4.44}$$

The situation for maps from a larger sphere to a smaller one is less simple. For all $k > 1$, $\pi_k(S^1) = 0$, but for $1 < n < k$ the group $\pi_k(S^n)$ is nontrivial, with a more complicated dependence on k and n.

We will also make use of some results concerning higher homotopy groups of Lie groups. First, a theorem due to Cartan [64] implies that

$$\pi_2(G) = 0, \quad G \text{ compact, connected,} \tag{4.45}$$

for any compact connected Lie group G. Next, recall that the group SU(2) is topologically equivalent to a three-sphere. Setting $n = 3$ in Eq. (4.43), we have

$$\pi_3(SU(2)) = Z. \tag{4.46}$$

SU(2) is a subgroup of every compact simple Lie group G, so every topologically nontrivial map from the three-sphere to SU(2) can also be viewed as a map from the three-sphere to G. Because G is a larger manifold, one might wonder if this map could be untwisted in G, even though it could not be untwisted in SU(2). The answer turns out to be no and we have

$$\pi_3(G) = Z, \quad G \text{ compact, connected, and simple.} \tag{4.47}$$

Finally, there is an important theorem concerning $\pi_2(G/H)$ that will prove to be very useful in our study of monopoles in the next chapter. It is analogous to the theorem for the fundamental group that was given in Eq. (4.16). One route [50] to obtaining this result is as follows. Let ϕ be a scalar field transforming under a gauge group G, and let H be the subgroup that leaves a particular vacuum value, ϕ_0, invariant. We are interested in maps from a two-sphere to the manifold of vacua, G/H. Any such map is given by a function $\phi(s,t)$, with $\phi = \phi_0$ when either s or t is equal to 0 or 1. Since all vacuum values of ϕ are related by gauge transformations, we can write

$$\phi(s,t) = g(s,t)\phi_0, \tag{4.48}$$

where $g(s,t)$ is an element (or rather a representation of an element) of G. However, $g(s,t)$ is not uniquely determined, since $g(s,t)h(s,t)\phi_0 = g(s,t)\phi_0$ for any element $h(s,t)$ of H.

Now consider the parameterization of the sphere given by Fig. 4.6. We can arbitrarily set $g = I$ at the beginning of each loop; i.e., $g(s,0) = I$. Also, because the $s = 0$ and $s = 1$ loops are simply points, $g(0,t) = g(1,t) = I$. Thus, $g = I$ on three sides of the square. On the fourth side, all that we know is that $g(s,1)\phi_0 = \phi_0$, which implies that

$$g(s,1) \equiv h(s) \tag{4.49}$$

is an element of H. The previous conditions imply that $h(0) = h(1) = I$, so $h(s)$ defines a loop in H. We thus have a correspondence between a map $\phi(s,t)$ from S^2 to G/H and a map $h(s)$ from S^1 to H.

Now suppose that there are two such maps, $\phi_1(s,t)$ and $\phi_2(s,t)$, that yield the same loop $h(s)$. If we write

$$\phi_1(s,t) = g_1(s,t)\phi_0,$$
$$\phi_2(s,t) = g_2(s,t)\phi_0, \tag{4.50}$$

then $g_1 = g_2$ everywhere on the boundary of the square. Hence, $g_3(s,t) \equiv g_2^{-1}(s,t)g_1(s,t)$ is equal to the identity everywhere on the boundary, and thus gives a map from the two-sphere to G. If this map is homotopic to the identity map, $g(s,t) = 1$, then ϕ_1 and ϕ_2 will be homotopic to each other.

However, we saw above that $\pi_2(G) = 0$ for any compact connected Lie group G. Therefore any map from the two-sphere to G, including $g_3(s,t)$, is homotopic to the identity map, and hence any two maps $\phi_1(s,t)$ and $\phi_2(s,t)$ that yield the same $h(s)$ are homotopic to each other. This clearly extends to the case where ϕ_1 and ϕ_2 yield different, but homotopic, loops $h_1(s)$ and $h_2(s)$. Thus, we have shown that there is a one-to-one mapping of elements of $\pi_2(G/H)$ to elements of $\pi_1(H)$.

Does this mapping yield all elements of $\pi_1(H)$? To answer this question, consider the series of loops $g_t(s) \equiv g(s,t)$. We have $g_0(s) = 1$ and $g_1(s) = h(s)$. Because $g(s,t)$ is a smooth function, this gives a homotopy connecting the loops $g_1(s)$ and the trivial loop $g_0(s)$. Hence, $h(s)$, *viewed as a loop in G*, is homotopic to the identity. Conversely, if we are given any $h(s)$ that is homotopic, as a loop in G, to the identity map, that homotopy defines a $g(s,t)$ and a $\phi(s,t)$. Thus, the mapping from $\pi_2(G/H)$ is onto the subgroup of $\pi_1(H)$ that is mapped onto the identity element of $\pi_1(G)$. If $\pi_1(G) = 0$, i.e., if G is simply connected, then all of $\pi_1(H)$ is mapped onto the single element of $\pi_1(G)$. The mapping from $\pi_2(G/H)$ to $\pi_1(H)$ is then both one-to-one and onto, so

$$\pi_2(G/H) = \pi_1(H), \quad G \text{ compact, connected, and simply connected.} \tag{4.51}$$

In the application of this result to the search for solitons in gauge theories G is always compact and connected, so it is only the last condition that is nontrivial. However, this too is always satisfied if G is taken to be the covering group of the Lie algebra, with H then being the appropriate subgroup of this covering group.

5

Magnetic monopoles with U(1) charges

The most important and best studied solitons in three spatial dimensions are magnetic monopoles, which arise when the spontaneous breaking of a gauge group G down to a subgroup H yields a nontrivial $\pi_2(G/H)$. Perhaps the most important of these are the groups that occur in grand unified theories. As we will see, magnetic monopoles are a generic prediction of all such theories.

In this chapter I will focus on the simplest case, that of SU(2) broken to U(1), which was first studied by 't Hooft [65] and Polyakov [2]. Monopoles associated with larger groups will be discussed in the next chapter.

5.1 Magnetic monopoles in electromagnetism

Before examining these solitons in detail, it is perhaps best to first review some basic properties of magnetic monopoles in electrodynamics.

It is a familiar fact that in the absence of sources Maxwell's equations display a duality symmetry between electric and magnetic fields. The four source-free equations can be combined into two complex equations,

$$0 = \boldsymbol{\nabla} \times (\mathbf{E} + i\mathbf{B}) - i\frac{\partial}{\partial t}(\mathbf{E} + i\mathbf{B}),$$

$$0 = \boldsymbol{\nabla} \cdot (\mathbf{E} + i\mathbf{B}), \tag{5.1}$$

that are invariant under the transformation

$$\mathbf{E} + i\mathbf{B} \to \mathbf{E}' + i\mathbf{B}' = e^{i\alpha}(\mathbf{E} + i\mathbf{B}). \tag{5.2}$$

This symmetry is lost when sources are present because only electric charges and currents appear. This naturally suggests the possibility of magnetically charged objects whose charge density and current would restore the symmetry. Such an object, known as a magnetic monopole, would be the source for a Coulomb magnetic field of the form

$$\mathbf{B} = \frac{Q_M}{4\pi r^2}\hat{\mathbf{r}} = \frac{g}{r^2}\hat{\mathbf{r}}. \tag{5.3}$$

In the rationalized natural units used here, Q_M is the magnetic charge, with the factor of 4π included so that its definition parallels that of the electric charge Q_E, which creates a Coulomb electric field

$$\mathbf{E} = \frac{Q_E}{4\pi r^2}\hat{\mathbf{r}}. \tag{5.4}$$

The auxiliary quantity $g = Q_M/4\pi$ has been introduced because it allows some equations to take on a simpler form. It should be noted that this usage is not universal in the literature, with some authors using definitions of electric and magnetic charges that differ by a factor of 4π.

In standard treatments of electrodynamics it is useful to write the electric and magnetic fields in terms of potentials. In particular, the fact that $\mathbf{B}$ is divergence-less means that it can be written as the curl of a vector potential, $\mathbf{B} = \boldsymbol{\nabla} \times \mathbf{A}$. If there are magnetic charges, $\mathbf{B}$ is no longer divergenceless, and so cannot be globally written as a curl. However, if the magnetic charges were point-like, or at least confined to a finite volume, one might try to write $\mathbf{B}$ as a curl in the regions where magnetic charges were absent.

Suppose, for example, that there were a point-like monopole at the origin with magnetic charge $Q_M = 4\pi g$. Let us define a vector potential with Cartesian components

$$A^{\mathrm{I}i} = -\epsilon^{ij3}\,\hat{r}^j\,\frac{g}{z+r} \tag{5.5}$$

or spherical components[1]

$$A_r^{\mathrm{I}} = A_\theta^{\mathrm{I}} = 0, \qquad A_\phi^{\mathrm{I}} = g(\cos\theta - 1). \tag{5.6}$$

This has a singularity, known as a Dirac string, along the negative z-axis. However, away from this singularity the curl of $\mathbf{A}^{\mathrm{I}}$ is equal to the Coulomb field of Eq. (5.3).

The location of the string seems somewhat arbitrary. Indeed, the potential

$$A^{\mathrm{II}i} = -\epsilon^{ij3}\,\hat{r}^j\,\frac{g}{z-r} \tag{5.7}$$

or, equivalently,

$$A_r^{\mathrm{II}} = A_\theta^{\mathrm{II}} = 0, \qquad A_\phi^{\mathrm{II}} = g(\cos\theta + 1), \tag{5.8}$$

that has a Dirac string along the positive z-axis, yields the same magnetic field, apart from singularities along the strings. [Note, for later reference, that $\mathbf{A}^{\mathrm{I}}(\mathbf{r}) = \mathbf{A}^{\mathrm{II}}(-\mathbf{r})$.] The difference between these two potentials,

---

[1] These are defined so that $A_x dx + A_y dy + A_z dz = A_r dr + A_\theta d\theta + A_\phi d\phi$.

$$A^{IIi} - A^{Ii} = g\,\epsilon^{ij3}\,\hat{r}^j\left(\frac{1}{z+r} - \frac{1}{z-r}\right)$$
$$= -\partial_i\left[2g\tan^{-1}(y/x)\right]$$
$$= -\partial_i(2g\phi), \tag{5.9}$$

is a gradient and so corresponds to a gauge transformation, albeit one that is singular along the initial and final strings. In fact, by an appropriate (but singular) gauge transformation the Dirac string can be moved to coincide with an arbitrary curve originating at the origin and running out to spatial infinity. It cannot, however, be gauged away.

In classical electrodynamics only the field strengths are measurable. The potentials are auxiliary quantities that, while useful for performing calculations, cannot be measured. There is therefore no reason to expect a singularity in $\mathbf{A}$ to have any consequences.

The situation is more complicated in the quantum theory, where the vector potential can affect the phase of a wavefunction. In particular, suppose that a particle with electric charge q travels around a closed curve C that encircles a region of nonzero magnetic flux. In the course of a full circuit the particle's wavefunction is multiplied by the gauge-invariant Aharonov–Bohm phase factor

$$U[C] = \exp\left(iq\int_C d\ell \cdot \mathbf{A}\right). \tag{5.10}$$

The particle is thus sensitive to the presence of the magnetic field, even if its wavefunction vanishes in the region where $\mathbf{B}$ is nonzero.

It was pointed out by Dirac [66] that because of this phenomenon the existence of even a single magnetic monopole could have remarkable consequences. Consider a magnetic monopole with its Dirac string oriented along the negative z-axis. Since the location of the string is arbitrary and gauge-dependent, it should not be observable. However, it would appear that the Aharonov–Bohm effect could provide a means to detect the string. Suppose one were to take a particle with electric charge q (and no magnetic charge), around a small closed loop C that encircles the string. The particle's wavefunction would then be multiplied by the phase factor in Eq. (5.10). If there were no string, the integral around the loop would vanish, giving $U = 1$, in the limit in which the loop was contracted to a point. If the string is to be unobservable, we must get the same result when C is contracted to an infinitesimal loop around the string.

The line integral is most easily evaluated using spherical coordinates. Choosing the loop C to be at a constant value of θ, and using Eq. (5.6), we have

$$q\int_C d\ell \cdot \mathbf{A} = q\int d\phi\, A_\phi = 2\pi q g(\cos\theta - 1). \tag{5.11}$$

If C is contracted so that it becomes an infinitesimal curve around the string, $\theta \to \pi$ and the line integral becomes $-4\pi q g$. Inserting this into Eq. (5.10), we see that for the Dirac string to be unobservable we must require that

$$e^{4\pi i q g} = 1 \tag{5.12}$$

or, equivalently, that

$$q g = \frac{n}{2} \tag{5.13}$$

with n an arbitrary integer. This equation, known as the Dirac quantization condition, has far-reaching implications. Since the particle with charge q could have been any electrically charged particle in the universe, Eq. (5.13) must be satisfied by every such charge (although not necessarily with the same integer n). Furthermore, since we would have obtained a similar constraint by considering a loop around the Dirac string of any other magnetic monopole, the quantization condition must be satisfied for every magnetic charge.

Thus, every possible choice of q and g must obey Eq. (5.13). The only way that this can be done is if all such charges are of the form

$$q = k\, q_{\min},$$
$$g = m\, g_{\min}, \tag{5.14}$$

where k and m are integers and

$$q_{\min}\, g_{\min} = \frac{1}{2}. \tag{5.15}$$

Thus, the detection of even a single magnetic monopole would provide an explanation for the observed quantization of electric charge.

In deriving this result I specified that the electrically charged particle did not carry any magnetic charge. If it were a dyon, a particle endowed with both electric and magnetic charges, it would have had a Dirac string of its own, which would get wound around the monopole's string as the dyon was carried around C. If the stationary monopole had only magnetic charge, the result would not be changed. However, there is an additional phase factor if both particles are dyons. It is straightforward to work out the result, but we can guess it already from a symmetry argument. The quantization condition must be invariant under the SO(2) duality transformation of Eq. (5.2), and it must reduce to our previous result if one of the particles has only magnetic charge. The only expression fitting these requirements is [67, 68]

$$q_1 g_2 - q_2 g_1 = \frac{n}{2}. \tag{5.16}$$

This is clearly satisfied if the electric and magnetic charges all obey Eqs. (5.14) and (5.15). However, it is also satisfied if all purely electric particles obey these equations and all magnetically charged particles are dyons with charges of the form

$$g = m\, g_{\min},$$
$$q = k\, q_{\min} + m\alpha, \tag{5.17}$$

where α is a fixed real number and m and k are integers that can vary from dyon to dyon. Thus, it is possible for dyons to have fractional, and even irrational, electric charge. We will see in Chap. 10 that this is actually required in some theories [69].

I have implicitly assumed that the U(1) of electromagnetism is the only unbroken gauge symmetry. If there are other unbroken symmetries, such as the SU(3) of QCD, then there can be additional contributions to the phase factor that must be taken into account. Thus, if the proton charge e were the smallest possible electric charge, Eq. (5.13) would imply that the smallest possible magnetic charge was $1/(2e)$. However, we know that there are quarks with charge $e/3$, which would seem to imply a minimum magnetic charge of $3/(2e)$. This is indeed true for monopoles carrying purely electromagnetic-type magnetic charge. However, if the monopole also carries color magnetic charge, the interaction with the quark's color electric charge can give an additional phase such that the U(1) magnetic charge of $1/(2e)$ is still allowed. I will return to this point, and obtain the general quantization conditions, in Sec. 6.1.

One might be uncomfortable with the use here of singular potentials. There is an alternative approach that dispenses with the Dirac string and provides a different derivation of the quantization condition [70]. The idea is to use different gauges in different parts of space, in a manner quite analogous to using several coordinate patches to describe a curved spacetime. With a magnetic monopole located at the origin, let us divide space into two regions. Region I is defined by $0 \le \theta < (\pi/2) + \Delta$, while Region II is given by $(\pi/2) - \Delta < \theta \le \pi$. For $r \ne 0$ the vector potential of Eq. (5.5) is nonsingular in Region I, while that of Eq. (5.7) is nonsingular in Region II. There remains a singularity at the origin, but this is physically meaningful, corresponding to the fact that the source of the magnetic field is a point-like, and thus singular, particle.

The one requirement on this construction is that in the overlap of Regions I and II the potentials be related by a gauge transformation. From Eq. (5.9), we see that the difference of the two potentials can be written as $\mathbf{A}_{\mathrm{II}} - \mathbf{A}_{\mathrm{I}} = -\boldsymbol{\nabla}\Lambda$, with $\Lambda = 2g\phi$, which is indeed a gauge transformation. Under this gauge transformation the wavefunction of a particle with electric charge q would transform according to

$$\psi \rightarrow \psi' = e^{iq\Lambda}\psi = e^{2iqg\phi}\psi. \qquad (5.18)$$

Requiring that the wavefunction remain single-valued, with $\psi'(r, \theta, \phi + 2\pi) = \psi'(r, \theta, \phi)$, gives Eq. (5.12), and so we recover Dirac's quantization condition.

Yet another way to obtain the quantization condition is from angular momentum considerations. If an electric charge moves in the Coulomb magnetic field of a stationary monopole, the spherical symmetry of the situation suggests that the angular momentum of the charged particle should be conserved. However, the Lorentz force equation,

$$m \frac{d\mathbf{v}}{dt} = q\mathbf{v} \times \mathbf{B} = qg \left(\frac{\mathbf{v} \times \hat{\mathbf{r}}}{r^2} \right), \qquad (5.19)$$

implies that

$$\frac{d}{dt}(m\mathbf{r} \times \mathbf{v}) = \frac{qg}{r}\hat{\mathbf{r}} \times (\mathbf{v} \times \hat{\mathbf{r}}) = qg\frac{d}{dt}\hat{\mathbf{r}}. \tag{5.20}$$

Thus, the conserved quantity is not the usual $m\mathbf{r} \times \mathbf{v}$, but rather

$$\mathbf{L} = m\mathbf{r} \times \mathbf{v} - qg\hat{\mathbf{r}}, \tag{5.21}$$

where $\hat{\mathbf{r}}$ is a unit vector pointing from the magnetic charge to the electric charge. Recalling that the component of the angular momentum in any given direction must be an integer or half-integer and that, in particular, $\mathbf{L} \cdot \hat{\mathbf{r}} = -qg$, we again recover the Dirac quantization condition.

It is notable that this new contribution to the angular momentum is present even if the monopole and charge are stationary. This result was known to J. J. Thomson, who calculated the angular momentum in electromagnetic fields due to a static electric charge and a static magnetic charge with separation $\mathbf{r}$ [71]. He obtained

$$\mathbf{L} = \int d^3r\, \mathbf{r} \times (\mathbf{E} \times \mathbf{B}) = -qg\hat{\mathbf{r}}, \tag{5.22}$$

in agreement with our previous results.

Thomson's calculation is readily extended to the case where the two objects are dyons, with charges q_j and g_j $(j = 1, 2)$. Because of the vector product, the only nonzero contributions to the integrand are from the terms containing the electric field of one dyon times the magnetic field of the other. This gives

$$\mathbf{L} = -(q_1 g_2 - q_2 g_1)\hat{\mathbf{r}}, \tag{5.23}$$

with $\hat{\mathbf{r}}$ pointing from dyon 2 to dyon 1. This immediately leads to the generalized quantization condition of Eq. (5.16).

To put these arguments on a firmer ground, we should check that the components of $\mathbf{L}$ satisfy the standard angular momentum commutation relations. Let us write

$$\mathbf{L} = \mathbf{r} \times \mathbf{\Pi} - qg\hat{\mathbf{r}}, \tag{5.24}$$

where

$$\mathbf{\Pi} \equiv \mathbf{p} - q\mathbf{A}(\mathbf{r}). \tag{5.25}$$

The Π_j obey

$$[\Pi_j, \Pi_k] = iq(\partial_j A^k - \partial_k A^j) = iqg\, \epsilon^{jkl}\frac{r^l}{r^3}, \tag{5.26}$$

where the second equality follows because we have a pure Coulomb magnetic field. Using this result, it is now a simple matter to obtain

$$[L_j, L_k] = i\epsilon^{jkl} L_l. \tag{5.27}$$

We must now find the possible eigenvalues of $\mathbf{L}^2$ and L_z. The algebra of raising and lowering operators implies that these can only be $l(l + 1)$, with l either an integer or an integer plus 1/2, and m, with m ranging from l to $-l$ in integer steps. The next step is to write $\mathbf{L}^2$ and L_z as differential operators and solve for their eigenfunctions, the spherical harmonics. In the absence of a vector potential, the fact that

$$L_z = -i\frac{\partial}{\partial\phi} \qquad (5.28)$$

implies that these must be of the form

$$Y_{lm}(\theta, \phi) = F_{lm}(\theta)e^{im\phi}. \qquad (5.29)$$

In order that these be single-valued m, and thus also l, must be integers. Because $\mathbf{L}$ involves $\mathbf{A}$ when a monopole is present, its representation as a differential operator, and therefore also the corresponding eigenfunctions, the monopole spherical harmonics [72, 73], must be gauge-dependent. Let us consider the two gauges in which the vector potential takes the forms $\mathbf{A}^{\mathrm{I}}$ and $\mathbf{A}^{\mathrm{II}}$ that were given in Eqs. (5.5) and (5.7). These will give us expressions for the monopole spherical harmonics that are nonsingular in Regions I and II, respectively, and that are related in the overlap region by the gauge transformation given in Eq. (5.18). Because the Dirac strings of these two potentials lie on the z-axis, the manifest axial symmetry is preserved and we can perform a similar separation of variables. We now have

$$L_z = -i\frac{\partial}{\partial\phi} \mp k, \qquad (5.30)$$

where $k = qg$ and the upper and lower signs refer to Regions I and II, respectively. Hence, the harmonics must be of the form

$$Y_{klm}^{\mathrm{I}} = F_{klm}(\theta)e^{i(m+k)\phi},$$

$$Y_{klm}^{\mathrm{II}} = F_{klm}(\theta)e^{i(m-k)\phi}. \qquad (5.31)$$

In order that these be single-valued, $m \pm k$ must both be integers. This requires that both k and m (and therefore also l) be multiples of 1/2, with k, l, and m either all integers or all integers plus 1/2. Furthermore, the fact that

$$\mathbf{L}^2 = (\mathbf{r} \times \mathbf{\Pi})^2 + k^2 \qquad (5.32)$$

implies classically that $\mathbf{L}^2 \geq k^2$; quantum mechanically, with $\mathbf{L}^2 = l(l+1)$, the bound becomes $l \geq |k|$.

The final step is to solve for $F_{klm}(\theta)$. The properly normalized result is [73]

$$F_{klm}(\theta) = N_{klm}(1 - \cos\theta)^{-(q+m)/2}(1 + \cos\theta)^{(q-m)/2}P_{l+m}^{(-q-m),(q-m)}(\cos\theta), \qquad (5.33)$$

where

$$N_{klm} = 2^m \left[\frac{(2l+1)}{4\pi}\frac{(l-m)!(l+m)!}{(l-q)!(l+q)!}\right]^{1/2} \qquad (5.34)$$

and the Jacobi polynomial

$$P_n^{\alpha,\beta}(x) = \frac{(-1)^n}{2^n n!}(1-x)^{-\alpha}(1+x)^{-\beta}\frac{d^n}{dx^n}\left[(1-x)^{\alpha+n}(1+x)^{\beta+n}\right]. \quad (5.35)$$

For example, the lowest $k = 1/2$ harmonics in Region I are

$$Y^{\mathrm{I}}_{\frac{1}{2},\frac{1}{2},\frac{1}{2}} = -\frac{1}{\sqrt{4\pi}}\sqrt{1-\cos\theta}\,e^{i\phi},$$

$$Y^{\mathrm{I}}_{\frac{1}{2},\frac{1}{2},-\frac{1}{2}} = \frac{1}{\sqrt{4\pi}}\sqrt{1+\cos\theta}. \quad (5.36)$$

These are both nonsingular at $\theta = 0$; the first is singular along the Dirac string of $\mathbf{A}^{\mathrm{I}}$, at $\theta = \pi$, but this lies outside Region I.

When dealing with fields with spin, it is useful to define harmonics that are eigenfunctions of the total angular momentum $\mathbf{J} = \mathbf{L} + \mathbf{S}$. In the absence of a monopole one would expand spin-1/2 fields in spinor harmonics with total angular momentum $j = 1/2, 3/2, \ldots$ and spin-1 fields in vector harmonics with total angular momentum $j = 0, 1, \ldots$. For the former there are two multiplets of harmonics for each j, and for the latter there are three multiplets for each $j \geq 1$.

In the presence of a monopole these are modified. For charged spin-1/2 fields we have spinor monopole harmonics $\mathcal{Y}^{(\lambda)}_{kjm}$ where $j = k-1/2, k+1/2, \ldots$. Except for $j = k - 1/2$, there are two multiplets of spinor harmonics for each value of j, distinguished by different values of λ. These can be chosen to be eigenfunctions of $\mathbf{L}^2$, with quantum numbers $l = k \pm 1/2$. For charged vector fields one can define vector monopole harmonics [74, 75] $\mathbf{C}^{(\lambda)}_{kjm}$. If $k = 1/2$, there are two multiplets with $j = 1/2$, and three each for $j \geq 3/2$. If $k \geq 1$, there is one multiplet with $j = k-1$, two with $j = k$, and three each for $j \geq k+1$. Note that a $j = 0$ vector monopole harmonic can only occur if $k = qg = 1$.

It may be helpful to conclude this section with a discussion of units. It is well known that many equations in electromagnetism take on different forms depending on the units that are used. The same is true of the Dirac quantization condition. Let us write the Coulomb field of a magnetic charge Q_M as

$$\mathbf{B} = \beta\frac{Q_M}{r^2}\hat{\mathbf{r}} \quad (5.37)$$

and the Lorentz force on an electric charge Q_E as

$$\mathbf{F} = Q_E(\mathbf{E} + \eta\mathbf{v}\times\mathbf{B}). \quad (5.38)$$

The Dirac quantization condition then takes the form

$$Q_E Q_M = \frac{\hbar}{\beta\eta}\frac{n}{2}. \quad (5.39)$$

In rationalized MKS or SI units, $\beta = 1/4\pi$ and $\eta = 1$, so

$$Q_E Q_M = 4\pi\hbar\frac{n}{2}, \quad \text{rationalized MKS.} \quad (5.40)$$

Setting $\hbar = 1$, to give the rationalized natural units used here, agrees with Eq. (5.13). In Gaussian units, $\beta = 1$ and $\eta = 1/c$, so

$$Q_E Q_M = \hbar c \frac{n}{2}, \quad \text{Gaussian.} \tag{5.41}$$

It should also be kept in mind that the definition of Q_E also varies between systems. Thus, the fine structure constant $\alpha \approx 1/137$ is given by $e^2/4\pi\epsilon_0\hbar c$, $e^2/4\pi$, and $e^2/\hbar c$ in rationalized MKS, rationalized natural, and Gaussian units, respectively. If we set $Q_E = e$, we have

$$\frac{Q_M^2}{4\pi\mu_0\hbar c} = \frac{n^2}{4}\left(\frac{4\pi\epsilon_0\hbar c}{e^2}\right) \approx \frac{137}{4}n^2, \quad \text{rationalized MKS,} \tag{5.42}$$

and

$$\frac{Q_M^2}{\hbar c} = \frac{n^2}{4}\left(\frac{\hbar c}{e^2}\right) \approx \frac{137}{4}n^2, \quad \text{Gaussian,} \tag{5.43}$$

with the relation $\mu_0\epsilon_0 = c^{-2}$ having been used in obtaining the first of these.

5.2 The 't Hooft–Polyakov monopole

The discussion in the previous section described the consequences that would follow from the existence of a magnetic monopole, but left completely open the question of whether such monopoles actually exist. It was therefore a remarkable development when 't Hooft [65] and Polyakov [2] showed that magnetic monopoles necessarily arise as solitons in certain gauge field theories.

To see how this comes about, let us consider how to find topological solitons in three spatial dimensions. Our experience with the kink and vortex solutions suggest that we look for theories where the space of vacua $\mathcal{M}$ has a nontrivial second homotopy group, so that the fields on the two-sphere at spatial infinity can take on configurations that cannot be deformed to a uniform vacuum solution [76].

The simplest possibility is a theory with SO(3) symmetry broken to U(1) by the vacuum expectation value of a triplet scalar field[2] ϕ. From Eq. (4.22) we see that $\mathcal{M} = \text{SO}(3)/\text{SO}(2) = S^2$ so, by Eq. (4.43), we have $\pi_2(\mathcal{M}) = Z$. Writing $\hat{\phi} = \phi/|\phi|$ and recalling Eq. (4.41)', we can define the integer topological charge

$$N_\phi = \frac{1}{8\pi}\epsilon^{ijk}\int dS^i\,\hat{\phi}\cdot\partial_j\hat{\phi}\times\partial_k\hat{\phi}. \tag{5.44}$$

A theory with global SO(3) symmetry will have solutions with nonzero topological charge but, like the global vortices of Sec. 3.1, these will have infinite energy. Let us therefore focus on a gauge theory with Lagrangian density

$$\mathcal{L} = \frac{1}{2}(D_\mu\phi)^2 - \frac{1}{4}\mathbf{F}_{\mu\nu}^2 - V(\phi), \tag{5.45}$$

---

[2] When discussing this theory it will often be convenient to suppress explicit SO(3) indices, and instead use bold-face vector notation.

where

$$D_\mu \phi = \partial_\mu \phi + e\mathbf{A}_\mu \times \phi, \tag{5.46}$$

$$\mathbf{F}_{\mu\nu} = \partial_\mu \mathbf{A}_\nu - \partial_\nu \mathbf{A}_\mu + e\mathbf{A}_\mu \times \mathbf{A}_\nu, \tag{5.47}$$

and the scalar field potential

$$V(\phi) = -\frac{\mu^2}{2}\phi^2 + \frac{\lambda}{4}(\phi^2)^2 + \frac{\lambda v^4}{4}. \tag{5.48}$$

This potential is minimized when

$$\phi^2 = \frac{\mu^2}{\lambda} \equiv v^2. \tag{5.49}$$

If we take

$$\langle \phi \rangle = (0, 0, v), \tag{5.50}$$

then the unbroken U(1) corresponds to SO(3) rotations about the 3-axis. It is convenient to use the language of electromagnetism[3] to describe this U(1). The spectrum of elementary particles then includes a massless photon corresponding to

$$A_\mu \equiv A_\mu^3, \tag{5.51}$$

two vector bosons with mass $m_W = ev$ and electric charges $\pm e$ that correspond to the complex vector field

$$W_\mu \equiv \frac{1}{\sqrt{2}}(A_\mu^1 + iA_\mu^2), \tag{5.52}$$

and a neutral Higgs scalar with mass $m_\phi = \sqrt{2}\,\mu$ corresponding to

$$\phi \equiv \phi^3. \tag{5.53}$$

For a finite energy configuration we require not just that $|\phi| \to v$ at spatial infinity, but also that $D_j\phi$ fall faster than $r^{-3/2}$. The latter condition implies that

$$0 = \frac{1}{8\pi}\epsilon^{ijk} \int dS^i \, \hat{\phi} \cdot D_j\hat{\phi} \times D_k\hat{\phi}, \tag{5.54}$$

where the integration is over the two-sphere at spatial infinity. Expanding the covariant derivatives, we obtain

---

[3] In fact, this theory was proposed by Georgi and Glashow as a possible theory of the electroweak interactions, with the unbroken U(1) being electromagnetism [77]. It was experimentally ruled out by the discovery of neutral weak currents.

$$0 = \frac{1}{8\pi}\epsilon^{ijk}\int dS^i \Big\{ \hat{\phi}\cdot\partial_j\hat{\phi}\times\partial_k\hat{\phi}$$
$$+ e\hat{\phi}\cdot\Big[\partial_j\hat{\phi}\times(\mathbf{A}_k\times\hat{\phi}) + (\mathbf{A}_j\times\hat{\phi})\times\partial_k\hat{\phi}\Big]$$
$$+ e^2\hat{\phi}\cdot(\mathbf{A}_j\times\hat{\phi})\times(\mathbf{A}_k\times\hat{\phi})\Big\}$$
$$= \frac{1}{8\pi}\epsilon^{ijk}\int dS^i \Big\{ \hat{\phi}\cdot\partial_j\hat{\phi}\times\partial_k\hat{\phi}$$
$$- e\partial_j\hat{\phi}\cdot\mathbf{A}_k + e\partial_k\hat{\phi}\cdot\mathbf{A}_j + e^2\mathbf{A}_j\times\mathbf{A}_k\cdot\hat{\phi}\Big\}$$
$$= \frac{1}{8\pi}\epsilon^{ijk}\int dS^i \Big\{ \hat{\phi}\cdot\partial_j\hat{\phi}\times\partial_k\hat{\phi} + e\hat{\phi}\cdot\mathbf{F}_{jk}\Big\}, \tag{5.55}$$

with an integration by parts being used in obtaining the last equality. If we define

$$\mathbf{B}_i = -\frac{1}{2}\epsilon^{ijk}\mathbf{F}_{jk} \tag{5.56}$$

and use Eq. (5.44), this equation can be rewritten as

$$N_\phi = \frac{e}{4\pi}\int dS^i\,\hat{\phi}\cdot\mathbf{B}_i. \tag{5.57}$$

The dot product with $\hat{\phi}$ picks out the component of $\mathbf{B}_i$ that lies in the unbroken U(1) subgroup. We can therefore identify this as the true electromagnetic magnetic field[4] and

$$Q_M = \int dS^i\,\hat{\phi}\cdot\mathbf{B}_i \tag{5.58}$$

as the magnetic charge. Hence, a soliton with nonzero winding number must be a magnetic monopole, with magnetic charge

$$Q_M = N_\phi\left(\frac{4\pi}{e}\right). \tag{5.59}$$

By the argument given below Eq. (4.42), the Higgs field of this soliton must have N_ϕ zeros.

Because the elementary charged particles in this theory have electric charges $\pm e$, the unit monopole, with $N_\phi = 1$, has twice the minimum magnetic charge allowed by the Dirac quantization condition. The absence of solitons with the minimum magnetic charge is explained by noting that we can add to the theory fields that transform according to a half-integer representation of the gauge group [which is then unambiguously SU(2), rather than SO(3)]. It is easy to arrange that the new fields leave the classical solutions unchanged (e.g., by choosing the new fields to be fermionic). After the symmetry breaking, these

---

[4] This identification is only valid far from the core of the soliton, where the massive gauge fields are exponentially small. At shorter distances, $\hat{\phi}\cdot\mathbf{B}_i$ is a combination of the electromagnetic field and a magnetic moment term due to the massive electrically charged vector fields. This will be discussed in more detail in the next section.

fields would produce particles with half-integer U(1) charges, with the minimum electric charges being $\pm e/2$, and so consistency with the quantization condition requires that the minimum soliton magnetic charge be $4\pi/e$.

Let us now try to find a solution with unit magnetic charge. This is most easily done by trying the ansatz

$$\phi^a = \hat{r}^a h(r),$$
$$A_i^a = \epsilon^{aim}\hat{r}^m \left[\frac{1 - u(r)}{er}\right],$$
$$A_0^a = 0. \tag{5.60}$$

This ansatz is spherically symmetric in the sense that it is invariant under a rotation combined with a global SU(2) gauge transformation. It is also invariant under a combined parity transformation that reverses the direction of the axes in both real space and in internal SU(2) space. (Imposing only spherical symmetry allows a more general ansatz. However, it turns out that this does not lead to any additional solutions.)

Either by directly substituting the ansatz into the static field equations, or by substituting it into the action and then varying with respect to h and u, one obtains

$$0 = h'' + \frac{2}{r}h' - \frac{2u^2 h}{r^2} + \lambda(v^2 - h^2)h, \tag{5.61}$$

$$0 = u'' - \frac{u(u^2 - 1)}{r^2} - e^2 u h^2. \tag{5.62}$$

Requiring that the fields be nonsingular at the origin gives the boundary conditions

$$h(0) = 0,$$
$$u(0) = 1, \tag{5.63}$$

while finiteness of the energy leads to

$$h(\infty) = v,$$
$$u(\infty) = 0. \tag{5.64}$$

Examining the field equations near $r = 0$, we find that $h(r) \sim r$ and $1 - u(r) \sim r^2$. At large distance, the coefficient functions u and h approach their asymptotic values exponentially fast, with

$$u(r) \sim e^{-evr} = e^{-m_w r} \tag{5.65}$$

and, by an analysis similar to that which led to Eq. (3.41),

$$v - h(r) \sim \begin{cases} e^{-\sqrt{2\lambda}vr} = e^{-m_\phi r}, & \lambda < 2e^2, \\ \\ \dfrac{e^{-2evr}}{r^2} = \dfrac{e^{-2mw}}{r^2}, & \lambda > 2e^2. \end{cases} \tag{5.66}$$

With our ansatz, the magnetic field is

$$B_i^a = -\frac{1}{2}\epsilon^{ijk}F_{jk}^a = \hat{r}^a\hat{r}^i\left(\frac{1-u^2}{er^2}\right) - (\delta^{ia} - \hat{r}^a\hat{r}^i)\frac{u'}{er}. \tag{5.67}$$

At large distance, with exponentially small contributions neglected, this reduces to

$$B_i^a = \hat{r}^a\hat{r}^i\frac{1}{er^2}. \tag{5.68}$$

This is parallel to ϕ in SU(2) space (and thus purely electromagnetic) and precisely of the form expected for the Coulomb field of a monopole with magnetic charge $4\pi/e$.

By rescalings similar to those in the analysis of the vortex solution of Sec. 3.3, one can show that the form of the solution depends, up to trivial rescalings, only on the ratio λ/e^2. Except in the somewhat singular, but very important, limiting case $\lambda/e^2 = 0$ (which we will study in Chap. 8), the field equations must be solved numerically. However, there is a simple heuristic argument that gives some semiquantitative understanding of the solutions.

We can think of the monopole as being composed of a core region, of radius R, in which all of the fields are nontrivial, and an exterior region, $r > R$, in which the fields take on their asymptotic form. If we ignore exponentially small terms, the only contribution to the energy in the exterior region is from the asymptotic magnetic field, which contributes

$$\begin{aligned}
E_{\text{Coulomb}} &= \int_{r>R} d^3r\left(\frac{1}{2}\mathbf{B}^2\right) \\
&= \frac{4\pi}{2e^2}\int_R^\infty \frac{dr}{r^2} \\
&= \frac{2\pi}{e^2R}.
\end{aligned} \tag{5.69}$$

For the core region, let us approximate the energy density as having a constant value ρ. If λ/e^2 is of order unity, the fact that ϕ has a zero at the center of the core suggests that $\rho \sim e^2v^4 \sim \lambda v^4$. The total mass of the monopole is then

$$M \approx E_{\text{core}} + E_{\text{Coulomb}}$$

$$\approx \frac{4\pi}{3}R^3\rho + \frac{2\pi}{e^2R}. \tag{5.70}$$

Choosing R to minimize this energy gives

$$R_{\text{core}} \approx \left(\frac{1}{2e^2\rho}\right)^{1/4} \approx \frac{1}{ev} \approx \frac{1}{m_W}, \tag{5.71}$$

which is consistent with the exponential falloff in Eqs. (5.65) and (5.66), and

$$M \approx \frac{8\pi v}{3e}. \tag{5.72}$$

Numerical integration of Eqs. (5.61) and (5.62) verifies our estimate of the core radius. Further, one finds that the monopole mass is

$$M = \frac{4\pi v}{e} f(\lambda/e^2), \tag{5.73}$$

where f is a monotonic function with $f(0) = 1$ and $f(\infty) \approx 1.787$, also consistent with our estimate [78].

5.3 Another gauge, another viewpoint

In the previous section we studied the 't Hooft–Polyakov monopole in what is often called the "hedgehog gauge", where the SU(2) orientation of the Higgs field is correlated with the location in space. It is instructive to also consider the theory in a gauge where the orientation of the Higgs field is uniform. The gauge transformation that connects the two gauges is necessarily singular, so that the vector potential acquires a Dirac string in this new "string gauge". In particular, if we apply the gauge transformation generated by the SU(2) element

$$U(\theta, \varphi) = e^{-i\varphi\tau_3/2} e^{i\theta\tau_2/2} e^{i\varphi\tau_3/2} \tag{5.74}$$

to the ansatz of Eq. (5.60), the Higgs field becomes

$$\phi^a = \delta^{a3} h(r), \tag{5.75}$$

as illustrated in Fig. 5.1. The unbroken U(1) subgroup now corresponds everywhere to the T_3 direction. We can therefore identify the electromagnetic vector potential

$$\mathcal{A}^i = -\mathcal{A}_i = -A_i^3 = -\epsilon^{ij3} \hat{r}^j \frac{1}{e} \frac{1}{z+r}. \tag{5.76}$$

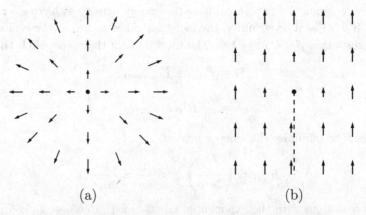

(a)   (b)

Fig. 5.1. The Higgs field of the monopole in the hedgehog (a) and string (b) gauges. The lengths and directions of the arrows indicate the magnitudes and gauge orientations of the fields, respectively. In (b), the dashed line indicates the position of the Dirac string.

This is just the vector potential of Eq. (5.5), with a Dirac string along the negative z-axis. This string singularity coincides with the singularity of $U(\theta, \varphi)$, which is ill-defined on the negative z-axis. The remaining components of A_i^a combine to form the field corresponding to the charged vector bosons,

$$W_i = \frac{1}{\sqrt{2}}(A_i^1 + iA_i^2) = \frac{u(r)}{er}v_i, \tag{5.77}$$

where

$$v_1 = \frac{i}{\sqrt{2}}\left[1 - e^{i\varphi}\cos\varphi(1 - \cos\theta)\right]$$

$$v_2 = -\frac{1}{\sqrt{2}}\left[1 + ie^{i\varphi}\sin\varphi(1 - \cos\theta)\right]$$

$$v_3 = -\frac{i}{\sqrt{2}}e^{i\varphi}\sin\theta. \tag{5.78}$$

Note that $\sum_j v_j{}^2 = 1$.

The Abelian field strength $\mathcal{F}_{ij}$ derived from $\mathcal{A}_i$ is not the same as F_{ij}^3, the component of $\mathbf{F}_{ij}$ parallel to $\hat{\phi}$. Instead, we have

$$\begin{aligned}\mathcal{F}_{ij} &= \partial_i \mathcal{A}_j - \partial_j \mathcal{A}_i \\ &= F_{ij}^3 - ie(W_i^* W_j - W_j^* W_i).\end{aligned} \tag{5.79}$$

The terms in parentheses on the second line represent the magnetic moment due to the spin-1 electrically charged vector fields. Outside the monopole core this term is exponentially small, and $\mathcal{F}_{ij} \approx \hat{\phi} \cdot \mathbf{F}_{ij}$, but inside the core the two quantities differ. The relation between the two is given by

$$\begin{aligned}\mathcal{F}_{ij} &\equiv \partial_i(\hat{\phi} \cdot \mathbf{A}_j) - \partial_j(\hat{\phi} \cdot \mathbf{A}_i) \\ &= \left[\hat{\phi} \cdot \mathbf{F}_{ij} - \frac{1}{e}\hat{\phi} \cdot (\mathbf{D}_i\hat{\phi}) \times (\mathbf{D}_j\hat{\phi})\right] + \frac{1}{e}\hat{\phi} \cdot \partial_i\hat{\phi} \times \partial_j\hat{\phi}.\end{aligned} \tag{5.80}$$

The expression in brackets on the second line is gauge invariant, and can be viewed as giving a definition of a U(1) field strength that is valid everywhere except at the zeros of ϕ [65]. This is possible because the subgroup that leaves $\phi(\mathbf{r})$ invariant is everywhere the same as the subgroup left unbroken by the vacuum. This is a special feature of our model with a single triplet Higgs field. One can construct both SU(2) theories with additional Higgs fields and theories with larger gauge groups where this is not true. In such cases there is no naturally preferred quantity to be identified as the field strength of the unbroken gauge group within the monopole core.

In the string gauge the Lagrangian can be written in terms of $\mathcal{A}_\mu$, W_μ, and $\phi \equiv \phi^3$ as

$$\begin{aligned}\mathcal{L} = &-\frac{1}{4}(\mathcal{F}_{\mu\nu} - M_{\mu\nu})^2 - \frac{1}{2}|\mathcal{D}_\mu W_\nu - \mathcal{D}_\nu W_\mu|^2 \\ &+ \frac{1}{2}(\partial_\mu\phi)^2 + e^2\phi^2|W_\mu|^2 - V(\phi),\end{aligned} \tag{5.81}$$

with the U(1) covariant derivative $\mathcal{D}_\mu = \partial_\mu + ie\mathcal{A}_\mu$ and the electromagnetic moment term $M_{\mu\nu} = -ie(W_\mu^* W_\nu - W_\nu^* W_\mu)$. When written in this form, the theory appears to be simply an Abelian gauge theory containing a charged vector field and a neutral scalar. Both the SU(2) gauge symmetry and the possibility of nontrivial topology are obscured. From this point of view, it might seem surprising that the theory should have a finite energy monopole solution. After all, one would expect the energy in the Coulomb field of a point monopole to diverge near the origin, just as would be the case for a point electric charge. However, from the first term in the Lagrangian we see that this divergence can be avoided if the $1/r^2$ behavior of the magnetic field is matched by a similar behavior in the magnetic moment, which is possible if $W_i \sim 1/r$ near the origin. It might then seem that such a singular W_i would give a divergent energy from the covariant curl term, but a calculation shows that the potentially dangerous angular derivative pieces actually cancel. Also, with $W_i \sim 1/r$, the W-mass term is greater than $V(\phi)$ near the origin. The energy density from the former term is minimized if ϕ vanishes, thus giving a physical explanation for the zero of the Higgs field at the origin [79].

5.4 Solutions with higher magnetic charge

In studying the U(1) vortex we saw that the rotationally invariant ansatz for the unit vortex could be rather easily generalized to obtain solutions of arbitrary vorticity. One might wonder if a similar generalization could be carried out in the monopole case. The answer is no. This was first demonstrated by imposing the requirement that the effect of an arbitrary rotation be canceled by a (possibly position-dependent) gauge transformation. After some tedious calculations it was shown that, except in the case of unit magnetic charge, this could not be done [80]. Not only are there no spherically symmetric solutions with multiple magnetic charge, but it is not even possible to write down a finite energy spherically symmetric configuration with greater than unit charge.

   This result can actually be obtained more easily by working in string gauge, with the Lagrangian written in the string gauge form of Eq. (5.81). In general, any field configuration can be expanded in terms of spherical harmonics, with the type of harmonic being appropriate to the properties of the field. A spherically symmetric configuration is one for which only $J = 0$ harmonics appear.

   In particular, the W_i field must be expanded in monopole vector harmonics. To have spherical symmetry, only the coefficient of the $J = 0$ harmonic can be nonzero. However, we saw in Sec. 5.1 that a $J = 0$ monopole vector harmonic only occurs if $k = qg = 1$. For all other values of k spherical symmetry requires that W_i vanish identically. If W_i vanishes, we have a free electromagnetic field, and the only spherically symmetric configuration with magnetic charge is the pure Coulomb field of a point monopole, whose energy diverges.

This leaves the possibility of static solutions with multiple magnetic charge that are not spherically symmetric. Static solutions comprising a number of well-separated unit monopoles are certainly excluded, because of the repulsive Coulomb forces between the individual monopoles. Although this argument does not rule out multicharged solutions with overlapping monopole cores, it seems unlikely that any exist.

An exception occurs in the BPS limit [81, 82] where $\lambda/e^2 = 0$. As we will see in Chap. 8, in this limit there is a long-range force mediated by the scalar field that exactly cancels the magnetic repulsion between static monopoles. As a result, there are multicharged solutions containing any number of monopoles, with almost arbitrary choices for the monopole positions [83, 84].

5.5 Zero modes and dyons

Let us now turn to the analysis of small fluctuations about the monopole, focusing in particular on the zero modes. As with the gauged U(1) vortex, we want to exclude the zero modes that simply correspond to local gauge transformations. We do this by imposing a background gauge condition. By steps analogous to those that led to Eq. (3.60), we find that this now takes the form

$$0 = \mathbf{D}_i \delta \mathbf{A}_i + e\boldsymbol{\phi} \times \delta\boldsymbol{\phi}. \tag{5.82}$$

As should be expected, there are three translational zero modes, which in background gauge are given by

$$\delta_k \mathbf{A}_j = \mathbf{F}_{kj}, \qquad \delta_k \boldsymbol{\phi} = \mathbf{D}_k \boldsymbol{\phi}. \tag{5.83}$$

These are handled by introducing collective coordinates z^k that specify the monopole position. Exciting these modes in a time-dependent fashion yields a moving monopole with nonzero linear momentum $\mathbf{P}$. As we did with the kink and the vortex, one can use a combination of virial theorems and Gauss's law to show that to leading order the resulting kinetic energy is $\mathbf{P}^2/(2M)$, where M is the monopole mass obtained from the static classical solution.

There is also a fourth zero mode. Its origin is most easily seen by working in the string gauge that was introduced in Sec. 5.3. The string gauge Lagrangian, Eq. (5.81), is clearly unchanged if we multiply W_i by a constant phase factor $e^{i\alpha}$; this is a manifestation of the unbroken U(1) symmetry. Because this is just a gauge transformation that takes one static solution to another physically equivalent solution, one might think that it should be excluded. Its significance becomes clear when we consider time-dependent excitations of this mode. This produces a nonzero value of the Noether charge corresponding to the U(1) symmetry. The electric charge is e times this Noether charge, and in terms of string gauge fields takes the form

$$Q_E = ie \int d^3x \left[W_j^* (\mathcal{D}_0 W_j - \mathcal{D}_j W_0) - W_j (\mathcal{D}_0 W_j^* - \mathcal{D}_j W_0^*) \right]. \tag{5.84}$$

Thus, excitation of this zero mode produces a dyon.[5]

When the theory is quantized, this zero mode is handled by introducing a phase collective coordinate with a finite range, 0 to 2π. The conjugate momentum must then be quantized in integer units. Since the electric charge is e times this momentum, we have

$$Q_E = ne. \tag{5.85}$$

Let us therefore start with our spherically symmetric solution in string gauge form and multiply W_i by a phase factor $e^{-i\omega t}$. Using Eq. (5.77), we find that the U(1) Gauss's law takes the form

$$0 = \mathcal{A}_0'' + \frac{2}{r}\mathcal{A}_0' - \frac{2u^2}{r^2}\left(\mathcal{A}_0 - \frac{\omega}{e}\right), \tag{5.86}$$

where, by analogy with Eq. (5.76), I have defined $\mathcal{A}_0 = A_0^3$ in string gauge.

Because $u(0) = 1$, we must require that $\mathcal{A}_0(0) = \omega/e$ if we want a nonsingular solution. Let us also set $\mathcal{A}_0(\infty) = 0$. Solving Eq. (5.86) then yields a solution for $\mathcal{A}_0$ that is proportional to ω. The back reaction of this $\mathcal{A}_0$ on W_i perturbs the initial monopole, but these effects are higher order and can be neglected for small ω; they are analogous to the Lorentz contraction of a moving soliton, which can be ignored for small velocity.

We thus have a time-dependent dyonic configuration that satisfies the field equations to leading order in ω. This can be transformed into a static solution by performing a U(1) gauge transformation

$$\mathcal{A}_\mu \to \tilde{\mathcal{A}}_\mu = \mathcal{A}_\mu - \partial_\mu \Lambda,$$

$$W_i \to \tilde{W}_i = e^{ie\Lambda} W_i, \tag{5.87}$$

with $\Lambda = (\omega/e)t$. This undoes the phase rotation of W_i and shifts $\mathcal{A}_0$ by a constant, so that $\tilde{\mathcal{A}}_0(0) = 0$ and $\tilde{\mathcal{A}}_0(\infty) = -\omega/e$.

We can now gauge-transform back into the hedgehog gauge. This takes us back to the ansatz of Eq. (5.60), except that now

$$A_0^a = \hat{r}^a \tilde{\mathcal{A}}_0 \equiv \hat{r}^a j(r). \tag{5.88}$$

The static field equations become[6]

$$0 = h'' + \frac{2}{r}h' - \frac{2u^2 h}{r^2} + \lambda(v^2 - h^2)h, \tag{5.89}$$

---

[5] A related approach that avoids the appearance of the singular Dirac string is obtained by working in axial gauge [85].

[6] One can obtain these equations directly by assuming a spherically symmetric static ansatz from the start; this is, in fact, how these solutions were first discovered by Julia and Zee [86]. However, this obscures the connection with the phase collective coordinate, which is essential for obtaining the quantization of the electric charge.

$$0 = u'' - \frac{u(u^2 - 1)}{r^2} - e^2 u(h^2 - j^2), \tag{5.90}$$

$$0 = j'' + \frac{2}{r} j' - \frac{2u^2 j}{r^2}. \tag{5.91}$$

The first of these is unchanged from the purely magnetic case, and the second differs only by a term proportional to j^2 that is $O(\omega^2)$. The boundary conditions for h and u are unchanged, while Eq. (5.88) gives $j(0) = 0$ and $j(\infty) = -\omega/e$.

With this static ansatz $\mathbf{F}_{0i}$ is parallel to ϕ at large distance, and the actual electric field is

$$\mathcal{F}_{0i} = -\partial_i j(r) = -\hat{r}^i j'(r). \tag{5.92}$$

By integrating Eq. (5.91), we find that

$$j'(r) = \frac{2}{r^2} \int_0^r ds \, u(s)^2 j(s). \tag{5.93}$$

The integrand falls exponentially fast outside the dyon core, so at large r we make a negligible error by setting the upper limit to infinity. But substituting our ansatz into Eq. (5.84) gives

$$Q_E = -8\pi \int_0^\infty ds \, u(s)^2 j(s). \tag{5.94}$$

Hence, at large distance we have

$$\mathcal{F}_{0i} = \hat{r}_i \frac{Q_E}{4\pi r^2}, \tag{5.95}$$

just as it should be for a Coulomb electric field.

The extra energy associated with the electric charge increases the mass of the dyon. To estimate this increase, let us assume that the charge is small enough that the deformations of $u(r)$ and $h(r)$ from their profiles in the uncharged monopole can be neglected. With our spherically symmetric ansatz, $\mathbf{D}_0\phi$ vanishes identically, so the additional energy comes entirely from the $\mathbf{F}_{0i}^2$ term. To leading order, this is

$$\begin{aligned}
\Delta M &= \frac{1}{2} \int d^3x \, \mathbf{F}_{0i}^2 \\
&= \frac{1}{2} \int d^3x \left[-\partial_i (\mathbf{A}_0 \cdot \mathbf{F}_{0i}) + \mathbf{A}_0 \cdot \mathbf{D}_i \mathbf{F}_{0i} \right] \\
&= -\frac{1}{2} j(\infty) Q_E = \frac{\omega}{2e} Q_E, \tag{5.96}
\end{aligned}$$

where the last line is obtained by using Gauss's law and the vanishing of $\mathbf{D}_0\phi$ to show that $\mathbf{D}_i\mathbf{F}_{0i} = 0$. To relate ω and Q_E, we rewrite Eq. (5.94) as

$$Q_E = \frac{8\pi\omega}{e} \int_0^\infty ds \, u(s)^2 \frac{j(s)}{j(\infty)} \equiv eI\omega. \tag{5.97}$$

Here I can be thought of as a moment of inertia associated with the phase rotation of the charged fields. (The factor of e is extracted because it is Q_E/e that is the actual conjugate momentum.) Because the integrand is of order unity inside the dyon core and vanishes rapidly outside the core radius $\sim 1/(ev)$, we have

$$I \sim \frac{1}{e^3 v} \tag{5.98}$$

and

$$\Delta M = \frac{1}{2} I \omega^2 = \frac{Q_E^2}{2e^2 I} \sim Q_E^2 ev \sim \frac{Q_E^2}{Q_M^2} M_{\text{mon}}. \tag{5.99}$$

Note that

$$\frac{d\Delta M}{dQ_E} \sim 2Q_E m_W, \tag{5.100}$$

so that the mass decrease from reducing the electric charge by one unit, with $Q_E \to Q_E - e$, is $\sim e Q_E m_W$. For weak coupling ($e \ll 1$) and Q_E not too large, this is less than the mass of a W, so the dyon is stable against losing its charge by emission of charged vector bosons. If instead eQ_E is of order unity, then we are well outside the range of validity of our approximation of treating $u(r)$ and $h(r)$ as unchanged.

5.6 Spin from isospin, fermions from bosons

We saw in Sec. 5.1 that, even when at rest, an electric charge in the presence of a magnetic monopole carries an additional angular momentum directed along the line joining their positions. One would then expect that a bound state of a charge and a monopole would possess an intrinsic spin arising from this angular momentum term. If the product of the electric and magnetic charges is the minimum allowed by the Dirac quantization condition, the radial component of the charge-monopole angular momentum is equal to $1/2$, which would seem to suggest that the bound state could have half-integer angular momentum even if its components were both spinless. One would thus obtain a fermion from bosonic components.

Although these arguments do not depend on the core structure of the monopole, the 't Hooft–Polyakov solution provides a setting for illustrating the idea in detail while avoiding any potential problems from singularities. If a soliton is not spherically symmetric, it will have rotational zero modes. As with other zero modes, these would be treated by introducing collective coordinates, which would specify the spatial orientation of the soliton. The variables conjugate to these would be the spin angular momenta, which become nonzero if the soliton undergoes a time-dependent rotation.

The 't Hooft–Polyakov monopole is rotationally invariant in the sense that the effects of a rotation can be completely compensated by a global gauge transformation. When written in the hedgehog gauge form of Eq. (5.60), it is invariant under the action of

$$\mathbf{J} = \mathbf{L} + \mathbf{T}, \tag{5.101}$$

where $\mathbf{L}$ represents a rotation of the spatial coordinates and $\mathbf{T}$ a global SU(2) ("isospin") gauge rotation. We can endow it with spin by giving it structure that breaks this rotational symmetry.

To do this, let us add to the theory a scalar field ψ that transforms under the doublet representation of the SU(2) isospin symmetry [87]. When this SU(2) is spontaneously broken to U(1), the components of ψ acquire U(1) charges $\pm 1/2$. Let us assume that, in addition to its couplings to the gauge field, ψ interacts with ϕ through a potential of the form

$$V(\phi, \psi) = -\frac{\mu^2}{2}\phi^2 + \frac{\lambda}{4}(\phi^2)^2 + [m^2 + h(\phi^2 - v^2)]\psi^\dagger\psi + g(\psi^\dagger\psi)^2, \tag{5.102}$$

with $v^2 = \mu^2/\lambda$, as previously, and g and h positive. The parameters can be chosen so that the minimum of the potential occurs when $\psi = 0$ and $\phi^2 = v^2$.

The topological arguments still guarantee the existence of a monopole solution. At large distance, ϕ again approaches a vacuum value, with $|\phi| \to v$, while ψ tends toward zero. However, we know that ϕ must vanish at the center of the monopole, so there can be a region where ϕ^2 is small enough that the coefficient of $\psi^\dagger\psi$ in the potential is negative, making a nonzero ψ energetically favored. Thus, with appropriate choices of couplings we expect to find a solution with nonzero $\psi(\mathbf{r})$. With A_μ^a and ϕ^a obeying the ansatz of Eq. (5.60), the symmetry of the theory implies that ψ must take the form

$$\psi = p(r)e^{i\boldsymbol{\alpha}\cdot\boldsymbol{\tau}/2}s, \tag{5.103}$$

where s is an arbitrary isospinor and $\boldsymbol{\alpha}$ an arbitrary three-vector, both independent of position. This is clearly not invariant under the action of $\mathbf{J}$, and so we have a soliton that is not rotationally invariant.[7] Furthermore, the fact that ψ has isospin $1/2$ implies that J will have half-integral eigenvalues. Hence, the monopole should now be a fermion.[8]

However, fermions are characterized not just by having half-integer spins, but also by their statistics. How do antisymmetric wavefunctions arise from bosonic components? To understand this, consider the case of two dyons, each of which is a bound state of an electrically charged particle and a magnetically charged particle, both spinless [89].

To start, recall that the Hamiltonian for a particle with electric charge q in combined electric and magnetic fields is

---

[7] The ansatz for A_i^a and ϕ^a can be maintained, despite the violation of rotational symmetry, because ψ couples to these fields through a density and current that are both spherically symmetric.

[8] An alternative approach [88] is to leave the classical soliton solution unchanged, but to add an isospinor scalar field whose normal modes in the presence of the monopole include a bound state. By arguments similar to those just outlined, a state with a single quantum in this mode will have half-integer angular momentum.

$$H_q = \frac{1}{2M_q} \left(\mathbf{p} - q\mathbf{A} \right)^2 + qA_0, \qquad (5.104)$$

where

$$\mathbf{B} = \boldsymbol{\nabla} \times \mathbf{A},$$

$$\mathbf{E} = -\boldsymbol{\nabla} A_0 - \frac{\partial \mathbf{A}}{\partial t}. \qquad (5.105)$$

From the duality symmetry of Maxwell's equations, we see that the analogous Hamiltonian for a particle with magnetic charge $4\pi g$ is (note the changed signs)

$$H_g = \frac{1}{2M_g} \left(\mathbf{p} + 4\pi g \tilde{\mathbf{A}} \right)^2 + 4\pi g \tilde{A}_0, \qquad (5.106)$$

where

$$\mathbf{E} = \boldsymbol{\nabla} \times \tilde{\mathbf{A}},$$
$$\mathbf{B} = -\boldsymbol{\nabla} \tilde{A}_0 + \frac{\partial \tilde{\mathbf{A}}}{\partial t}. \qquad (5.107)$$

Just like the vector potential $\mathbf{A}$ generated by a monopole, the dual vector potential $\tilde{\mathbf{A}}$ generated by an electric charge has a string singularity that is undetectable if the Dirac quantization condition is satisfied. As with the Dirac vector potential, the location of the dual string singularity is arbitrary and can be changed by a gauge transformation, with a corresponding change in the phase of the wavefunctions of monopoles feeling that potential. However, some choices are particularly convenient. Let us suppose that we have an electric charge q_1 at $\mathbf{r}_1$ and a magnetic charge $4\pi g_2$ at $\mathbf{r}_2$. The velocities of these two particles are

$$\mathbf{v}_1 = \frac{1}{M_1} \left[\mathbf{p}_1 - q_1 \mathbf{A}^{(2)}(\mathbf{r}_1) \right],$$

$$\mathbf{v}_2 = \frac{1}{M_2} \left[\mathbf{p}_2 + 4\pi g_2 \tilde{\mathbf{A}}^{(1)}(\mathbf{r}_2) \right], \qquad (5.108)$$

where the superscripts in parentheses indicate the source of the potential. We want the sum of the canonical momenta of the two particles to vanish in the center-of-mass frame, where $M_1\mathbf{v}_1 + M_2\mathbf{v}_2 = 0$. This will be the case if

$$q_1 \mathbf{A}^{(2)}(\mathbf{r}_1) = 4\pi g_2 \tilde{\mathbf{A}}^{(1)}(\mathbf{r}_2). \qquad (5.109)$$

This constraint, together with the requirement that the potentials generate the proper magnetic and electric fields, is satisfied by taking

$$\mathbf{A}^{(2)}(\mathbf{r}) = g_2 \mathbf{f}(\mathbf{r} - \mathbf{r}_2),$$

$$\tilde{\mathbf{A}}^{(1)}(\mathbf{r}) = \left(\frac{q_1}{4\pi} \right) \mathbf{f}(\mathbf{r}_1 - \mathbf{r}), \qquad (5.110)$$

where, apart from a factor of the magnetic charge,

$$\mathbf{f}(\mathbf{s}) \cdot d\mathbf{s} = (1 - \cos\theta)d\phi \tag{5.111}$$

is the Dirac potential given in Eq. (5.5). Equation (5.9) and the comments preceding it tell us that

$$\mathbf{f}(-\mathbf{s}) = \mathbf{f}(\mathbf{s}) - 2\boldsymbol{\nabla}\phi. \tag{5.112}$$

If we ignore the internal degrees of freedom of the dyons, our two-dyon system is described by a two-particle wavefunction $\psi_B(\mathbf{r}_1, \mathbf{r}_2)$ whose time evolution is governed by the Hamiltonian

$$
\begin{aligned}
H_B &= \frac{1}{2M_1}\left[\mathbf{p}_1 - q_1\mathbf{A}^{(2)}(\mathbf{r}_1) + 4\pi g_1\tilde{\mathbf{A}}^{(2)}(\mathbf{r}_1)\right]^2 \\
&\quad + \frac{1}{2M_2}\left[\mathbf{p}_2 - q_2\mathbf{A}^{(1)}(\mathbf{r}_2) + 4\pi g_2\tilde{\mathbf{A}}^{(1)}(\mathbf{r}_2)\right]^2 + V_{\text{Coul}} \\
&= \frac{1}{2M_1}\left[\mathbf{p}_1 - q_1 g_2\mathbf{f}(\mathbf{r}_1 - \mathbf{r}_2) + g_1 q_2\mathbf{f}(\mathbf{r}_2 - \mathbf{r}_1)\right]^2 \\
&\quad + \frac{1}{2M_2}\left[\mathbf{p}_2 - q_2 g_1\mathbf{f}(\mathbf{r}_2 - \mathbf{r}_1) + g_2 q_1\mathbf{f}(\mathbf{r}_1 - \mathbf{r}_2)\right]^2 + V_{\text{Coul}}, \tag{5.113}
\end{aligned}
$$

where

$$V_{\text{Coul}} = \frac{q_1 q_2 + (4\pi)^2 g_1 g_2}{4\pi|\mathbf{r}_1 - \mathbf{r}_2|}. \tag{5.114}$$

Now let us take the dyons to be identical, with $M_1 = M_2 = M$, $q_1 = q_2 = q$, and $g_1 = g_2 = g$. Because they are composed of spinless particles, their wavefunction should be symmetric, with $\psi_B(\mathbf{r}_1, \mathbf{r}_2) = \psi_B(\mathbf{r}_2, \mathbf{r}_1)$. By making use of Eq. (5.112), the Hamiltonian can be put in the form

$$H_B = \frac{1}{2M}\left[\mathbf{p}_1 - 2qg\boldsymbol{\nabla}\phi_{12}\right]^2 + \frac{1}{2M}\left[\mathbf{p}_2 - 2qg\boldsymbol{\nabla}\phi_{21}\right]^2 + V_{\text{Coul}}, \tag{5.115}$$

where ϕ_{12} is the azimuthal angle defined by the vector $\mathbf{r}_1 - \mathbf{r}_2$.

The form of this Hamiltonian is hardly transparent. It can be put into a simpler form, with manifest rotational invariance, by a gauge transformation that takes $\psi_B(\mathbf{r}_1, \mathbf{r}_2)$ to

$$\psi_S(\mathbf{r}_1, \mathbf{r}_2) = e^{2iqg\phi_{12}}\psi_B(\mathbf{r}_1, \mathbf{r}_2) \tag{5.116}$$

and at the same time takes H_B to

$$H_S = \frac{\mathbf{p}_1^2}{2M} + \frac{\mathbf{p}_2^2}{2M} + V_{\text{Coul}}. \tag{5.117}$$

It is now evident, as it was not before, that the interaction between the two dyons is a purely Coulomb one. This could have been anticipated by noting that a duality rotation of the form of Eq. (5.2) can simultaneously transform both dyons to purely electric charges.

Because $\phi_{21} = \phi_{12} + \pi$, under interchange of $\mathbf{r}_1$ and $\mathbf{r}_2$ we have

$$\psi_S(\mathbf{r}_2, \mathbf{r}_1) = e^{2iqg\pi} \psi_S(\mathbf{r}_1, \mathbf{r}_2). \qquad (5.118)$$

The wavefunction is thus symmetric or antisymmetric, depending on whether qg is an integer or a half-integer. Our gauge transformation has not only simplified the Hamiltonian, but it has also given our states the bosonic or fermionic character appropriate to the dyon spins.

5.7 Fermions and monopoles

We have seen in previous chapters that interesting effects can arise from the interaction of fermion fields with topological solitons. In two spacetime dimensions there are fermion zero modes about the kink that lead to degenerate states with fractional fermion number. Jackiw and Rebbi showed that a similar phenomenon occurs with magnetic monopoles [14].

To see this, let us add to the Lagrangian of Eq. (5.45) a fermionic SU(2) "isospin" multiplet Ψ_m. This field is governed by the fermion Lagrangian

$$\mathcal{L}_{\text{fermion}} = i\bar{\Psi}_n \gamma^\mu (D_\mu \Psi)_n - G\bar{\Psi}_m T^a_{mn} \Psi_n \Phi^a, \qquad (5.119)$$

where the T^a are the SU(2) generators in the appropriate representation, and acquires a mass solely from its interaction with the Higgs field. It is convenient to work in a basis where the Dirac matrices are

$$\gamma^0 = \begin{pmatrix} 0 & -iI \\ iI & 0 \end{pmatrix}, \qquad \gamma^j = \begin{pmatrix} -i\sigma^j & 0 \\ 0 & i\sigma^j \end{pmatrix}, \qquad (5.120)$$

and to decompose Ψ_m into a pair of two-component spinors,

$$\Psi_m = \begin{pmatrix} \psi^+_m \\ \psi^-_m \end{pmatrix}. \qquad (5.121)$$

In a gauge where the Higgs and gauge fields are given by the hedgehog ansatz of Eq. (5.60), the Dirac equation for a zero-frequency mode can be written as

$$0 = \left\{ \delta_{mn} \mathbf{S} \cdot \boldsymbol{\nabla} + i\left[\frac{1 - u(r)}{r}\right] T^a_{mn} (\mathbf{S} \times \hat{\mathbf{r}})^a \mp \frac{G}{2} h(r) \hat{r}^a T^a_{mn} \right\} \psi^\pm_m, \qquad (5.122)$$

with $\mathbf{S} = \boldsymbol{\sigma}/2$.

In the fixed monopole background we can write the conserved fermion angular momentum as

$$\mathbf{J} = \mathbf{L} + \mathbf{S} + \mathbf{T}. \qquad (5.123)$$

The simplest case to consider is that of an isodoublet spinor. The most likely candidate for a zero-frequency mode is one with $J = 0$. In general, such modes will be linear combinations of fields with orbital angular momentum $L = 0$ and

$L = 1$. However, let us seek a solution that is purely $L = 0$, implying that it is invariant under $\mathbf{S} + \mathbf{T}$. Such a solution must be of the form

$$\psi^\pm_{\alpha m} = f_\pm(r)\epsilon_{\alpha m},\qquad\qquad(5.124)$$

where α and m are spin and isospin indices, respectively. Equation (5.122) then reduces to

$$0 = \left[\frac{\partial}{\partial r} - \frac{1 - u(r)}{r} \pm \frac{1}{2}Gh(r)\right]f_\pm,\qquad\qquad(5.125)$$

which is solved by

$$f_\pm(r) = C\exp\left\{\int_0^r dr'\left[\frac{1 - u(r')}{r'} \mp \frac{1}{2}Gh(r')\right]\right\}.\qquad(5.126)$$

This gives a normalizable zero mode only for the upper sign. The analysis of [14] shows that this is in fact the only zero mode. Hence, there are two degenerate monopole states, with fermion number $\pm\frac{1}{2}$.

The analogous Dirac equation for an isotriplet fermion field was also analyzed in [14]. In this case there are two zero modes, and hence four degenerate states, a result that will be of importance in Sec. 8.7. Results for a much wider class of Dirac equations can be obtained from an index theorem of Callias [90].

Studying the scattering of fermions off a monopole reveals further unusual effects. Already at the classical level, angular momentum considerations suggest that this might be the case. Thus, consider a classical charged particle moving on a straight-line trajectory toward a point monopole. The particle's angular momentum includes the anomalous charge-monopole angular momentum directed along the line from the charge to the monopole. If the charge were to pass through the monopole, the direction of this term would be reversed. For the total angular momentum to be conserved, either the charge or the helicity of the particle would have to undergo a sudden change.

This effect is seen quantum mechanically in the scattering state solutions of the Dirac equation for a charged massless fermion in the presence of a point-like monopole. One would expect to find eight states in the $J = 0$ sector: two helicities for each choice of particle or antiparticle and incoming or outgoing. In actuality, only half of these are present. For the positively charged particle there is an incoming state with negative helicity and an outgoing state with positive helicity; the helicities are reversed for the antiparticle. Furthermore, these solutions are all singular at the origin, a symptom of the fact that the Hamiltonian corresponding to this Dirac equation is not self-adjoint; this can be corrected by imposing a boundary condition at the origin that determines the matching of incoming and outgoing states.

One would expect that replacing the point monopole with a nonsingular 't Hooft–Polyakov monopole would uniquely determine the boundary condition. Assuming that the theory respects the conservation of electric charge, two possibilities need to be considered. One is that an incoming particle or antiparticle flips its helicity and scatters outward with its original charge. The other is that an incoming particle deposits charge on the monopole and then leaves as an antiparticle with the original helicity.

The estimate in Eq. (5.99) of the energy associated with the dyon charge seems to suggest that the latter possibility is energetically forbidden in low-energy scattering. This would indeed be the case within the context of the first-quantized theory. In the full quantum field theory the situation is more subtle. The monopole is surrounded by a fermion field condensate that is quite analogous to the vacuum polarization cloud surrounding an electric charge. The radius of this cloud is roughly equal to the inverse of the fermion mass, and so for a light fermion is much greater than the monopole core radius. Spreading any excess charge over this larger region would entail a much smaller Coulomb energy, and so is not obviously forbidden.

Rubakov [91] and Callan [92, 93] argued that a good approximation could be obtained by restricting one's attention to the $J = 0$ fermion sector. Not only is this the sector where the mismatch of states is seen, but at low energy the angular momentum barrier prevents fermions with nonzero J from reaching the monopole core. Restricting to this sector effectively reduces the problem to a (1+1)-dimensional field theory, with the one-dimensional space actually a half-line, $0 \leq r < \infty$. We saw in Sec. 2.7 that the fermionic massive Thirring model was equivalent to the bosonic sine-Gordon model, with the correspondence between the fermion and boson fields given by Eq. (2.124). By a similar "bosonization" the $J = 0$ fermionic theory on the half-line can be recast as an equivalent bosonic theory that is more easily analyzed.

More specifically, let us focus on the theory with an isodoublet fermion field that was considered at the beginning of this section. After the spontaneous breaking of the SU(2) symmetry, there are two species of fermions. For one the particle has U(1) charge $e/2$ and the antiparticle has charge $-e/2$, while for the other species the charges are reversed. Although the naïve expectation would be for eight incoming and eight outgoing $J = 0$ modes, only half of these are actually present.

Bosonizing the theory leads to a theory with two scalar fields, each with a sine-Gordon-like self-interaction, and with a $1/r^2$ Coulomb interaction between the two fields. There are kink and antikink solutions for each of the fields. Each of these four solitons can move either inward or outward, giving a total of eight states, in one-to-one correspondence with the fermionic scattering states. Analysis of the bosonic theory leads to the conclusion that in low-energy fermion-monopole scattering charge is not deposited on either the monopole or its surrounding fermionic condensate, and that instead the fermion chirality is reversed. Importantly, the cross-section for this scattering process is not suppressed by factors of the monopole mass or radius, but is instead roughly given by the square of the fermion Compton wavelength.

The origin of the nonconservation of chiral charge in this scattering process can be traced to the Adler–Bell–Jackiw anomaly that gives an anomalous divergence of the chiral current proportional to $F_{\mu\nu}\tilde{F}^{\mu\nu} = 4\mathbf{E} \cdot \mathbf{B}$. We will see in Chap. 11 that instanton effects can also lead to nonconservation of chiral charge via the anomaly. However, these are quantum tunneling processes and so entail an exponential suppression. The striking feature here is that the presence of a monopole removes the need for tunneling so that there is no such suppression. Indeed, using the fact that $\mathbf{E} \cdot \mathbf{B}$ is nonzero about a dyon, it can be shown that a highly charged massive dyon will emit light fermions in a chirality-violating manner [94, 95].

6

Magnetic monopoles in larger gauge groups

Having studied in some detail the monopole that arises when SU(2) is broken to U(1), let us now consider the possibilities that occur with larger, and possibly more realistic, gauge groups. Perhaps the most important of these are the groups that occur in grand unified theories. As we will see, magnetic monopoles are a generic prediction of all such theories.

6.1 Larger gauge groups—the external view

Although our primary interest is in monopoles with nonsingular cores, which arise when a gauge group G is spontaneously broken to a subgroup H with $\pi_2(G/H)$ nontrivial, I will begin by focusing on the long-range field, the part outside the monopole core, and determine the possible magnetic charges. This is essentially an extension of Dirac's analysis to the case where the unbroken gauge group H is larger than U(1) and, in general, non-Abelian. Like Dirac's analysis, this analysis applies equally well to singular point monopoles and to nonsingular solitons. Also as with the Dirac analysis, we will find that in some cases there are allowed magnetic charges for which there are no nonsingular monopoles.

Let us therefore start by assuming that we have an unbroken gauge group H, and that the magnetic components F_{ij} of the field strength have long-range tails that fall as $1/r^2$. Let us further assume that the vector potential can be expanded in inverse powers of r, with its Cartesian components falling as $1/r$ or faster, and that[1]

$$A_i = \frac{f_i(\theta, \phi)}{r} + O(1/r^2), \tag{6.1}$$

where A_i and f_i should be understood to be matrices representing elements of the Lie algebra of H. The $1/r^2$ terms do not contribute to the $1/r^2$ part of the

---

[1] In order to simplify the notation, the gauge fields in this section have been rescaled so as to absorb the gauge coupling constant.

magnetic field, and so will henceforth be omitted. If we now go over to spherical components, these assumptions imply that A_θ and A_ϕ are independent of r at large distance. Finally, we will restrict ourselves to static fields with no electric charge, and with $A_0 = 0$.

The first step is to choose a gauge where $A_r = 0$ outside a sphere of radius R. (Excluding the region $r < R$ avoids any issues of singularities at the origin, and has no effect on the fields at large distance.) This can be done by means of a gauge transformation with

$$U^{-1}(r, \theta, \phi) = P \exp\left[\frac{i}{g} \int_R^r dr' \, A_r(r', \theta, \phi)\right], \qquad (6.2)$$

with the P denoting path ordering. With this gauge choice and our dropping of the $1/r^2$ terms in A_i, the only nonzero component of the field strength is $F_{\theta\phi}$, which is independent of r.

Next, we set $A_\theta = 0$ in this asymptotic region by means of an r-independent gauge transformation obtained in a similar fashion by integrating along arcs of fixed ϕ starting at the north pole and moving downward. The field equation

$$0 = D_a(\sqrt{g} \, F^{a\phi}) \qquad (6.3)$$

then implies that A_ϕ, the only nonzero component of the gauge potential, is of the form

$$A_\phi = C(\phi) + \frac{Q_M(\phi)}{4\pi} \cos\theta. \qquad (6.4)$$

To avoid a singularity at the north pole, $\theta = 0$, we must set $C(\phi) = -Q_M(\phi)/4\pi$. The field equation

$$0 = D_a(\sqrt{g} \, F^{a\theta}) \qquad (6.5)$$

then tells us that Q_M is independent of ϕ, so that

$$A_\phi = \frac{1}{4\pi} Q_M(\cos\theta - 1) \qquad (6.6)$$

and (returning to Cartesian components)

$$B_i = -\frac{1}{2}\epsilon^{ijk} F_{jk} = \frac{Q_M}{4\pi} \frac{\hat{r}_i}{r^2}. \qquad (6.7)$$

This looks very much like the U(1) case, with a Coulomb magnetic field and a Dirac vector potential with its string along the negative z-axis. There are two crucial differences. First, the U(1) charge was simply a number, whereas now Q_M is a matrix in the appropriate representation of the Lie algebra. Second, Q_M, like B_i itself, is not gauge invariant, but only gauge covariant. Had we made other gauge choices, it could even have varied with angle.

Despite these differences, the derivation of the quantization condition goes through pretty much as before. Requiring that the potential of Eq. (6.6) be related by a single-valued gauge transformation to an equivalent potential with the Dirac string along the positive z-axis gives the condition [96–98]

$$e^{iQ_M} = I. \tag{6.8}$$

The corresponding condition in the U(1) case, Eq. (5.12), was required to hold for all electric charges q that are present in the theory. Similarly, Eq. (6.8) must hold with Q_M a matrix in any representation of the Lie algebra that actually occurs in the theory.

Let us start by considering the case where the unbroken gauge symmetry has the algebra of SU(2), with the generators taken to be the T_a ($a = 1, 2, 3$) with the standard normalization. Since any element in the Lie algebra can be rotated by a global gauge transformation to be proportional to T_3, there is no loss of generality in writing

$$Q_M = 4\pi k T_3. \tag{6.9}$$

The quantization condition then becomes

$$e^{4\pi k T_3} = I, \tag{6.10}$$

so that for every eigenvalue t_3 we must have

$$2k t_3 = n, \tag{6.11}$$

with n an integer.

If all of the fields in the theory transform under integer "spin" representations of SU(2), then the t_3 will all be integers and k can be either an integer or an integer plus 1/2. If there are also fields transforming under half-integer representations, then k must be an integer.

These conditions can be rephrased by being more careful about specifying the gauge group, keeping in mind that SO(3) and SU(2) share the same Lie algebra. The allowed eigenvalues of T_3 are known as weights. These are all integers for SO(3), and either integers or half-integers for SU(2). If the theory only has fields in integer representations, and thus only integer weights, then the gauge group can be taken to be either SO(3) or SU(2); let us choose the former. If there are also fields transforming under half-integer representations, then the gauge group is unambiguously SU(2). What we have found is that if the "electric" gauge group is SU(2), then the "magnetic weight" k must be a weight of the "magnetic group" SO(3). Conversely, if the electric group is SO(3), then the allowed magnetic weights are those of SU(2).

Configurations with different values of k are not all physically distinct. Because a global gauge transformation can reverse the sign of T_3, monopoles with magnetic weights k and $-k$ are gauge equivalent and can be continuously transformed one into the other. (It is possible to continuously connect these two discrete values because for the interpolating configurations Q_M is not simply a multiple of T_3.)

This leads to a curious phenomenon. Consider a configuration with two monopoles, each with magnetic charge $Q_M = 2\pi T_3$, held fixed at some separation much larger than their core size. Outside the monopole cores there is an obvious solution of the field equations in which the magnetic field is equal to $2\pi T_3$ times the ordinary electrostatic field corresponding to two like-sign unit charges. However, changing the sign of one of the charges by a gauge rotation leads us to another quasi-Abelian solution, of lower energy, formed from the electrostatic field of two charges with opposite sign. If the field and charges are initially in the former solution, even the slightest perturbation should be enough to cause them to transform into the latter, radiating off the excess energy in the process.

In this process, the total charge (i.e., that seen at spatial infinity) will have changed from $4\pi T_3$ to 0. By considering configurations with more monopoles, we can see that configurations with total charges differing by any integer multiple of $4\pi T_3$ can be continuously connected (although the intermediate configurations may not be solutions of the static field equations). Hence, two configurations can be continuously deformed into one another if they both have integer, or both have half-integer magnetic weights. What if one has an integer and the other a half-integer magnetic weight?

One can define a topological quantity that shows that these cannot be connected [99]. Consider a sphere at very large radius, where the gauge fields can be assumed to take on their asymptotic form. This sphere can be covered by a family of loops, each of which begins and ends at the north pole. One such family consists of loops $C(\tau)$ that go from the north pole ($\theta = 0$) to the south pole ($\theta = \pi$) along the arc $\phi = 0$ and then back along the arc $\phi = 2\pi\tau$. As τ ranges from 0 to 1, these cover the sphere.

Each such loop defines a group element

$$h(\tau) = P \exp \left[-i \int_{C(\tau)} d\boldsymbol{\ell} \cdot \mathbf{A} \right], \tag{6.12}$$

where A_i is understood to be a matrix and the P indicates path ordering. This Wilson loop is gauge invariant, so we can evaluate it in a gauge where $A_\theta = 0$ on the sphere. For nonzero Q_M there is a gauge singularity at the south pole, with $A_\phi \neq 0$. Including an infinitesimal section around this singularity, we can view the Wilson loop as being composed of three parts: (1) $0 \leq \theta \leq \pi$, with $\phi = 0$; (2) $0 \leq \phi \leq 2\pi\tau$, with $\theta = \pi$; and (3) $0 \leq \theta \leq \pi$, with $\phi = 2\pi\tau$. Since $A_\theta = 0$, the first and last segments do not contribute to the line integral. With A_ϕ given by Eq. (6.6), the integral is straightforward to evaluate, and we have

$$h(\tau) = \exp \left[i\tau Q_M \right]. \tag{6.13}$$

Clearly $h(0) = I$. By the quantization condition of Eq. (6.8), we also have $h(1) = I$. Thus, $h(\tau)$ for $0 \leq \tau \leq 1$ defines a loop in the (electric) gauge group H and so can be assigned to an element of $\pi_1(H)$. If H is the simply connected

SU(2), then π_1 has only a single element, and the topological invariant is trivial. If instead $H = \text{SO}(3)$, then π_1 has two elements, one arising when k is an integer and the other when it is a half-integer. Hence, a configuration with integer magnetic weight cannot be continuously deformed into one with a half-integer magnetic weight.

Now note that the energy in the long-range Coulomb field of a monopole is proportional to $\text{tr}\, Q_M^2$. Since configurations with the same topological charge can have different values for this quantity, one might wonder if solutions with higher values were unstable and could decay to solutions with lower values of $\text{tr}\, Q_M^2$. This instability is not completely obvious, both because there might be an energy barrier along the path joining the two solutions, and because we have not considered the fields in the core region. These concerns are dispelled by detailed analysis of small fluctuations about a pure Coulomb solution, as was done by Brandt and Neri [100] and by Coleman [101].[2] They showed that, even if one imposes the restriction that the field be held fixed inside some sphere of radius R, there is always such an instability. Stable solutions must have the minimum value of $\text{tr}\, Q_M^2$ consistent with their topological charge. In the present case, monopoles can only be stable if $k = \pm 1/2$; monopoles with integer magnetic weights are all unstable.

The language of roots and weights[3] is ideally suited for generalizing this discussion to the case of an arbitrary semisimple unbroken gauge group H [97, 98]. The Cartan subalgebra can be chosen to include any given element of the Lie algebra.[4] Hence, there is no loss of generality in taking Q_M to be a linear combination of the generators H_a of the Cartan subalgebra and writing

$$Q_M = 4\pi \mathbf{k} \cdot \mathbf{H}. \tag{6.14}$$

The components k_a are called the magnetic weights of the monopole. In a basis where the H_a, and hence Q_M, are simultaneously diagonalized, the diagonal elements of Q_M are of the form $4\pi i \mathbf{k} \cdot \mathbf{w}$, where $\mathbf{w}$ is a weight vector of the representation in which the generators are being expressed. The quantization condition of Eq. (6.8) becomes the requirement that for every weight $\mathbf{w}$ that appears in the theory

$$2\mathbf{k} \cdot \mathbf{w} = n \tag{6.15}$$

for some integer n. If the group is SU(2), Eqs. (6.14) and (6.15) reduce to our previous results, Eqs. (6.9) and (6.11), respectively.

---

[2] Their analysis does not apply in the BPS limit, in which the Higgs field also has a long-range tail.

[3] Root and weight vectors are reviewed in Appendix A.

[4] For the important case of SU(N), where the Cartan subalgebra can be taken to be the elements that are diagonal in the fundamental representation, this is just the statement that any Hermitian matrix can be diagonalized.

Any weight $\mathbf{w}$ and root $\boldsymbol{\alpha}$ must satisfy

$$\frac{2\mathbf{w} \cdot \boldsymbol{\alpha}}{\alpha^2} = N, \tag{6.16}$$

with N an integer. Hence one solution of the quantization condition is given by

$$\mathbf{k} = \sum n_\alpha \frac{\boldsymbol{\alpha}}{\alpha^2} = \sum n_\alpha \boldsymbol{\alpha}^*, \tag{6.17}$$

with the n_α integers; i.e., by taking $\mathbf{k}$ to be an element of the root lattice of a dual group H^v whose roots are the dual roots $\boldsymbol{\alpha}^* = \boldsymbol{\alpha}/\alpha^2$. If H is simple and the $\boldsymbol{\alpha}^*$ differ from the $\boldsymbol{\alpha}$ only by an overall rescaling, then H and H^v share the same Lie algebra, although they will not in general be the same group. For a semisimple H, equality of the two Lie algebras only requires that the rescaling be the same within each simple factor.

If H is the universal covering group of the Lie algebra, so that all representations appear, Eq. (6.17) gives the only solution for $\mathbf{k}$. If not, there are additional solutions. For example, if H is the adjoint group, whose representations all have weights lying on the root lattice, then any weight $\mathbf{w}$ can be written as an integral sum of the roots $\boldsymbol{\alpha}$. Applying Eq. (6.15) to the adjoint representation, whose weights are equal to the roots, yields the requirement that

$$2\mathbf{k} \cdot \boldsymbol{\alpha} = \frac{2\mathbf{k} \cdot \boldsymbol{\alpha}^*}{\alpha^{*2}} = N', \tag{6.18}$$

be an integer. This is the same as requiring that $\mathbf{k}$ be a weight of the Lie algebra of H^v.

Thus, just as with the case of SU(2), we have an electric group whose weights correspond to the representations of the elementary fields and a magnetic group whose weights correspond to the allowed magnetic charges. The former has roots $\boldsymbol{\alpha}$, while their duals $\boldsymbol{\alpha}^*$ are the roots of the latter. The larger one group is, the smaller the other must be. If one is the universal covering group of its algebra, the other is the adjoint group.[5]

We saw for the case of SU(2) that solutions with magnetic weights k and $-k$ are gauge-equivalent. In the general case, magnetic weight vectors $\mathbf{k}$ and $\mathbf{k}'$ that are related by Weyl reflections lead to physically equivalent solutions.

For SU(2) we also found that any two configurations with integer k or any two with half-integer k could be continuously deformed into one another. On the other hand, a configuration with integer k and one with half-integer k could not be continuously connected, a fact that could be verified by noting that the loop defined by $h(\tau)$ associated them with different elements of $\pi_1(H)$.

The generalization of this result can be expressed in terms of sublattices of the weight lattice. The weight lattice of the simply connected covering group can be

---

[5] The possible intermediate cases are discussed in [98].

decomposed into sublattices such that any two weights in the same sublattice differ by an integer sum of root vectors, while the difference between weights in different sublattices is never of this form. Just as in the case of $H = SU(2)$, with its two sublattices, these sublattices are in one-to-one correspondence with the elements of $\pi_1(H)$. Configurations with magnetic weights in the same sublattice can be continuously deformed into one another, while those with weights in different sublattices cannot.

Finally, the Brandt–Neri–Coleman analysis generalizes immediately to larger groups, and shows that stable solutions must have the minimum value of $\operatorname{tr} Q_M^2$ consistent with their topological charge. In particular, solutions with nonzero magnetic weights lying in the root lattice, which includes the origin, are all unstable.

The discussion thus far has assumed that H is semisimple. If this assumption is relaxed, then the U(1) electric and magnetic charges also contribute to the quantization condition. For simplicity, let us assume that H contains a single U(1) factor with generator $T_{U(1)}$. The magnetic charge will then have a non-Abelian component, given by Eq. (6.14), and an Abelian component equal to $4\pi k_{U(1)} T_{U(1)}$. If there is an electrically charged particle with non-Abelian electric weight $\mathbf{w}$ and U(1) electric charge $q_{U(1)}$, then Eq. (6.15) is replaced by

$$2\mathbf{k} \cdot \mathbf{w} + 2k_{U(1)} q_{U(1)} = n, \tag{6.19}$$

with an obvious generalization if there are multiple U(1) factors. The theory always includes the gauge bosons of the semisimple part of H, which carry no U(1) charge and have weights in the root lattice. Imposing the quantization condition using these shows that $\mathbf{k}$ must lie in the magnetic weight lattice.

Let us consider an explicit example. In the real world, there is an unbroken SU(3)×U(1) gauge group, with the first factor corresponding to QCD and the latter to electromagnetism. Experimentally, there is a correlation between the color and electromagnetic charges, in that particles invariant under SU(3) or transforming under zero triality representations have integer U(1) charges ne, while particles corresponding to triality ± 1 representations (e.g., quarks and antiquarks, respectively), have fractional electric charges of the form $(N \pm \frac{1}{3})e$.

If the monopole's magnetic charge has no SU(3) component, we recover the original Dirac quantization condition

$$2k_{U(1)} q_{U(1)} = n. \tag{6.20}$$

The same is true if the monopole has a magnetic weight in the dual root lattice, since then $2\mathbf{k} \cdot \mathbf{w}$ is an integer for any $\mathbf{w}$. For either case the existence of down quarks with U(1) electric charge $-e/3$ implies that the minimum U(1) magnetic charge is $4\pi k_{U(1)} = 6\pi/e$.

Now suppose that the magnetic weight $\mathbf{k}$ of the monopole lies in the triality 1 sublattice. Applying Eq. (6.19) with an electrically charged particle of SU(3) triality 0 leads again to Eq. (6.20). Because all triality 0 particles have integer electric charges, this quantization condition is satisfied if the magnetic charge is a multiple of $2\pi/e$. However, we must also consider the electrically charged particles with nonzero triality and fractional U(1) charge. For any triality ± 1 representation with weight $\mathbf{w}$ the product $2\mathbf{k} \cdot \mathbf{w}$ is of the form $n_1 \pm \frac{2}{3}$, and the U(1) electric charge is $(n_2 \pm \frac{1}{3})e$. Substituting these into Eq. (6.19) leads to the requirement that $2k_{U(1)}e$ be an integer, so that the U(1) component of the magnetic charge is still of the form

$$4\pi k_{U(1)} = \frac{2\pi n}{e}. \tag{6.21}$$

The net effect of all of this is that if a color singlet electron, with charge $-e$, goes around the Dirac string of this monopole, it acquires a phase of 2π and the string is unobservable, as required. If a quark is carried around the same path, the U(1) charges give a phase that is less than 2π, but the deficit is made up by the phase from the SU(3) magnetic and electric charges.

6.2 Larger gauge groups—topology

The analysis in the previous section focused on the long-range fields, extending Dirac's analysis to determine what magnetic charges are allowed by the requirement that the Dirac string be unobservable. However, not every allowed magnetic charge can be realized in a nonsingular soliton. For such a soliton to arise and be topologically stable, the manifold of vacuum solutions, $\mathcal{M} = G/H$, must have a nontrivial second homotopy group. In this section I will illustrate some of the possible behaviors. In these examples Eq. (4.51), which reduces the calculation of $\pi_2(G/H)$ to the calculation of $\pi_1(H)$, will be of considerable help.

6.2.1 SU(3) broken to SU(2)×U(1)

Consider a $G = $ SU(3) gauge theory with gauge coupling g and an octet Higgs field that can be viewed as a traceless 3×3 Hermitian matrix. Such a matrix can always be diagonalized, and so can be characterized by its eigenvalues. If the scalar field potential is such that the Higgs vacuum expectation value has three unequal eigenvalues, the symmetry is broken to U(1)×U(1). Let us focus instead on the other possibility, with two equal eigenvalues and a vacuum expectation value

$$\phi_0 = \text{diag} (2b, -b, -b). \tag{6.22}$$

The generators of the unbroken symmetry can then be taken to be

$$T_8 = \text{diag} \left(\frac{1}{\sqrt{3}}, -\frac{1}{2\sqrt{3}}, -\frac{1}{2\sqrt{3}} \right) \tag{6.23}$$

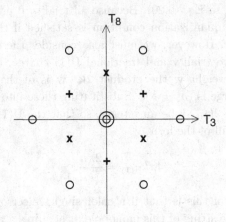

Fig. 6.1. The weights of the octet (open circles), triplet (+'s), and antitriplet (×'s) representations of SU(3). Note that the octet has a pair of weights at $T_3 = T_8 = 0$.

together with the three generators

$$T_a = \begin{pmatrix} 0 & 0 \\ 0 & \frac{\tau_a}{2} \end{pmatrix}, \qquad a = 1, 2, 3, \tag{6.24}$$

obtained by embedding the Pauli matrices in the lower right 2×2 block. These generate an SU(2) × U(1) algebra, but because $e^{2\pi i T_a} e^{2\sqrt{3}\pi i T_8} = I$, the actual unbroken symmetry group is $H = [\text{SU}(2) \times \text{U}(1)]/Z_2$. The generators of the Cartan subalgebra can be chosen to be $H_1 = T_3$ and $H_2 = T_8$.

After the symmetry breaking the spectrum of states from the octet fields includes massive scalars and massless vectors in the singlet and triplet representations of SU(2), all with $T_8 = 0$, and two SU(2) doublets of massive vectors with $T_8 = \pm\frac{\sqrt{3}}{2}$. The corresponding weight vectors are shown in Fig. 6.1. Also shown are the weights that would appear if the theory also included fields transforming under the triplet and antitriplet representations of the original SU(3).

Nonsingular monopoles correspond to nontrivial elements of $\pi_2(G/H)$. Because SU(3) is simply connected we can make use of Eq. (4.51), which gives

$$\pi_2(G/H) = \pi_1(H) = \pi_1\{[\text{SU}(2) \times \text{U}(1)]/Z_2\} = Z, \tag{6.25}$$

with the Z arising from the U(1). Hence, there are topologically stable nonsingular monopoles carrying U(1) magnetic charge. We will see that they can also carry SU(2) magnetic charge.

A nonsingular spherically symmetric configuration with unit magnetic charge can be obtained by embedding the 't Hooft–Polyakov ansatz of Eq. (5.60) in the SU(2) subgroup lying in the upper left 2×2 block of the SU(3) matrices; i.e., the subgroup generated by $\frac{1}{2}\lambda_1$, $\frac{1}{2}\lambda_2$, and $\frac{1}{2}\lambda_3$, where the λ_a are the Gell-Mann matrices defined in Eq. (A.3). This by itself would not give the correct eigenvalues for ϕ at spatial infinity, so a term proportional to λ_8 must be added. Thus, we have [102, 103]

$$\phi = \frac{1}{2}\sum_{a=1}^{3}\hat{r}^a\lambda_a h(r) + \frac{1}{2}\lambda_8 j(r),$$

$$A_i = \frac{1}{2}\sum_{a=1}^{3}\epsilon^{iam}\hat{r}^m\lambda_a\left[\frac{1-u(r)}{gr}\right]. \tag{6.26}$$

Requiring that ϕ be nonsingular at the origin and be gauge equivalent to ϕ_0 at spatial infinity gives the boundary conditions

$$h(0) = 0, \quad j'(0) = 0,$$
$$h(\infty) = 3b, \quad j(\infty) = \sqrt{3}\,b \tag{6.27}$$

for the scalar field, while for the gauge field we have $u(0) = 1$ and $u(\infty) = 0$, as in the SU(2) monopole. Solving the field equations with this ansatz and these boundary conditions gives a monopole with a mass

$$M_{\text{mon}} \approx \frac{4\pi(3b)}{g}. \tag{6.28}$$

According to the analysis of the previous section, the asymptotic magnetic field should take the Coulomb form of Eq. (6.7), where Q_M can be gauge rotated to

$$Q_M = \frac{4\pi}{g}(k_1 H_1 + k_2 H_2) = \frac{4\pi}{g}(k_1 T_3 + k_2 T_8). \tag{6.29}$$

Indeed, the monopole solution that follows from the ansatz of Eq. (6.26) has a magnetic charge that along the positive z-axis is given by

$$Q_M = \frac{4\pi}{g}\,\text{diag}\left(\frac{1}{2}, -\frac{1}{2}, 0\right), \qquad k_1 = -\frac{1}{2}, \quad k_2 = \frac{\sqrt{3}}{2}. \tag{6.30}$$

Our ansatz was obtained by making use of the SU(2) subgroup lying in the upper left 2×2 block of the SU(3) matrices. We could equally well have used the SU(2) subgroup defined by the four corner elements of the SU(3) matrices. In this case the magnetic field along the positive z-axis would have given

$$Q_M = \frac{4\pi}{g}\,\text{diag}\left(\frac{1}{2}, 0, -\frac{1}{2}\right), \qquad k_1 = \frac{1}{2}, \quad k_2 = \frac{\sqrt{3}}{2}. \tag{6.31}$$

The two ansatzes are related by a global gauge rotation in the unbroken SU(2).

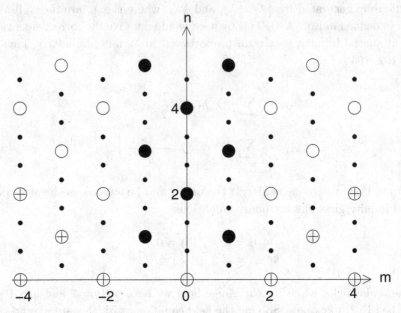

Fig. 6.2. Allowed magnetic weights for SU(3) broken to SU(2)×U(1). Here $m = 2k_1$ and $n = 2k_2/\sqrt{3}$, with the k_j defined as in Eq. (6.29). The large circles represent magnetic weights that are consistent with any representation for the electrically charged particles, while the weights denoted by small circles are only allowed if the electrically charged particles are all in triality zero representations. Only the former can be obtained from configurations containing collections of nonsingular monopoles ($n = 1$) and antimonopoles ($n = -1$). Of these, the large solid circles represent weights that are stable by the Brandt–Neri–Coleman analysis; these can all be obtained from configurations containing only monopoles. The large open circles can also be obtained using only monopoles, but are unstable by this analysis. The large circles with crosses require assemblies of monopoles and antimonopoles, and are also unstable. The pattern of weights for $n < 0$ is similar, with the roles of monopoles and antimonopoles interchanged.

By assembling a number of these monopoles, using various combinations of the two forms, we can construct configurations with $k_1 = m/2$ and $k_2 = n\sqrt{3}/2$, where m and n are either both even or both odd integers and $|m| \leq n$. These points are indicated in Fig. 6.2. It is n, from the coefficient of the U(1) generator, that is the conserved topological charge. Configurations with different values of m can be deformed into one another and, by the results of Brandt, Neri, and Coleman, will reduce their long-range non-Abelian components until $k_1 = \pm\frac{1}{2}$ or 0. If the only fields in the theory are the SU(3) adjoint Higgs and gauge fields, then the generalized Dirac quantization condition allows a larger set of charges, which are also shown in Fig. 6.2. The absence of nonsingular solitons with these charges can be understood by noting that they would be forbidden if fields with nonzero triality were added to the theory.

6.2.2 A Z_2 monopole

Let us again consider an SU(3) gauge theory with gauge coupling g, but this time with a Higgs field S that transforms according to the **6** representation of SU(3) [104]. This can be viewed as a symmetric 3×3 matrix that transforms as

$$S \to USU^T, \tag{6.32}$$

where U is an SU(3) matrix and a superscript T denotes the transpose. If the Higgs potential is minimized by

$$S_0 = \sigma \begin{pmatrix} 1 & 0 & 0 \\ 0 & 1 & 0 \\ 0 & 0 & 1 \end{pmatrix}, \tag{6.33}$$

the unbroken symmetry is the SO(3) subgroup generated by the antisymmetric matrices λ_2, λ_5, and λ_7. Because the triplet of the original SU(3) transforms as a vector under SO(3), only integer-spin representations of SO(3) appear, confirming that the unbroken group really is SO(3), and not SU(2).

Making use of Eq. (4.51), we find that the second homotopy group of the vacuum manifold is

$$\pi_2[\mathrm{SU}(3)/\mathrm{SO}(3)] = \pi_1[\mathrm{SO}(3)] = Z_2. \tag{6.34}$$

Thus, we should expect to find monopoles with Z_2 topological charges. A combination of two of these should be topologically trivial, and so the monopole should be its own antiparticle.

To see more explicitly how this can occur, let us start with a singular string gauge configuration in which $S = S_0$ is spatially uniform at large distance while the gauge potential has the Dirac form

$$A_r = A_\theta = 0, \quad A_\phi = \frac{n}{2g}(\cos\theta - 1)\lambda_2, \tag{6.35}$$

with the Dirac string also along the negative z-axis. This has a magnetic charge

$$Q_M = \frac{4\pi}{g}\frac{n}{2}\lambda_2. \tag{6.36}$$

If we take λ_2 to be the single generator of the Cartan subalgebra, this corresponds to magnetic weight $k = n/2$.

This string singularity can be removed by a gauge transformation generated by the gauge function

$$U_n(\theta, \varphi) = e^{in\lambda_2\varphi/2}e^{i\lambda_3\theta/2}e^{-in\lambda_2\varphi/2}, \tag{6.37}$$

which is singular along the negative z-axis. [See the analogous transformation given by Eq. (5.74).] In particular, the asymptotic Higgs field at $r = \infty$ becomes

$$S_n = U_n S_0 U_n^T$$

$$= \begin{pmatrix} \cos\theta + i\sin\theta\cos(n\varphi) & -i\sin\theta\sin(n\varphi) & 0 \\ -i\sin\theta\sin(n\varphi) & \cos\theta - i\sin\theta\cos(n\varphi) & 0 \\ 0 & 0 & 1 \end{pmatrix}. \quad (6.38)$$

The covariant derivatives of S were rapidly vanishing at large distance before the gauge transformation, and so must also be afterwards. Hence, we have a finite energy configuration, and the usual arguments show that for $n = \pm 1$ there are actual static solutions with this asymptotic behavior. These would be expected to have a mass $\sim 4\pi\sigma/g$.

A Z_2 monopole should be its own antiparticle. We can see that this is so by noting that

$$S_{-1} = e^{i\pi\lambda_5} S_1 \left(e^{i\pi\lambda_5}\right)^T, \quad (6.39)$$

so that the $n = 1$ monopole and $n = -1$ antimonopole Higgs fields (and in fact their vector potentials also) are gauge equivalent. Furthermore, let us define a unitary matrix

$$V_n(\theta, \varphi) = e^{i\lambda_5\theta} e^{in\lambda_2\varphi/2} e^{-i\lambda_5\theta} e^{-i\lambda_3\theta/2} e^{-in\lambda_2\varphi/2}. \quad (6.40)$$

This is multiple-valued for odd n, because $V_n(\theta, \varphi) \neq V_n(\theta, \varphi + 2\pi)$. For even n, on the other hand, it is nonsingular and single-valued, with $V_n(0, \varphi) = I$, and has the property that

$$V_n S_n V_n^T = S_0. \quad (6.41)$$

Hence, any configuration with even n is equivalent by a smooth gauge transformation to one with $n = 0$. Because $\pi_2(\mathrm{SU}(3)) = 0$, this gauge transformation at $r = \infty$ can be smoothly deformed to the identity, thus giving a homotopy connecting S_n and S_0. Note that Eqs. (6.39) and (6.41) both required gauge transformations involving matrices that went outside the 2×2 block that contains the twisting of S_n.

6.2.3 A light doubly charged monopole

Let us now add an SU(3) triplet Higgs field ψ to the model of the previous example [104]. Let us also assume that ψ has a nonzero vacuum expectation value, and that in the vacuum the two Higgs fields are (up to a gauge transformation)

$$\begin{aligned} S_0^{ab} &= \sigma\delta^{ab}, \\ \psi_0^a &= v\delta^{a3}. \end{aligned} \quad (6.42)$$

The unbroken gauge group is now U(1), so the relevant homotopy group is

$$\pi_2[\mathrm{SU}(3)/\mathrm{U}(1)] = \pi_1[\mathrm{U}(1)] = Z. \quad (6.43)$$

The topological charge on the monopoles is now an ordinary additive integer charge. Choosing the asymptotic Higgs fields so that $S = S_n(\theta, \varphi)$ as in Eq. (6.38) and $\psi = \psi_0$ gives a configuration with topological charge n.

Ordinarily, we would expect to have a static particle-like solution only for $n = \pm 1$, with all larger values of n corresponding to multimonopole configurations. A new feature appears if $v \ll \sigma$, so that the symmetry breaking can be viewed as a two-step process

$$SU(3) \xrightarrow[S]{} SO(3) \xrightarrow[\psi]{} U(1). \tag{6.44}$$

At the first stage we obtain a Z_2 monopole with mass $M_1 \sim \sigma/g^2$. This remains a solution, with only slight modifications, at the second stage. However, the transformation of Eq. (6.40), which turned the monopole into an antimonopole, is no longer possible, because λ_5 is not a generator of the unbroken group.

Now consider an $n = 2$ configuration. With just the breaking to SO(3), this would be topologically trivial and could be unwound by applying the V_2 of Eq. (6.40). With the breaking to U(1), it has topological charge 2 and is topologically nontrivial. We can still use V_2 to unwind S, but this would have the effect of twisting ψ. However, because the mass scale associated with ψ is much less than that associated with S, shifting the winding from S to ψ reduces the energy considerably, thus allowing us to obtain a charge 2 monopole with $M_2 \sim v/g^2 \ll M_1$.

All configurations with $n \geq 3$ presumably relax to multimonopole solutions.

6.2.4 Electroweak monopoles?

The previous examples in this section were illustrative, but not of direct phenomenological significance. Let us now consider the standard electroweak model, with SU(2)×U(1) broken to U(1) by a complex doublet Higgs field. Because the full gauge group is not simply connected, we cannot use Eq. (4.51) to determine $\pi_2(\mathcal{M}) = \pi_2(G/H)$. This is no problem, because we showed in Sec. 4.5 that the space of vacua for this theory is a three-sphere [see Eq. (4.26)]. Since $\pi_2(S^3) = 0$, there are no topologically stable monopoles in the Weinberg–Salam model.

6.3 Monopoles in grand unified theories

The idea of a grand unified theory (GUT) whose spontaneous breakdown leads to the observed gauge symmetries of the Standard Model remains an attractive possibility. Various implementations of this idea, often with several stages of symmetry breaking, have been proposed. By definition, all begin with a simple gauge group G that is ultimately broken down to the SU(3)×U(1) of QCD and electromagnetism. If we take G to be the covering group of the Lie algebra, Eq. (4.51) tells us that

$$\pi_2(G/H) = \pi_1(H) = \pi_1[SU(3) \times U(1)] = Z, \tag{6.45}$$

with the Z arising from the unbroken electromagnetic U(1). Thus any grand unified theory must contain topologically stable magnetic monopoles. Their mass will be of the order of $4\pi v/e$, where v is the vacuum expectation value of the Higgs field responsible for the symmetry breaking that first gives rise to nontrivial topology. Since this is typically a GUT scale of roughly 10^{16} GeV, these monopoles will be supermassive.

Let us consider two important examples.

6.3.1 SU(5) monopoles

The prototypical grand unified theory is based on an SU(5) gauge group, with a gauge coupling g, that is broken in two stages,

$$\text{SU}(5) \xrightarrow[\Phi]{} \text{SU}(3) \times \text{SU}(2) \times \text{U}(1) \xrightarrow[\chi]{} \text{SU}(3) \times \text{U}(1). \qquad (6.46)$$

The first breaking is due to an adjoint representation Higgs field Φ that acquires a GUT-scale vacuum expectation value, while the second is due to a fundamental representation Higgs field χ that includes the Weinberg–Salam doublet with an electroweak scale vacuum expectation value.

Because the fermions fall into the $\bar{\mathbf{5}}$ and $\mathbf{10}$ representations, the initial group is indeed the covering group, SU(5), and not a factor group. By arguments similar to those for the SU(3) example of Sec. 6.2.1, the final unbroken subgroup is actually $[\text{SU}(3) \times \text{U}(1)]/Z_3$, with the factoring by Z_3 explaining the observed correlation between SU(3) triality and fractional electric charge.

To start, let us focus on the first breaking and set $\chi = 0$. The scalar field potential can then be chosen so that Φ has a vacuum expectation value of the form

$$\Phi_0 = \begin{pmatrix} v & 0 & 0 & 0 & 0 \\ 0 & v & 0 & 0 & 0 \\ 0 & 0 & v & 0 & 0 \\ 0 & 0 & 0 & -\frac{3}{2}v & 0 \\ 0 & 0 & 0 & 0 & -\frac{3}{2}v \end{pmatrix}. \qquad (6.47)$$

The generators of the unbroken symmetry then take on a block diagonal form, with SU(3) generators $\lambda_a/2$ lying in the upper left 3×3 block, SU(2) generators $\tau_a/2$ in the lower right 2×2 block, and the U(1) generator being

$$T_{\text{U}(1)} = \frac{1}{\sqrt{15}} \text{diag}\left(1, 1, 1, -\frac{3}{2}, -\frac{3}{2}\right). \qquad (6.48)$$

Twelve of the SU(5) gauge bosons acquire a mass $M_X = \sqrt{25/8}\, gv$ at this stage of symmetry breaking.

Because $\pi_1[\text{SU}(3) \times \text{SU}(2) \times \text{U}(1)] = Z$, monopoles already appear at this first stage of symmetry breaking. Classical solutions can be obtained by following a strategy similar to that used for the SU(3) example of Sec. 6.2.1 [105]. We choose an ansatz such that the twisting of the Higgs field lies entirely within a 2×2 SU(2) subgroup corresponding to the intersections of columns and rows 1, 2, or 3 with columns and rows 4 or 5, and then add diagonal components to Φ to ensure the correct eigenvalues at spatial infinity. Choosing, for example, the subgroup defined by rows and columns 3 and 4 gives a Higgs field ansatz of the form

$$\Phi = \begin{pmatrix} a(r) & 0 & 0 & 0 \\ 0 & a(r) & 0 & 0 \\ 0 & 0 & h(r)\hat{\mathbf{r}} \cdot \boldsymbol{\tau} + b(r)I_2 & 0 \\ 0 & 0 & 0 & -2[a(r)+b(r)] \end{pmatrix}. \tag{6.49}$$

The nonzero components of A_i all lie within the chosen SU(2), and lead to an asymptotic magnetic field with magnetic charge along the positive z-axis

$$\begin{aligned} Q_M &= \frac{4\pi}{g} \operatorname{diag}\left(0,0,\frac{1}{2},-\frac{1}{2},0\right) \\ &= \frac{4\pi}{g}\left[\operatorname{diag}\left(\frac{1}{6},\frac{1}{6},\frac{1}{6},-\frac{1}{4},-\frac{1}{4}\right) + \operatorname{diag}\left(-\frac{1}{6},-\frac{1}{6},\frac{1}{3},0,0\right)\right. \\ &\qquad \left. + \operatorname{diag}\left(0,0,0,-\frac{1}{4},\frac{1}{4}\right)\right], \end{aligned} \tag{6.50}$$

where the second equality shows the decomposition into U(1), SU(3), and SU(2) components, respectively. The classical energy of this monopole is approximately $4\pi M_X/g^2$, and its core radius is of the order of M_X^{-1}.

The electroweak symmetry breaking is driven by the vacuum expectation value of χ, which is at a mass scale 14 or so orders of magnitude lower than the GUT scale. The effects of this symmetry breaking only become significant at length scales of order M_W^{-1}, so the corrections to the monopole core structure and mass are negligible. However, at distances much larger than M_W^{-1} the Coulomb magnetic field must lie within the unbroken gauge group. Thus, whatever the orientation of the SU(2) magnetic field near the core, at large distances Q_M must be rotated so that it is a linear combination of an SU(3) charge and the electromagnetic charge generator

$$Q_{\text{em}} = \operatorname{diag}\left(\frac{1}{3},\frac{1}{3},\frac{1}{3},-1,0\right). \tag{6.51}$$

The normalization of Q_{em} is such that $e = \sqrt{3/8}\,g$ (evaluated at the GUT scale). Taking this into account, we find that the minimally charged monopole, with a core profile given by Eq. (6.49), has an electromagnetic magnetic charge $2\pi/e$.

In Eq. (6.49) the choice of the SU(2) subgroup that contained the twisting of Φ was somewhat arbitrary. We could, for example, have used the first and fifth rows and columns, leading to a solution with asymptotic magnetic charge

$$Q'_M = \frac{4\pi}{g} \operatorname{diag}\left(\frac{1}{2},0,0,0,-\frac{1}{2}\right). \tag{6.52}$$

Now consider a configuration composed of two monopoles, one with charge Q_M and one with Q'_M. At distances that are small compared to the electroweak scale, the Coulomb interaction between the two is proportional to $\text{tr}\,(Q_M Q'_M) = 0$. What has happened is that their long-range U(1) repulsion has been exactly canceled by the attractive SU(2) and SU(3) forces. The interaction between the two is then determined by the Yukawa forces mediated by the massive Higgs and gauge bosons. By proper choice of the masses of the bosons, one can arrange for the net effect of these to be attractive, giving a stable monopole with two units of U(1) magnetic charge. In fact, by this mechanism one can also obtain solutions with three, four, and six units of U(1) magnetic charge [106].

6.3.2 SO(10) monopoles

A second widely studied model is based on SO(10). One possible symmetry breaking pattern is commonly written as

$$\text{SO}(10) \underset{\phi_1}{\to} \text{SU}(4) \times \text{SU}(2) \times \text{SU}(2) \underset{\phi_2}{\to} \text{SU}(3) \times \text{SU}(2) \times \text{U}(1) \underset{\phi_3}{\to} \text{SU}(3) \times \text{U}(1). \quad (6.53)$$

This is correct as far as the Lie algebras go, but to get the homotopy right we need to be careful about specifying the groups.

To start, we note that the Standard Model quarks and leptons of one generation, together with a right-handed neutrino, fill out a 16-component SO(10) spinor. With a spinor representation present, the original gauge group G is unambiguously the covering group, Spin(10). In the first stage of symmetry breaking, a Higgs field ϕ_1 transforming under the 54-dimensional traceless symmetric tensor representation obtains a vacuum expectation value of order v_1 that breaks this symmetry down to a subgroup H_1 that is locally SO(6) × SO(4) = SU(4) × SU(2) × SU(2). Under this breaking the fundamental spinor of SO(10) decomposes into $(\mathbf{4}, \mathbf{1}, \mathbf{2}) + (\bar{\mathbf{4}}, \mathbf{2}, \mathbf{1})$. Here the **4** and $\bar{\mathbf{4}}$ are conjugate SO(6) spinors [or, equivalently, the fundamental and antifundamental representations of SU(4)], while the **2**'s are SU(2) spinors. A rotation by 2π multiplies a spinor by -1, so simultaneous rotations by 2π in the SO(6) and SO(4) subgroups give two factors of -1 and thus act as the identity on the fermions. Hence,

$$H_1 = [\text{SU}(4) \times \text{SU}(2) \times \text{SU}(2)]/Z_2, \quad (6.54)$$

which is not simply connected. Thus, we have

$$\pi_2(G/H_1) = \pi_1(H_1) = Z_2, \quad (6.55)$$

which means that there is a monopole carrying a Z_2 charge with a mass of order v_1/g. Like the Z_2 monopole of Sec. 6.2.2, this monopole is its own antiparticle.

At the next stage of symmetry breaking an SO(10) spinor Higgs field ϕ_2 gets a vacuum expectation value v_2 that breaks the symmetry down to

$$H_2 = [\text{SU}(3) \times \text{SU}(2) \times \text{U}(1)]/Z_6. \quad (6.56)$$

The crucial point here is the appearance of the U(1) factor, so that

$$\pi_2(G/H_2) = \pi_1(H_2) = Z \qquad (6.57)$$

and we now have monopoles with ordinary additive charges. If $v_2 \ll v_1$, the situation is essentially the same as in the example of Sec. 6.2.3. The Z_2 monopole that appeared at the first stage remains, but now with a unit Z charge. In addition, there is a new monopole, with two units of magnetic charge, associated with a nontrivial winding of ϕ_2 but a topologically trivial ϕ_1. This monopole has a core size of order $(gv_2)^{-1}$ and a mass $M_2 \sim v_2/g$ that can be several orders of magnitude smaller than that of the unit monopole [107].

Both monopoles survive the final stage of symmetry breaking with negligible corrections to their masses.

6.4 Chromodyons

We have seen that when a soliton is not invariant under a symmetry of the theory, the spectrum of fluctuations about the soliton includes a zero mode that requires the introduction of a collective coordinate z. Exciting this mode in a time-dependent fashion gives a nonzero conjugate momentum $p = I\dot{z}$, where I can be thought of as a generalized moment of inertia, and leads to a tower of excited states with energies $p^2/2I$ above the ground state. Thus, any soliton breaks translation invariance, and solitons with time-dependent position collective coordinates have nonzero linear momentum; here I is simply the soliton mass. If there is an unbroken U(1) internal symmetry that acts nontrivially on a soliton, then a time-dependent phase rotation gives the soliton a U(1) charge Q; for the case of the 't Hooft–Polyakov monopole, this yields the dyons studied in Sec. 5.5.

The GUT monopole solutions described in the previous section have Coulomb magnetic fields with nonzero components in the unbroken color SU(3). These are acted upon by the SU(3) generators, and so one might expect to obtain monopoles with SU(3) electric-type charges—chromodyons—from solutions that rotate in the internal SU(3) space.

Matters are not so simple. The first indication that there might be a problem is the slow falloff of the zero modes. A magnetic field falling as $1/r^2$ corresponds to a vector potential falling as $1/r$. Acting on such a field, an SU(3) transformation that was nontrivial at spatial infinity and did not commute with the magnetic charge would produce an infinitesimal transformation δA_j that also fell as $1/r$, making the resulting zero mode non-normalizable. One might naïvely view this as corresponding to an infinite moment of inertia, and conclude that the tower of chromodyon states collapses to a set of degenerate states. However, in a gauge theory one must proceed more carefully, making sure that the Gauss's law constraints are satisfied, as was done for the U(1) dyon in Sec. 5.5. This entails finding an A_0 that satisfies Gauss's law and that has a $1/r$ behavior consistent with the chromodyonic charge. It turns out that this cannot be done[6] [108].

The underlying explanation for these difficulties is that the long-range non-Abelian fields of the monopole create a topological obstruction that makes it impossible to define a set of generators for the unbroken gauge group that is nonsingular everywhere on the sphere at spatial infinity. Without these generators, one cannot define the global gauge rotations that would give rise to the chromodyons [109–114].

To see this more explicitly [109], consider the example of Sec. 6.2.1, where the unbroken gauge group is SU(2)×U(1). At spatial infinity the adjoint Higgs field is

$$\phi(\theta, \varphi) = b \begin{pmatrix} \frac{1}{2} + \frac{3}{2}\hat{\mathbf{r}} \cdot \boldsymbol{\tau} & 0 \\ 0 & -1 \end{pmatrix} = b\, U^{-1} \begin{pmatrix} 2 & 0 & 0 \\ 0 & -1 & 0 \\ 0 & 0 & -1 \end{pmatrix} U, \qquad (6.58)$$

where U is a 3×3 matrix with the block diagonal form

$$U(\theta, \varphi) = \begin{pmatrix} \mathcal{U} & 0 \\ 0 & 1 \end{pmatrix}, \qquad (6.59)$$

with $\mathcal{U}$ being the 2×2 SU(2) matrix given in Eq. (5.74).

At the north pole, $\theta = 0$, the unbroken SU(2) corresponds to the lower right 2×2 block, and a standard choice for the U(1) generator T_0 and the SU(2) generators T_k is

$$T_0 = \begin{pmatrix} 2 & 0 & 0 \\ 0 & -1 & 0 \\ 0 & 0 & -1 \end{pmatrix}, \qquad T_1 = \frac{1}{2}\begin{pmatrix} 0 & 0 & 0 \\ 0 & 0 & 1 \\ 0 & 1 & 0 \end{pmatrix},$$

$$T_2 = \frac{1}{2}\begin{pmatrix} 0 & 0 & 0 \\ 0 & 0 & -i \\ 0 & i & 0 \end{pmatrix}, \qquad T_3 = \frac{1}{2}\begin{pmatrix} 0 & 0 & 0 \\ 0 & 1 & 0 \\ 0 & 0 & -1 \end{pmatrix}. \qquad (6.60)$$

By acting on these with U, we can obtain a set that commutes with $\phi(\theta, \varphi)$ and has the correct commutation relations. Two of these,

$$T_0(\theta, \varphi) = \begin{pmatrix} \frac{1}{2} + \frac{3}{2}\cos\theta & \frac{3}{2}\sin\theta e^{-i\varphi} & 0 \\ \frac{3}{2}\sin\theta e^{i\varphi} & \frac{1}{2} - \frac{3}{2}\cos\theta & 0 \\ 0 & 0 & -1 \end{pmatrix} \qquad (6.61)$$

and

$$T_3(\theta, \varphi) = \frac{1}{2}\begin{pmatrix} \frac{1}{2} - \frac{1}{2}\cos\theta & -\frac{1}{2}\sin\theta e^{-i\varphi} & 0 \\ -\frac{1}{2}\sin\theta e^{i\varphi} & \frac{1}{2} + \frac{1}{2}\cos\theta & 0 \\ 0 & 0 & -1 \end{pmatrix}, \qquad (6.62)$$

---

[6] One can also work in $A_0 = 0$ gauge, in which case Gauss's law must be imposed as an additional constraint. In this approach, the construction of the chromodyon only goes through if there is a zero mode corresponding to a gauge transformation with a gauge function Λ that is nonzero at spatial infinity and satisfies $D_k D_k \Lambda + g^2[\phi, [\phi, \Lambda]] = 0$. This equation has no solutions if Λ does not commute with the magnetic charge [108].

are well defined everywhere, but the other two,

$$T_1(\theta, \varphi) = \frac{1}{2}\begin{pmatrix} 0 & 0 & -\sin(\theta/2)e^{-i\varphi} \\ 0 & 0 & \cos(\theta/2) \\ -\sin(\theta/2)e^{i\varphi} & \cos(\theta/2) & 0 \end{pmatrix} \quad (6.63)$$

and

$$T_2(\theta, \varphi) = \frac{1}{2}\begin{pmatrix} 0 & 0 & i\sin(\theta/2)e^{-i\varphi} \\ 0 & 0 & -i\cos(\theta/2) \\ -i\sin(\theta/2)e^{i\varphi} & i\cos(\theta/2) & 0 \end{pmatrix}, \quad (6.64)$$

are singular at $\theta = \pi$, the south pole.[7]

In this example, the generators that commuted with the magnetic charge (T_0 and T_3) could be defined globally. It was only the two that did not commute with Q_M that failed to be well defined. This can be understood as follows. One way to define a global gauge rotation is to choose a Lie algebra element Ω at one point P on a sphere at large r and then use parallel transport to obtain Ω at any other point P' on the sphere. This only works if the result of the parallel transport is independent of the path from P to P'. This in turn requires that the surface integral of $[\mathbf{B}, \Omega]$ over the area between any two such paths vanishes. In the limit of infinite radius only the $1/r^2$ part of $\mathbf{B}$, i.e., the magnetic charge, contributes to this integral. Hence, only the generators of the subgroup that commutes with the magnetic charge are well defined.

This suggests a loophole that might allow chromodyons to exist. Consider a monopole with a purely Abelian magnetic charge in a theory with an unbroken non-Abelian subgroup. The Abelian magnetic charge would not be an obstacle to defining global color transformations. Although these would have no effect on the asymptotic magnetic field, there might well be fields nearer the core that were not invariant under color transformation. The corresponding zero modes would be normalizable, and would provide the basis for constructing a chromodyonic solution.

A monopole of just this sort can be constructed in a gauge theory with SO(5) broken to SU(2)×U(1). However, numerical study of the classical evolution of the chromodyon solution reveals that the rate of rotation in color space decreases with time, corresponding to a loss of color charge [115]. This is apparently due to radiation of energy and color charge via the massless gauge boson field, with all indications being that the radiation continues until the charge has been completely lost. Thus, even when there are no topological obstacles to their existence, chromodyons appear to be dynamically unstable.

---

[7] An alternative approach is to use two patches, with sets of generators that are nonsingular in the upper and lower hemispheres, respectively. Consistency then requires that the two sets be related by a nonsingular gauge transformation in the overlap region. Again, this turns out to be impossible.

6.5 The Callan–Rubakov effect

We saw in Sec. 5.7 that the scattering of massless fermions off an 't Hooft–Polyakov monopole has some unusual aspects. In the $J = 0$ sector one finds only half of the expected incoming states and half of the expected outgoing states; this is ultimately a consequence of the extra charge-monopole contribution to the angular momentum. The matching of incoming to outgoing states requires that either the fermion chirality or the fermion electric charge must change. The analyses of Rubakov [91] and of Callan [92, 93] showed that the former is the case, and that no electric charge is deposited either on the monopole or on the surrounding fermion condensate.

An analogous effect occurs with monopoles in larger gauge groups, in particular those that arise in grand unified theories [91, 93, 116–118]. The new feature here is that the incoming and outgoing states that are paired have different baryon and lepton numbers. To be specific, let us consider a monopole in the SU(5) theory that has a magnetic charge given by Eq. (6.50). This is essentially an embedding of the SU(2) monopole in the subgroup corresponding to the third and fourth rows and columns.

Each family of fermion fields in the SU(5) model can be assembled into two multiplets of Weyl fields. The first family, which can be treated as approximately massless, contains an antifundamental $\bar{\mathbf{5}}$ representation,

$$\psi = (d_1^c, d_2^c, d_3^c, e^-, \nu)_L^t, \tag{6.65}$$

and a symmetric tensor $\mathbf{10}$ representation,

$$\chi = \frac{1}{\sqrt{2}} \begin{pmatrix} 0 & u_3^c & -u_2^c & -u_1 & -d_1 \\ -u_3^c & 0 & u_1^c & -u_2 & -d_2 \\ u_2^c & -u_1^c & 0 & -u_3 & -d_3 \\ u_1 & u_2 & u_3 & 0 & -e^+ \\ d_1 & d_2 & d_3 & e^+ & 0 \end{pmatrix}_L . \tag{6.66}$$

(Here subscripts are SU(3) color indices and a superscript c denotes charge conjugation and the d's should be understood as the CKM-rotated mixtures.) When these SU(5) multiplets are decomposed into representations of the SU(2) defined by the monopole embedding, we find four doublets,

$$\begin{pmatrix} e^+ \\ d_3 \end{pmatrix}_L, \quad \begin{pmatrix} d_3^c \\ e^- \end{pmatrix}_L, \quad \begin{pmatrix} u_1 \\ u_2^c \end{pmatrix}_L, \quad \begin{pmatrix} u_2 \\ -u_1^c \end{pmatrix}_L . \tag{6.67}$$

In the $J = 0$ sector, the upper components of the doublets only appear as incoming waves, and the lower ones as outgoing waves.

As with the SU(2) theory considered in Sec. 5.7, the analysis of the system is most easily done by bosonizing the theory. In the previous case there were two Dirac, or four Weyl, fermion fields, leading to two scalar fields. Now, with eight Weyl fermion fields, we have four scalar fields, but the analysis is otherwise similar.

The monopole is surrounded by a fermion condensate formed from the doublets listed above. The energy of the monopole-fermion system is minimized by requiring vanishing charge under all components of the unbroken gauge group. A set of particles from the relevant doublets that meets this criterion is the electrically neutral, color singlet combination $e^- u_1 u_2 d_3$ (or the corresponding set of antiparticles). The analysis of the bosonized theory shows that the ground state of the system is a superposition of states with arbitrary numbers of this set of fermions. This allows scattering processes that effectively add or subtract particles in this combination. An example is $u_1 + \text{Monopole} \rightarrow \bar{u}_2 + \bar{d}_3 + e^+ + \text{Monopole}$, a process that violates the conservation of both baryon number B and lepton number L (but not of $B - L$). With the initial u being a valence quark in a proton, this process could lead to the monopole-catalyzed decay of a proton to a positron plus a $\pi^+ \pi^-$ pair or to a positron plus a photon.

The possibility of a baryon number violating process is not surprising, since it is well known that the SU(5) theory allows proton decay. The striking feature is that there is no suppression by factors of the masses of the superheavy gauge bosons, or of the monopole core size. Instead, the cross-section is essentially geometric, and so is expected to be of typical strong interaction size.

This analysis in the SU(5) theory depended in a detailed manner on the way in which the light fermions transformed under the SU(2) defined by the magnetic charge. With a different embedding of the magnetic charge in the GUT gauge group, this catalysis of baryon number violation might not occur. This is confirmed by a detailed examination of a number of theories. The monopoles in the SO(10) theory considered in Sec. 6.3.2 are important examples [119]. The heavier, singly charged monopoles that arise at the first stage of symmetry breaking catalyze baryon number violation, but the lighter, doubly charged ones do not.

7

Cosmological implications and experimental bounds

In the previous chapters solitons were examined largely as theoretical constructs. Let us now address the question of whether they exist as actual physical objects. Condensed matter systems with structures analogous to kinks and vortices certainly exist and have been well studied; some of these have already been briefly mentioned. However, there has been as yet no confirmed experimental or observational evidence of a soliton in a relativistic quantum field theory. The natural question, then, is what conclusions can be drawn from this. The most plausible source of domain walls and strings is as relics surviving from the early universe. The same is true of magnetic monopoles if, as in grand unified theories, their masses are far beyond the reach of possible accelerator experiments. As we will see, all of these could have been produced during the course of symmetry-breaking cosmological phase transitions. Comparison of the expected production rates with the present-day bounds on the abundances of these objects yields important constraints on the underlying field theories and cosmological scenarios.

7.1 Brief overview of big bang cosmology

There is strong evidence, both from the spatial distribution of galaxies and, especially, observations of the cosmic microwave background radiation, that the universe (or at least the part accessible to our observations) possesses a high degree of spatial homogeneity and isotropy. Any homogeneous and isotropic spacetime can be described by the Robertson–Walker metric, which can be written as

$$ds^2 = dt^2 - a(t)^2 \left(\frac{dr^2}{1 - kr^2} + r^2 d\theta^2 + r^2 \sin\theta^2 d\varphi^2 \right). \qquad (7.1)$$

Here k indicates the nature of the spatial slices. It has three possible values, yielding flat Euclidean space ($k = 0$), a three-dimensional sphere ($k = 1$), or a three-dimensional hyperboloid ($k = -1$). These are referred to as flat, closed, and open universes, respectively. In an open or a closed universe the scale factor $a(t)$ is the time-dependent curvature radius. For a flat universe the overall scale of a is arbitrary, but the ratio of its values at two different times is a measure of the cosmic expansion and is physically meaningful.

The coordinates used here are comoving coordinates. A worldline with fixed r, θ, and φ is a geodesic, with t measuring the proper time along the worldline. One can view $a(t)$ as being a conversion factor between a comoving coordinate distance and a physical distance. Two comoving objects separated by a proper distance $\ell_{\text{phys}} = a(t)\ell_{\text{coord}}$ recede from one another with a velocity

$$\frac{d\ell_{\text{phys}}}{dt} = \dot{a}\ell_{\text{coord}} = \frac{\dot{a}}{a}\ell_{\text{phys}} \equiv H\ell_{\text{phys}}, \tag{7.2}$$

where overdots denote time derivatives and

$$H = \frac{\dot{a}}{a} \tag{7.3}$$

is the Hubble parameter.

Homogeneity and isotropy imply that the energy–momentum tensor $T_{\mu\nu}$ can be expressed in terms of just two functions of t, the energy density $\rho(t)$ and the pressure $p(t)$. Einstein's equations then imply the Friedmann equation,

$$\left(\frac{\dot{a}}{a}\right)^2 = \frac{8\pi\rho}{3M_{\text{Pl}}^2} - \frac{k}{a^2}, \tag{7.4}$$

where the Planck mass is related to Newton's constant by $M_{\text{Pl}} = G_N^{-1/2} = 1.2 \times 10^{19}$ GeV.

The fact that $T_{\mu\nu}$ is covariantly conserved gives the equation

$$\dot{\rho} = 3H(\rho + p). \tag{7.5}$$

Given an equation of state, this determines the evolution of ρ as the universe expands. The contents of the universe today can be classified into three components which, because their mutual interactions are relatively weak today, separately obey Eq. (7.5). Nonrelativistic matter, including both ordinary matter (baryons and electrons) and the dark matter, is essentially pressureless, and so obeys

$$\rho_{\text{matt}} \sim a^{-3}, \tag{7.6}$$

which can be understood as conservation of particle number. Massless radiation (e.g., photons), with $p = \frac{1}{3}\rho$, obeys

$$\rho_{\text{rad}} \sim a^{-4}. \tag{7.7}$$

Finally, current observations are consistent with the dark energy being a cosmological constant with $\rho_\Lambda = -p = $ constant.

Today, the dark energy dominates, with ρ_Λ roughly three times ρ_{matt}, and ρ_{rad} much smaller. These densities, together with the current value of H, determine the magnitude of the k/a^2 curvature term in the Friedmann equation. We can then work backward to trace the evolution of the universe at earlier times. Doing so, we see that while the dark energy makes the greatest contribution to ρ today, nonrelativistic matter was dominant before that, and at earlier times (those in which we will be most interested here) the universe was in a radiation-dominated regime. We also find that the curvature term in the Friedmann equation, which makes only a small contribution today, was completely negligible at earlier times, so we can safely set $k = 0$ in our considerations. It then follows that $a \sim t^{2/3}$ during the matter-dominated era and $a \sim t^{1/2}$ during the radiation-dominated era.

The microwave background radiation today has a Planck spectrum corresponding to a temperature $T = 2.7$ K. Because the interactions of this radiation with matter (and with itself) are negligible today, this is only a nominal temperature, characterizing the spectrum of the microwave background, and not a measure of a system in thermal equilibrium.[1] However, at the higher densities of the early universe, the matter and radiation interacted rapidly enough to maintain the universe in true thermal equilibrium. The temperature fell as the universe expanded, but the expansion was slow and smooth enough that it can be treated as an adiabatic process, with entropy conserved and the entropy density S obeying

$$a^3 S = \text{constant}. \tag{7.8}$$

In a radiation-dominated era the energy and entropy densities (in units with Boltzmann's constant equal to unity) are

$$\rho = \frac{\pi^2}{30} \mathcal{N} T^4, \tag{7.9}$$

$$S = \frac{2\pi^2}{45} \mathcal{N} T^3. \tag{7.10}$$

Here $\mathcal{N} = N_b + \frac{7}{8} N_f$, where N_b and N_f are the numbers of effectively massless bosonic and fermionic degrees of freedom. These count the number of spin states of particles with mass much less than T, and so are approximately stepwise constant, with a step downward each time the temperature falls below another particle mass.[2] Equations (7.8) and (7.10) imply that aT is constant between such thresholds.

---

[1] Even in the absence of interactions, the redshifting of massless radiation is such that an initially thermal distribution maintains the Planck form, but with T varying as $1/a$. This is the case here.

[2] Note that $\mathcal{N}$ was at least of the order of 10^2 at early times; for temperatures above the electroweak scale the standard model particles alone give $\mathcal{N} = 106.75$.

Substituting Eq. (7.9) into Eq. (7.4) and using the fact that the expansion is adiabatic leads to a differential equation for $T(t)$ in the radiation-dominated era. Its solution is

$$T = \left(\frac{45}{16\pi^3}\right)^{1/4} \mathcal{N}^{-1/4} \sqrt{\frac{M_{\mathrm{Pl}}}{t}} = 0.55 \,\mathcal{N}^{-1/4} \sqrt{\frac{M_{\mathrm{Pl}}}{t}}. \tag{7.11}$$

The integration constant here has been chosen so that $t = 0$ is the time at which the temperature and energy density diverge and the scale factor $a \to 0$. Of course, these are only formal statements, since the Friedmann–Robertson–Walker approximation must break down at sufficiently high T, and certainly cannot be trusted if T is Planckian in size.

Causality considerations will be of particular importance for us. Consider a light signal emitted from $r = 0$ at time t_0. At a later time t, it will have traveled a coordinate distance (assuming a flat universe)

$$\ell_{\mathrm{coord}}(t_0, t) = \int_{t_0}^{t} \frac{dt'}{a(t')} \tag{7.12}$$

that corresponds to a physical distance

$$\ell_{\mathrm{phys}}(t_0, t) = a(t) \int_{t_0}^{t} \frac{dt'}{a(t')}. \tag{7.13}$$

Setting $t_0 = 0$ gives the size

$$d_H(t) = a(t) \int_{0}^{t} \frac{dt'}{a(t')} \tag{7.14}$$

of what is called the particle horizon. If two objects are separated by more than twice this horizon distance, their past light cones have no points in common, and the objects are causally disconnected. For a flat radiation-dominated universe, we find

$$d_H = 2t = 0.60 \mathcal{N}^{-1/2} \frac{M_{\mathrm{Pl}}}{T^2}. \tag{7.15}$$

An analogous calculation gives $d_H = 3t$ for a matter-dominated universe.[3]

7.2 Symmetry restoration and cosmological phase transitions

It is a common phenomenon that symmetries that are spontaneously broken at low temperature are restored at high temperature. The magnetization of a ferromagnet, which spontaneously breaks the rotational symmetry of the Hamiltonian,

---

[3] It is believed that there was an earlier era of cosmological inflation during which the universe, or at least our portion of it, expanded exponentially fast [120]. As a result of this, the actual horizon distances would be vastly larger than the expressions above; indeed, this is precisely how inflation explains the homogeneity of the presently observed portion of the universe. However, the expressions given here are the appropriate ones for determining the maximum causal influence of events occurring in post-inflationary times.

disappears above the Curie temperature. The crystal structure of a solid breaks both translational and rotational symmetry, but these are restored if the crystal is heated to its melting temperature.

A similar high-temperature symmetry restoration can occur in a quantum field theory [121–124]. To understand this, consider a weakly coupled theory in which a complex scalar field ϕ interacts with a massless fermion field ψ and an Abelian gauge field A_μ, giving them masses $G|\phi|$ and $g|\phi|$, respectively. At zero temperature the equilibrium value of ϕ is determined by minimizing the energy density of a uniform configuration, $V(\phi)$. [More precisely, one should find the minimum of the effective potential, $V_{\text{eff}}(\phi)$, which includes the higher-order quantum corrections to the tree-level potential [125].] At a finite temperature T the quantity to be minimized is the free energy density, usually expressed as a finite temperature effective potential $V_{\text{eff}}(\phi, T)$.

With weak coupling, the various particle species can be treated as essentially ideal gases. The free energy density is then given, to a first approximation, by the sum of the zero-temperature energy density $V(\phi)$ and the ideal-gas free energies of the various particle species. The latter depend on the masses of the particles, which in turn depend on ϕ. For $M \ll T$, the free energy density per spin degree of freedom of an ideal gas of bosons with mass M is

$$F = -\frac{\pi^2}{90}T^4 + \frac{M^2}{24}T^2 + \cdots, \tag{7.16}$$

while for fermions of mass M we have

$$F = -\frac{7\pi^2}{720}T^4 + \frac{M^2}{48}T^2 + \cdots. \tag{7.17}$$

Hence, if the tree-level potential is

$$V(\phi) = -\mu^2|\phi|^2 + \frac{1}{2}\lambda|\phi|^4, \tag{7.18}$$

with $\mu^2 > 0$, the finite temperature effective potential for small $|\phi|$ is [123]

$$V_{\text{eff}}(\phi, T) = -\frac{\pi^2}{90}\mathcal{N}T^4 + (-\mu^2 + \sigma T^2)|\phi|^2 + O(|\phi|^3), \tag{7.19}$$

where

$$\sigma = \frac{1}{8}g^2 + \frac{1}{12}G^2 + \frac{1}{3}\lambda \tag{7.20}$$

reflects the contributions to the free energy from the ϕ-dependent part of the masses of A_μ, ψ, and ϕ itself, with the coefficients including factors for the various possible polarizations.

At zero temperature $\phi = 0$ is a local maximum of the potential and the symmetry is spontaneously broken. For $T > T^* = \sqrt{\mu^2/\sigma}$, the coefficient of $|\phi|^2$ is positive, and $\phi = 0$ is a local minimum of the effective potential. For sufficiently

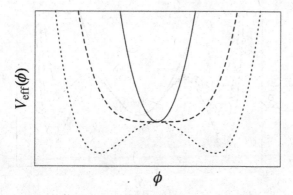

Fig. 7.1. Evolution of the shape of the finite temperature effective potential for a second-order phase transition. The curves correspond to $T > T_c$ (solid line), $T = T_c$ (dashed line), and $T < T_c$ (dotted line). Arbitrary constants have been added to make the curves coincide at $\phi = 0$.

large T this is always the global minimum, but to determine whether this is the case for $T \sim T^*$ we need to know the behavior of the effective potential at all values of ϕ. For weak coupling, this can be done by diagrammatic methods that sum all one loop vacuum graphs in the presence of a spatially uniform background ϕ [122]. The case of strong coupling is more difficult to address analytically, but can often be studied by numerical lattice field theory methods.

Generically, there are two possible behaviors, depending on whether or not the equilibrium value of $\langle\phi\rangle$ is continuous at the critical temperature. The former case, called a second-order transition, is illustrated in Fig. 7.1. In this example the origin goes directly from being the global minimum to being a local maximum at T^*, which is therefore the critical temperature T_c. As the universe cools below the critical temperature, $\langle\phi\rangle$ increases from zero until it reaches its zero-temperature value. If this is the case for the example of Eq. (7.19) and $g^2 \sim G^2 \sim \lambda$,

$$T_c = \sqrt{\frac{\mu^2}{\sigma}} \sim \sqrt{\frac{\mu^2}{\lambda}} \sim \langle\phi\rangle_0, \qquad (7.21)$$

where $\langle\phi\rangle_0$ is the zero-temperature vacuum expectation value.

The other possibility, a first-order transition, is illustrated in Fig. 7.2. In this example, at very high temperature $\phi = 0$ is the only minimum of V_{eff}. An asymmetric local minimum appears at T_1, becomes degenerate with the symmetric minimum at $T_c < T_1$, and is the global minimum for $T < T_c$. In the example outlined above the symmetric minimum disappears at T^*, but for other theories it may persist down to $T = 0$. In a first-order transition the low-temperature phase does not emerge smoothly from the high-temperature phase. Instead, the transition proceeds by the nucleation of bubbles of the low-temperature phase, a process which we will examine further in Chap. 12. Assuming that the nucleation

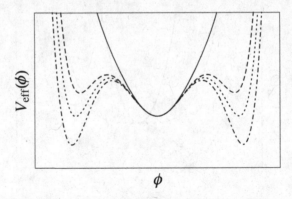

Fig. 7.2. Evolution of the shape of the finite temperature effective potential for a first-order phase transition. The curves correspond to $T \gg T_c$ (solid line), $T > T_c$ (dashed line), $T = T_c$ (dotted line), and $T < T_c$ (dot-dashed line). Arbitrary constants have been added to make the curves coincide at $\phi = 0$.

rate is large compared to the rate at which the universe is cooling, these bubbles expand and eventually merge to form a uniform low-temperature phase.[4]

In particular, it is believed that at $T \sim 10^2$ MeV the universe went from a QCD phase with manifest chiral symmetry and unconfined quarks, to the present confining phase with broken chiral symmetry. Earlier, at a higher temperature, $T_c \sim 10^2$ GeV, there was an electroweak transition from a phase in which the SU(2)×U(1) symmetry was manifest, to the current low-temperature phase in which this symmetry is spontaneously broken. If there is a grand unified theory, there would have been at least one, and possibly more, transitions corresponding to the breaking of the GUT symmetry at still earlier times.

7.3 The Kibble mechanism

Let us consider a phase transition from a symmetric phase characterized by a vanishing scalar field ϕ to an asymmetric phase in which the effective potential has degenerate minima at nonzero values of ϕ. (For simplicity, I will refer to these as vacuum values of ϕ, but it should be kept in mind that the minima of the effective potential at finite T are not necessarily the same as those at $T = 0$.) If the transition is second-order, then as the universe cools past the critical temperature ϕ will become nonzero and move toward one of its vacuum values. Because the vacua are all physically equivalent, all vacuum values are equally probable. Although it would be energetically favorable for the same vacuum to

---

[4] If the nucleation rate is small relative to the cosmological expansion, the universe enters a regime of extreme supercooling, and the transition is never globally completed [126, 127]. For the present discussion I will assume that this is not the case, and that the transition is completed.

be chosen everywhere, the choice can only be uniform over a finite distance, leading to a system of domains characterized by a correlation length ξ.

If the transition is first order, then the vacuum can be uniform within a single bubble (at least until it collides with another bubble), but the choices in different bubbles will be uncorrelated. When the bubbles coalesce, a domain structure again appears, with the characteristic bubble size at coalescence playing the role of ξ.

Once the transition is completed, the dynamics will tend to smooth out the variations in the field at the domain boundaries. However, as pointed out by Kibble, there can be topological obstructions that prevent this, leading to the creation of topological defects [128].

Perhaps the simplest case to visualize is that with a discrete symmetry leading to two distinct vacua, with $\langle \phi \rangle = \pm v$. Once the fields have settled down after the transition there will be regions of positive $\langle \phi \rangle$ and ones with negative $\langle \phi \rangle$, with domain walls—(3+1)-dimensional generalizations of the kink—along the boundaries between them. Any region with volume more than a few times ξ^3 would be expected to have at least one domain wall traversing it.

A second possibility is that the phase transition corresponds to the breaking of a symmetry group G to a subgroup H, with $\pi_1(G/H)$ being nontrivial. For definiteness, consider a theory where a complex scalar field ϕ develops a nonzero vacuum expectation value. Figure 7.3 shows a caricature of the domain structure along a two-dimensional spatial slice just after the transition is completed, with the arrows indicating the phase of ϕ in the various domains. The field dynamics will tend to align the phases of neighboring domains. However, this relaxation to a uniform phase cannot be complete, because there will inevitably be some domain junctions, such as the ones shown in Fig. 7.4, with a net vorticity that

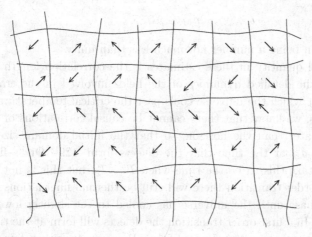

Fig. 7.3. Domain structure shortly after a phase transition in which a U(1) symmetry is broken by a nonzero complex scalar field. The arrows indicate the phase of the field.

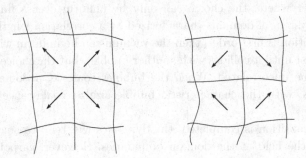

Fig. 7.4. Domain junctions leading to the formation of a vortex (left) or an antivortex (right).

cannot be smoothed away. Instead, these will lead to the formation of topological strings that appear as a vortices on this two-dimensional slice. The number of strings per unit area will be roughly

$$n_V \sim p_V \xi^{-2}, \tag{7.22}$$

where p_V measures the probability of nontrivial vorticity arising at a domain junction from the random phases in the adjacent domains. The value of p_V depends on the details of the theory, but it cannot be much less than order unity; a reasonable estimate is $p_V \sim 1/10$.

Finally, if the transition corresponds to a symmetry breaking with nontrivial $\pi_2(G/H)$, there can be point defects at domain junctions, leading to point-like solitons such as magnetic monopoles. By arguments analogous to those for the strings, we see that the initial density of these will be

$$n_M \sim p_M \xi^{-3}, \tag{7.23}$$

with p_M again being a number not much less than unity.

The crucial quantity in these estimates is the correlation length ξ. Its value depends on the detailed dynamics of the fields involved in the transition and on the rate at which the universe cools past the critical temperature. Whatever the dynamics, we know that there cannot be causal correlations of the field on distances greater than the horizon, so that the initial domain size cannot be greater than d_H at the time that the defects form [129]. One's first thought might be to take this to be the time when $T = T_c$, but this is not quite right. In a second-order transition there will still be thermal fluctuations back to the symmetric phase until the universe has cooled to the slightly lower Ginzburg temperature. In a first-order transition the defects will form at the time that the bubbles coalesce to complete the transition. Because of supercooling, this will be at a temperature somewhat less than the critical temperature. If we denote the temperature at which the defects form by $\tilde{T}_c$, then in the radiation-dominated era the causality bound on the initial correlation length is [129]

$$\xi \le d_H(\tilde{T}_c) \sim \frac{M_{\text{Pl}}}{\tilde{T}_c^2}. \tag{7.24}$$

This upper bound on ξ implies lower bounds on the initial densities of the various topological defects. It must be stressed that in many, if not all, cases the actual value of ξ will be much less than the horizon distance, so that these lower bounds may vastly underestimate the actual initial densities. They will, however, be sufficient for our purposes.

One might object to applying these arguments in a gauge theory, because they have been phrased in terms of the group orientation of the scalar field, which is not a gauge-invariant quantity. This is easily remedied. To give a concrete example, consider the production of monopoles in a theory where a triplet Higgs field breaks SU(2) to U(1). Now consider a spherical surface of radius $L \gg \xi$. According to Eq. (4.41), the topological charge contained within this surface is

$$N_\phi = \frac{1}{8\pi}\epsilon^{ijk} \int dS^i \, \hat{\phi} \cdot \partial_j \hat{\phi} \times \partial_k \hat{\phi}. \tag{7.25}$$

Using Eq. (5.55), this can be rewritten as a surface integral with a gauge-invariant integrand,

$$N_\phi = \frac{1}{8\pi}\epsilon^{ijk} \int dS_i \left[\hat{\phi} \cdot (\mathbf{D}_j\hat{\phi} \times \mathbf{D}_k\hat{\phi} - e\mathbf{F}_{jk}) \right]. \tag{7.26}$$

We can analyze this integral by arguments similar to those we have been using. At any point on the integration surface either sign for the integrand is equally likely, and so we would expect the sign to be correlated only over distances of order ξ. Hence, the integral should be viewed as a sum over roughly $(L/\xi)^2$ patches, with signs assigned randomly in each patch. The difference in the numbers N_+ and N_- of monopoles and antimonopoles should then be of the order of the square root of the number of patches,

$$|N_+ - N_-| \sim \frac{L}{\xi}. \tag{7.27}$$

Because our previous arguments showed that the total number of monopoles and antimonopoles was

$$N_+ + N_- \sim \left(\frac{L}{\xi}\right)^3, \tag{7.28}$$

we might have expected that $|N_+ - N_-| \sim (L/\xi)^{3/2}$. The fact that this quantity is only linear in L indicates a correlation between the positions of the monopoles and the antimonopoles.

7.4 Gravitational and cosmological consequences of domain walls and strings

Both domain walls and cosmic strings would be recognized primarily by their gravitational effects. We will see that the effects of the walls are disastrous except at very low mass scales, thus placing very stringent conditions on theories with spontaneous breaking of discrete symmetries. Strings, on the other hand, could quite plausibly be detected.

One would expect a planar domain wall to have large and obvious gravitational effects. In Newtonian gravity such a wall would give rise to an attractive force that was independent of distance. However, the general relativistic analysis leads to rather different results. Let us focus on a planar domain wall that is described by the extension of a one-dimensional scalar field kink solution to three dimensions. If the wall is in the x-y plane, we have

$$\phi(x, y, z) = \phi_{\text{kink}}(z). \tag{7.29}$$

The energy–momentum tensor is

$$T_{\mu\nu} = \partial_\mu \phi \, \partial_\nu \phi - g_{\mu\nu} \mathcal{L}, \tag{7.30}$$

so that

$$T_{00} = -T_{11} = -T_{22} = \frac{1}{2}\phi'(z)^2 + V(\phi(z)),$$
$$T_{33} = \frac{1}{2}\phi'(z)^2 - V(\phi(z)), \tag{7.31}$$

where the prime indicates a derivative with respect to z. Integrating across the thickness of the wall, we find negative pressures (i.e., positive tensions) in the x- and y-directions with magnitudes equal to the energy density per unit area, while the net pressure in the z-direction vanishes by virtue of the virial identity in Eq. (2.20). These negative pressures have a repulsive effect that is twice the attractive effect of the energy density, so that an observer at the wall sees test particles moving away from the wall with constant acceleration. However, the spacetime away from the wall is actually flat [assuming that $V(\phi)$ vanishes at its minima], very much like a higher-dimensional analogue of Rindler spacetime. In fact, one can find coordinates in which the metric on one side of the wall is that of a portion of Minkowski spacetime surrounded by a spherical wall that collapses to a minimum size and then expands, always with a constant outward acceleration [130–132].

More relevant for cosmology is the effect, not of a single wall, but of the network of domain walls produced via the Kibble mechanism during a phase transition where a discrete symmetry is broken. This network will evolve as the universe cools. The general tendency will be for walls to become more planar and for closed walls that enclose finite domains to contract until the domain has disappeared. The average size of the domains that remain will increase with time.

Even without a detailed study of the dynamics of this evolution, we can obtain a useful constraint just from causality. The same arguments that tell us that the correlation length just after the phase transition must be less than the horizon length at that time imply that the characteristic domain size at any later time cannot be greater than the horizon length at that time. It follows that at any time t the total domain wall area per unit volume must be at least of the order of $1/d_H(t)$. The energy density from domain walls is therefore

$$\rho_{\text{wall}} \gtrsim \frac{\sigma}{d_H(t)}, \tag{7.32}$$

where σ is the wall mass density per unit area; from our analysis of the kink solutions we know that $\sigma \sim m^3/\lambda$, where m and λ are the mass and coupling constant associated with the fields underlying the domain wall.

In both the radiation- and matter-dominated regimes the horizon distance is proportional to t, implying that the wall energy density falls more slowly than those of radiation and matter, and could potentially come to dominate them. However, the subsequent evolution of a wall-dominated universe is sufficiently different as to be clearly in conflict with cosmological observations. Excluding this possibility places an upper limit on σ [133]. For example, requiring that horizon-crossing walls not dominate the energy density today implies that [48, 134]

$$\sigma \lesssim (100\,\text{MeV})^3. \tag{7.33}$$

A stronger bound, $\sigma \lesssim (1\,\text{MeV})^3$, is obtained by considering the effects of walls on the anisotropy of the cosmic microwave background [48, 133].

Hence, any domain walls surviving to the present must be associated with very low-energy physics that has so far escaped discovery. Domain walls with a higher energy scale could have existed in the past, but only if they later ceased to be stable. This instability could result from a later phase transition that changed the vacuum structure of the theory, or it could happen if the discrete symmetry whose breaking led to the domain wall was only approximate. In the latter case domain walls would persist while the energy difference between the vacua on either side was negligible compared to the cosmic temperature, but at sufficiently low temperature the pressure from the lower-energy true vacuum would cause the regions of higher-energy vacuum to shrink and the walls to disappear.

More detailed discussions of the evolution of networks of cosmic domain walls and of their observational consequences can be found in [48, 134].

The situation with strings is rather different. By analogy with the argument for domain walls, a minimum expectation is that there should be at least one string crossing the visible universe. For strings arising in gauge theories, where the energy density is concentrated in a narrow core, the gravitational effects are hardly as dramatic as that of a domain wall. For a straight solitonic string only one component of the tension is nonzero, with the same magnitude as the energy density, so a nearby test particle feels neither attraction or repulsion. There is, however, a conical singularity at the string. This has a lensing effect, so that one signal of such a string would be double images of galaxies located behind it.

A major focus has been on the study of the network of strings that would emerge from a suitable phase transition. It was suggested that with an appropriate choice of the symmetry-breaking scale these could have served as the seeds for the density inhomogeneities that grew and evolved to the structure that we see in the universe today. As a result, considerable efforts, both analytic and numerical, have been devoted to the study of the problem. However, it has now become clear that while strings can lead to inhomogeneities of roughly the right magnitude, they cannot reproduce the detailed features of the cosmic microwave background spectrum. These are instead much better fit by inhomogeneities arising from slow-roll inflation, although the possibility of a small contribution from strings is not ruled out. These considerations place an upper bound on the energy per unit length μ of the string. Some recent studies [135, 136] quote bounds of roughly

$$G\mu \lesssim 7 \times 10^{-7}, \tag{7.34}$$

corresponding to a symmetry-breaking scale no higher than 10^{15} GeV or so. Bounds of a roughly similar range are obtained by considerations of the effects of the gravitational radiation from the strings.

A comprehensive discussion of cosmic strings is given in [48]. Some more recent reviews are [137, 138].

7.5 Evolution of the primordial monopole abundance

Magnetic monopoles are produced by the Kibble mechanism with an initial density given by Eq. (7.23). Assuming weak gauge coupling, the monopole mass is greater than the critical temperature of the transition where they are formed. They are therefore nonrelativistic, and their initial abundance is considerably greater than what it would be in thermal equilibrium. Monopoles disappear through monopole–antimonopole annihilation although, as we will see, the dilution of the monopole density by the cosmic expansion eventually brings an end to this annihilation.

Monopole–antimonopole annihilation in the early universe was studied by Zeldovich and Khlopov [139] and, with a particular emphasis on GUT monopoles, by Preskill [140]. It is perhaps best viewed as a two-step process in which the monopole and antimonopole are first captured into a Coulomb bound state, and then subsequently move down to lower bound states and eventually annihilate. It is the capture process that limits the annihilation rate. This is a purely electromagnetic process, and so does not depend on the details of the monopole's non-Abelian core.

The essential requirement for capture is that the initially free monopole and antimonopole, each with mass m and with magnetic charges $\pm Q_M$, lose enough of their initial kinetic energies that they can form a bound state. At high temperatures they are moving in a plasma of relativistic charged particles. They undergo Brownian motion with a mean free path

$$\ell \sim \frac{1}{CT}\sqrt{\frac{m}{T}}, \tag{7.35}$$

where $C \sim (1-5)\mathcal{N}_c$ if the number of charged degrees of freedom is in the range $1 \lesssim \mathcal{N}_c \lesssim 100$. As long as this is less than the capture radius,

$$r_c \sim \frac{Q_M^2}{4\pi T}, \tag{7.36}$$

where the negative Coulomb potential energy of the pair becomes comparable to their thermal kinetic energy, the drag forces exerted by the plasma can dissipate enough energy for the pair to be captured. However, once the universe has cooled below the temperature $T_1 \sim (4\pi)^2 m/(C^2 Q_M^4)$ where $\ell = r_c$, capture is only possible if an initially unbound monopole–antimonopole pair loses enough energy through bremsstrahlung to become bound.

In either temperature regime the time derivative of the monopole density n_M can be written as

$$\dot{n}_M = -D n_M^2 - 3\frac{\dot{a}}{a} n_M, \tag{7.37}$$

where D represents the effects of the annihilation processes and the last term is the effect of the cosmic expansion.

In the high-temperature regime the Coulomb attraction felt by a monopole at a distance r from an antimonopole is opposed by the drag forces from the plasma, with the net effect being a drift velocity

$$v_{\text{Drift}} \sim \frac{Q_M^2}{4\pi} \frac{1}{CT^2 r^2} \tag{7.38}$$

toward the antimonopole. If the typical monopole separation is $d \sim n_M^{-1/3}$, the capture time is

$$\tau \sim \frac{d}{v_{\text{Drift}}} \sim \frac{4\pi}{Q_M^2} \frac{CT^2}{n_M} \tag{7.39}$$

and

$$D \sim \frac{1}{\tau n_M} \sim \frac{Q_M^2}{4\pi} \frac{1}{CT^2}. \tag{7.40}$$

In the low-temperature regime, with an initial monopole thermal velocity $\sim \sqrt{T/m}$, the cross-section for radiative capture via bremsstrahlung emission is

$$\sigma_{\text{rad}} \sim \left(\frac{Q_M^2}{4\pi T}\right)^2 \left(\frac{T}{m}\right)^{3/5}, \tag{7.41}$$

giving

$$D \sim v\sigma_{\text{rad}} \sim \left(\frac{Q_M^2}{4\pi m}\right)^2 \left(\frac{m}{T}\right)^{9/10}. \tag{7.42}$$

With the above two expressions for D, we can now solve Eq. (7.37) to obtain the evolution of the monopole density. However, it is better to separate the effects of annihilation and expansion by working instead with the monopole-to-entropy ratio

$$r = \frac{n_M}{S}. \tag{7.43}$$

If the expansion is adiabatic, Eq. (7.8) implies that

$$\dot{r} = -DSr^2. \tag{7.44}$$

It is convenient to express r as a function of temperature rather than time. In a radiation-dominated regime with the number of massless degrees of freedom constant, so that $\dot{T}/T = -\dot{a}/a$,

$$\frac{dr}{dT} = -\frac{1}{T} DSr^2 = \left(\frac{\pi \mathcal{N}}{45}\right)^{1/2} DM_{\text{Pl}}\, r^2. \tag{7.45}$$

Integrating this gives

$$r(T) = \left[\frac{1}{r_{\text{init}}} + \frac{1}{r_*(T)}\right]^{-1}, \tag{7.46}$$

where r_{init} is the initial monopole to entropy ratio and

$$r_*(T) = \left[\left(\frac{\pi \mathcal{N}}{45}\right)^{1/2} M_{\text{Pl}} \int_T^{T_c} dT'\, D(T')\right]^{-1}. \tag{7.47}$$

When the high- and low-temperature expressions for $D(T)$ are substituted into Eq. (7.47), the integration divides into two regimes, both dominated by the region near T_1. Thus at temperatures below T_1

$$r_* \approx \left(\frac{45}{\pi \mathcal{N}}\right)^{1/2} \left[C\left(\frac{Q_M^2}{4\pi}\right)^3 + 10 C^{-1/5}\left(\frac{Q_M^2}{4\pi}\right)^{9/5}\right]^{-1} \frac{m}{M_{\text{Pl}}}. \tag{7.48}$$

For $C \approx 10^2$, $\mathcal{N} \approx 10^2$, and $Q_M = 2\pi/e$, and with

$$m_{17} \equiv \frac{m}{10^{17}\,\text{GeV}}, \tag{7.49}$$

this gives

$$r_* \approx 10^{-10}\, m_{17}. \tag{7.50}$$

Under the same assumptions Eqs. (7.23) and (7.24) give a lower bound

$$r_{\text{init}} \gtrsim p\, \frac{1}{S(\tilde{T}_c)\,[d_H(\tilde{T}_c)]^3} \approx \left(\frac{\tilde{T}_c}{M_{\text{Pl}}}\right)^3. \tag{7.51}$$

If $\tilde{T}_c$ is about an order of magnitude smaller than the monopole mass, this gives

$$r_{\text{init}} \gtrsim 10^{-9}\, m_{17}^3. \tag{7.52}$$

From Eq. (7.46), we see that at large times r will be given by the lesser of Eqs. (7.50) and (7.52).

7.6 Observational bounds and the primordial monopole problem

Let us now compare the predictions for the monopole-to-entropy ratio from the previous section to the various observational bounds on the current monopole density. The simplest of these is obtained by noting that r is related to the monopole fraction of the critical density, Ω_M, by

$$r \approx 10^{-27} m_{17}^{-1} \Omega_M. \tag{7.53}$$

Even in the rather implausible case that monopoles were to account for all of the mass usually attributed to dark matter, so that $\Omega_M \approx .25$, this exceeds both r_* and r_{init} unless $m \lesssim 10^{12}$ GeV.

Other density bounds follow from the limits on the monopole flux F in our galaxy. If the monopoles are uniformly distributed throughout the universe, $n_M = F/v$, where v is the typical monopole velocity. If instead they cluster with the galaxies, the average value of n_M would be up to five orders of magnitude smaller for a given value of F.

If there were no galactic magnetic field, monopoles in the galaxy would have typical velocities on the order of $10^{-3}c$, which is both the virial velocity in the galaxy and its peculiar velocity with respect to the rest frame of the cosmic microwave background. However, our galaxy does have a magnetic field, with a magnitude of approximately 3×10^{-6} gauss, that is coherent over distances of the order of 10^{23} cm. A monopole with magnetic charge $2\pi/e$ would be accelerated by this field to a velocity

$$v_{\text{mag}} \sim \begin{cases} c, & m \lesssim 10^{11} \text{ GeV}, \\ 10^{-3} m_{17}^{-1/2} c, & m \gtrsim 10^{11} \text{ GeV}. \end{cases} \tag{7.54}$$

Hence, monopoles with masses less than about 10^{17} GeV will be accelerated sufficiently to be ejected from the galaxy, and thus certainly do not cluster with our galaxy.

The acceleration of these monopoles drains energy from the galactic field. Requiring that the rate of this loss be small compared to the time scale on which the field can be regenerated (roughly 10^8 yrs) gives the Parker bound [141, 142]

$$F < F_{\text{Parker}} = \begin{cases} 10^{-15} \text{ cm}^{-2}\text{sr}^{-1}\text{sec}^{-1}, & m \lesssim 10^{17}\text{GeV}, \\ 10^{-15} m_{17} \text{ cm}^{-2}\text{sr}^{-1}\text{sec}^{-1}, & m \gtrsim 10^{17}\text{GeV}. \end{cases} \tag{7.55}$$

The two cases here reflect the fact that while the lighter monopoles are carried along the magnetic field lines, the heavier ones experience only small deflections by the field. Applying similar reasoning to an earlier seed field from which the present galactic field developed gives the somewhat stronger bound [143]

$$F < \left[m_{17} + (3 \times 10^{-6}) \right] 10^{-16} \text{ cm}^{-2}\text{sr}^{-1}\text{sec}^{-1}. \tag{7.56}$$

Reasoning along these lines can also be applied to the magnetic fields in galactic clusters, giving a bound which, although less certain, is about two orders of magnitude tighter than the Parker bound [144].

There are also limits from direct searches for monopoles in cosmic rays. For monopoles with $v > 10^{-4}c$, the MACRO experiment [145] places an upper bound of about $10^{-16}\,\mathrm{cm}^{-2}\mathrm{sr}^{-1}\mathrm{sec}^{-1}$. Somewhat stronger bounds have been obtained by other experiments, but these are limited to monopoles with higher velocities.

Even more stringent bounds apply for GUT monopoles that catalyze baryon number violation via the Callan–Rubakov effect. The essential idea is that such monopoles would be captured by compact astrophysical objects. They would then catalyze baryon decay, with the energy released in the decay leading to an increase in the luminosity of the object. A variety of bounds have been obtained by considering neutron stars [146–150], white dwarfs [151], and Jovian planets [152]. These depend on the details of the astrophysical scenario, such as whether monopoles captured by a progenitor star survive its collapse to a white dwarf or neutron star, and on the degree to which monopole–antimonopole annihilation reduces the accumulated density in the object. The bounds obtained in this manner lie in the range

$$F\left(\frac{\sigma_{\Delta B} v}{10^{-27}\,\mathrm{cm}^2}\right) \lesssim (10^{-18} - 10^{-25})\,\mathrm{cm}^{-2}\mathrm{sr}^{-1}\mathrm{sec}^{-1}, \qquad (7.57)$$

where $\sigma_{\Delta B}$ is the cross-section for catalysis of baryon number violation.

For a GUT monopole mass of 10^{17} GeV, with the monopoles not clustering with the galaxies, we have upper bounds on r of 10^{-26} from both the mass density and Parker bounds, 10^{-27} from direct observation, and perhaps as low as 10^{-36} for monopoles that catalyze baryon number violation. These range from 15 to 25 orders of magnitude below the predictions of Eqs. (7.50) and (7.52). Even taking into account the uncertainties in the various estimates involved, there is a very clear conflict between the cosmological predictions and the observational bounds. This conflict persists for monopole masses down to about 10^{12} GeV and to even lower masses if the monopoles catalyze baryon number violation.

This poses a serious problem for any grand unified theory. All such theories necessarily predict the existence of superheavy monopoles, and any plausible unification scale predicts that at least one species of these has a mass well above 10^{12} GeV. [Lighter multiply-charged monopoles, such as the SO(10) example discussed in Sec. 6.3.2, could have masses this low.] One might therefore decide to simply abandon all such theories. However, the idea of unification is sufficiently attractive as to motivate attempts to find a resolution of this primordial monopole problem that is consistent with grand unification.

The most attractive solution to this problem is based on the inflationary universe scenario, in which the universe undergoes a period of exponential expansion followed by a reheating process in which vacuum energy is converted to particles with a thermal distribution. (Indeed, it was consideration of the primordial monopole problem that led Guth to the idea of inflation [120].) If the inflation takes place after monopoles have been produced, any pre-existing monopoles will be diluted by an exponential factor. As long as the reheating after inflation does not raise the temperature of the universe above the critical temperature of the GUT phase transition, the present-day value of r will be unobservably small.

Although inflation is the most widely accepted solution, there is an alternative proposal that is of interest, if only for illustrative purposes, that was put forth by Langacker and Pi [153]. Consider, for example, a scenario with the following series of phase transitions:[5]

$$SU(5) \rightarrow SU(3) \times SU(2) \times U(1) \rightarrow SU(3) \rightarrow SU(3) \times U(1). \qquad (7.58)$$

Monopoles are formed at the first transition, when an $SU(5)$ adjoint field ϕ gets a nonzero vacuum expectation value, because $\pi_2[SU(5)/SU(3) \times SU(2) \times U(1)]$ is nontrivial. At the next transition, the breaking of the $U(1)$ symmetry leads to the formation of strings. The monopoles cannot survive as free objects after this transition, because $\pi_2[SU(5)/SU(3)] = 0$. Instead, they become bound to strings that have a monopole and an antimonopole at opposite ends. Because the energy of a string is proportional to its length, the monopole–antimonopole pair are drawn to each other by a constant force, leading to a rapid and efficient annihilation.

Free monopoles can exist again once the $U(1)$ symmetry is restored. One might therefore expect that the horizon bound together with the Kibble argument would produce monopoles with a minimum density of roughly one per horizon volume. (Since the critical temperature for this transition must be of the order of the electroweak scale, this would not conflict with observation.) This reasoning is incorrect. When the fields settle down after the first transition, the only constraint on $\phi(x)$ is that it be continuous. This allows the configuration of ϕ on a sphere with radius $\gg \xi$ to correspond to any element of $\pi_2[SU(5)/SU(3) \times SU(2) \times U(1)]$. After the second transition, when an additional field ψ also becomes nonzero, the requirement is that both ϕ and ψ be continuous. If there are constraints on the relative orientation of ϕ and ψ, these may eliminate the ϕ configurations corresponding to nontrivial elements of $\pi_2[SU(5)/SU(3) \times SU(2) \times U(1)]$, so that monopoles do not reappear when ψ becomes zero again.

---

[5] The final transition may seem a bit odd, since the low-temperature phase has higher symmetry than the high-temperature one. This is possible in a theory with several scalar fields. Recall that the effect of finite temperature is to add an effective scalar field mass σT^2. For the example considered in Sec. 7.2, σ was given in Eq. (7.20). In order that $V(\phi)$ be bounded from below, λ, and hence the right-hand side of Eq. (7.20), must be positive. In the more general case, with arbitrary numbers of scalar, spinor, and gauge fields, the contributions to σ from the gauge and Yukawa couplings remain positive. However, it is possible to have scalar quartic couplings of both signs that give a net negative contribution to σ while still keeping $V(\phi)$ bounded from below. With a negative σ, this unconventional ordering of symmetries is possible.

To put this more formally [154], let us suppose that we have a theory with two fields such that $\langle\phi\rangle$ by itself breaks G to H, while the combined effect of $\langle\phi\rangle$ and $\langle\psi\rangle$ is to break G to a subgroup $\hat{H} \subset H$. Every configuration of ϕ corresponds to an element of $\pi_2(G/H)$, which is assumed to be nontrivial. Every combined configuration of ϕ and ψ corresponds to an element of $\pi_2(G/\hat{H})$. Considering only the ϕ in such a combined configuration gives an element of $\pi_2(G/H)$, thus giving a map from $\pi_2(G/\hat{H})$ into $\pi_2(G/H)$. However, this map need not be onto, and it could even be the case that all elements $\pi_2(G/\hat{H})$ map to the identity element of $\pi_2(G/H)$.

If we start with a symmetric phase with $\langle\phi\rangle = \langle\psi\rangle = 0$ and then go directly to one with $\langle\phi\rangle \neq 0$, $\langle\psi\rangle = 0$, and unbroken symmetry H, then configurations corresponding to all elements of $\pi_2(G/H)$ can be created. On the other hand, if the breaking is from G to $\hat{H}$ to H, then in the final state only those elements of $\pi_2(G/H)$ that lie in the image of $\pi_2(G/\hat{H})$ can arise; if this image is just the identity element of $\pi_2(G/H)$, then no monopoles are created by the Kibble mechanism.

8

BPS solitons, supersymmetry, and duality

In addition to the direct prediction of physical objects, the study of solitons has also led to deeper insights into the structure of quantum field theories. We have already seen an early example of this, the discovery of the duality between the sine-Gordon and massive Thirring models [33]. A particularly fruitful direction was inspired by a special limit for the study of magnetic monopoles that was first proposed by Bogomolny [81] and by Prasad and Sommerfield [82]. It turns out that this BPS limit can be generalized and extended to many other systems, and that it has deep connections with supersymmetry and dualities.

Although there are a variety of solitons displaying BPS properties, in this chapter I will focus primarily on BPS magnetic monopoles. However, I will also use as examples a special case of the U(1) vortices that were studied in Chap. 3, as well as kink solutions in two spacetime dimensions.

The discussion in this chapter is necessarily limited. A review covering many of these topics in greater depth is [155].

8.1 The BPS limit as a limit of couplings

In Sec. 5.2 we considered magnetic monopoles in an SU(2) gauge theory with a triplet Higgs field governed by the potential

$$V(\phi) = -\frac{\mu^2}{2}\phi^2 + \frac{\lambda}{4}(\phi^2)^2 + \frac{\lambda v^4}{4}.$$

(8.1)

The ansatz

$$\phi^a = \hat{r}^a h(r),$$

$$A_i^a = \epsilon^{aim}\hat{r}^m \left[\frac{1 - u(r)}{er}\right],$$

$$A_0^a = \hat{r}^a j(r),$$

(8.2)

led to the static field equations

$$0 = h'' + \frac{2}{r} h' - \frac{2u^2 h}{r^2} + \lambda(v^2 - h^2)h,$$

$$0 = u'' - \frac{u(u^2 - 1)}{r^2} - e^2 u(h^2 - j^2),$$

$$0 = j'' + \frac{2}{r} j' - \frac{2u^2 j}{r^2}. \tag{8.3}$$

Three parameters enter these equations. Two of these, the Higgs field vacuum expectation value v and the gauge coupling e, can be eliminated from the equations by rescaling h and r. The resulting equations then depend only on the combination λ/e^2. Although these equations cannot be solved analytically for arbitrary values of λ/e^2, one might hope that they would be more tractable for some special values. In particular, Prasad and Sommerfield [82] proposed considering the case $\lambda/e^2 = 0$, obtained by taking a limit in which μ^2 and λ are taken to zero, but with the ratio $\mu^2/\lambda = v^2$ held fixed and the boundary condition $h(\infty) = v$ maintained. In this BPS limit the last term in the first of Eqs. (8.3) is absent, and by trial and error one can find the solution

$$h(r) = v \coth(evr) - \frac{1}{er},$$

$$u(r) = \frac{evr}{\sinh(evr)},$$

$$j(r) = 0. \tag{8.4}$$

This gives a monopole with $Q_M = 4\pi/e$ and mass

$$M = \frac{4\pi v}{e} = Q_M v. \tag{8.5}$$

This solution carries no electric charge. However, by noting that h and j obey identical equations when $\lambda = 0$, we see that dyon solutions can be obtained by a simple rescaling of the $Q_E = 0$ solution. For any real number γ there is a solution

$$h(r) = v \coth(e\hat{v}r) - \frac{\cosh\gamma}{er},$$

$$u(r) = \frac{e\hat{v}r}{\sinh(e\hat{v}r)},$$

$$j(r) = -v \tanh\gamma \coth(e\hat{v}r) + \frac{\sinh\gamma}{er}, \tag{8.6}$$

where $\hat{v} = v/\cosh\gamma$. The solution has magnetic charge $Q_M = 4\pi/e$, while from the gradient of the long-range tail of $j(r)$ we find that the electric charge is

$$Q_E = \frac{4\pi}{e} \sinh\gamma. \tag{8.7}$$

The mass of this dyon is

$$M = \frac{4\pi v}{e} \cosh\gamma = \sqrt{Q_M^2 + Q_E^2}\, v. \tag{8.8}$$

For these solutions $h(r)$ falls as $1/r$, in contrast to its usual exponential falloff. This is a consequence of the fact that the Higgs mass vanishes when μ^2 is taken to 0. Because this scalar field is now massless, it mediates a long-range force whose consequences will become evident presently.

8.2 Energy bounds

There are further striking features of this limit. With an eye to generalization to larger gauge groups later in this chapter, let us write the triplet Higgs field Φ as an element of the Lie algebra. The Lagrangian density for an arbitrary potential then takes the form

$$\mathcal{L} = -\frac{1}{2}\text{tr}\, F_{\mu\nu}^2 + \text{tr}\,(D_\mu\Phi)^2 - V(\Phi), \tag{8.9}$$

with

$$F_{\mu\nu} = \partial_\mu A_\nu - \partial_\nu A_\mu - ie[A_\mu, A_\nu], \tag{8.10}$$

$$D_\mu\Phi = \partial_\mu\Phi - ie[A_\mu, \Phi]. \tag{8.11}$$

Let us assume that Φ is nonzero, with magnitude v at the minima of V, and that V is everywhere positive.

It is convenient to define

$$\mathcal{Q}_M = 2\int d^2 S^i\, \text{tr}\,(\Phi B_i),$$

$$\mathcal{Q}_E = 2\int d^2 S^i\, \text{tr}\,(\Phi E_i), \tag{8.12}$$

where $B_i = -\frac{1}{2}\epsilon^{ijk}F_{jk}$, $E_i = F_{0i}$, and the integrals are over the surface at $r = \infty$. These quantities are, up to an overall rescaling, the magnetic and electric charges, with $\mathcal{Q}_M = vQ_M$ and $\mathcal{Q}_E = vQ_E$.

The energy functional can be written as

$$\begin{aligned}
E &= \int d^3x\, \left[\text{tr}\, E_i^2 + \text{tr}\, B_i^2 + \text{tr}\,(D_0\Phi)^2 + \text{tr}\,(D_i\Phi)^2 + V(\Phi)\right]\\
&= \int d^3x\, \left[\text{tr}\,(B_i \mp \cos\alpha D_i\Phi)^2 + \text{tr}\,(E_i \mp \sin\alpha D_i\Phi)^2\right.\\
&\quad \left. + \text{tr}\,(D_0\Phi)^2 + V(\Phi)\right]\\
&\quad \pm 2\int d^3x\,\left[\cos\alpha\,\text{tr}\,(B_i\, D_i\Phi) + \sin\alpha\,\text{tr}\,(E_i\, D_i\Phi)\right], \tag{8.13}
\end{aligned}$$

where α is an arbitrary angle. The terms in the last integral can be rewritten by integrating by parts. With the aid of the Bianchi identity $D_iB_i = 0$ and Gauss's law,

$$D_iE_i = ie[\Phi, D_0\Phi], \tag{8.14}$$

this gives

$$2\int d^3x\,[\text{tr}\,(B_i\, D_i\Phi)] = 2\int d^3x\,[\partial_i\,\text{tr}\,(B_i\Phi) - \text{tr}\,(\Phi D_i B_i)] = \mathcal{Q}_M,$$

$$\begin{aligned}
2\int d^3x\,[\text{tr}\,(E_i\, D_i\Phi)] &= 2\int d^3x\,[\partial_i\,\text{tr}\,(E_i\Phi) - \text{tr}\,(\Phi D_i E_i)]\\
&= \mathcal{Q}_E - 2ie\int d^3x\,\text{tr}\,(\Phi[\Phi, D_0\Phi]) = \mathcal{Q}_E. \tag{8.15}
\end{aligned}$$

Substituting these results into Eq. (8.13), we obtain

$$E = \int d^3x \left[\text{tr}\,(B_i \mp \cos\alpha D_i\Phi)^2 + \text{tr}\,(E_i \mp \sin\alpha D_i\Phi)^2 \right.$$
$$\left. + \text{tr}\,(D_0\Phi)^2 + V(\Phi) \right]$$
$$\pm \cos\alpha\, \mathcal{Q}_M \pm \sin\alpha\, \mathcal{Q}_E$$

$$\geq \pm \cos\alpha\, \mathcal{Q}_M \pm \sin\alpha\, \mathcal{Q}_E. \tag{8.16}$$

This inequality holds for either choice of sign and for any value of α. For fixed values of the charges, the most stringent inequality,

$$E \geq \sqrt{\mathcal{Q}_M^2 + \mathcal{Q}_E^2}, \tag{8.17}$$

is obtained by setting $\alpha = \tan^{-1}(\mathcal{Q}_E/\mathcal{Q}_M)$ and choosing the upper or lower signs according to whether $\mathcal{Q}_M$ is positive or negative. Without loss of generality, let us assume that $\mathcal{Q}_M > 0$ and take the upper signs.

If $V(\Phi)$ is nonzero, the lower bound of the energy can never be attained by a nontrivial configuration. However, let us go to the BPS limit in which V is taken to zero but the asymptotic boundary conditions on Φ are maintained. With V removed, the remaining terms in the integrand in Eq. (8.16) are all perfect squares. They will all vanish, and the lower bound on the energy will be achieved, if [81, 156]

$$B_i = \cos\alpha\, D_i\Phi,$$
$$E_i = \sin\alpha\, D_i\Phi,$$
$$D_0\Phi = 0. \tag{8.18}$$

In particular, for vanishing electric charge $\alpha = 0$ and

$$B_i = D_i\Phi. \tag{8.19}$$

This is often referred to as the Bogomolny equation. We now use the fact that configurations that minimize the energy for fixed values of the charges are solutions of the full equations of motion, provided that they satisfy the Gauss's law constraint. Solutions of Eq. (8.18) do indeed satisfy this constraint, since for these

$$D_i E_i - ie[\Phi, D_0\Phi] = \tan\alpha\, D_i B_i - 0 = 0. \tag{8.20}$$

Hence, solutions of the first-order equations are in fact static solutions of the full field theory.[1] This can be verified explicitly by differentiating Eq. (8.18) and substituting the results into the second-order Euler–Lagrange equations.

---

[1] These are not the only static solutions of the theory. The existence of nontrivial solutions that do not obey Eq. (8.18) was demonstrated in [157, 158]. However, these correspond to saddle points of the energy, and are therefore not stable.

For the particular case of unit magnetic charge in the SU(2) theory, substitution of the ansatz of Eq. (8.2) into Eq. (8.18) yields

$$h' = \frac{1 - u^2}{er^2} \sec \alpha,$$
$$u' = -euh \cos \alpha,$$
$$j = -h \sin \alpha, \tag{8.21}$$

which are indeed satisfied by the solution in Eq. (8.6).

We have several remarkable results here. First, we have found a set of first-order equations whose solutions are also solutions of the full set of second-order field equations. Second, the energy of these solutions is determined completely by their charges. In particular, for solutions carrying only magnetic charges the energy is strictly proportional to that charge. Finally, the fact that these solutions minimize the energy for fixed charges guarantees their stability against decay to some other solution with the same magnetic and electric charges.

One might perhaps wonder whether these results are a consequence of the rather unusual limiting process that took the scalar field potential to zero but retained the boundary conditions implied by that potential. However, there are other systems with similar behavior that do not require any such problematic limiting procedure. An example is the (2+1)-dimensional U(1) gauge theory with a single complex field that was studied in Sec. 3.2. For static configurations with vanishing A_0, the energy functional implied by the Lagrangian density of Eq. (3.24) can be written as

$$E = \int d^2x \left[\frac{1}{2}|D_j\phi|^2 + \frac{1}{4}F_{jk}^2 + \frac{\lambda}{4}(|\phi|^2 - v^2)^2 \right], \tag{8.22}$$

with $D_j\phi = \partial_j\phi + ieA_j\phi$ and $F_{jk} = \partial_j A_k - \partial_k A_j$. Now let us choose $\lambda = e^2/2$, which, it may be recalled, is the value that makes the vector and scalar masses equal and corresponds to the boundary between type I and type II superconductors.

The energy can then be written as [41, 81]

$$E = \frac{1}{2} \int d^2x \left[|D_1\phi|^2 + |D_2\phi|^2 + F_{12}^2 + \frac{e^2}{4}(|\phi|^2 - v^2)^2 \right]$$
$$= \frac{1}{2} \int d^2x \left\{ |D_1\phi \mp iD_2\phi|^2 + \left[F_{12} \pm \frac{e}{2}(|\phi|^2 - v^2) \right]^2 \right.$$
$$\left. \pm i\left[(D_1\phi)^*D_2\phi - (D_2\phi)^*D_1\phi \right] \mp eF_{12}(|\phi|^2 - v^2) \right\}$$
$$= \frac{1}{2} \int d^2x \left\{ |D_1\phi \mp iD_2\phi|^2 + \left[F_{12} \pm \frac{e}{2}(|\phi|^2 - v^2) \right]^2 \right\}$$
$$\pm \frac{ev^2}{2} \int d^2x F_{12}, \tag{8.23}$$

where the third line is obtained by an integration by parts for which the surface term vanishes. Recalling Eq. (3.31), we recognize the last integral as being $2\pi/e$ times the vorticity n and see that we again have a lower bound on the energy in terms of the topological charge. For positive n we take the upper signs and have

$$E \geq \pi v^2 n, \tag{8.24}$$

with equality if

$$0 = D_1\phi - iD_2\phi,$$

$$0 = F_{12} + \frac{e}{2}(|\phi|^2 - v^2). \tag{8.25}$$

By arguments similar to those for the BPS monopole, solutions of these first-order equations are also static solutions to the full set of second-order field equations. In contrast with the monopole case, these equations cannot be solved analytically. However, as we saw in Sec. 3.4, it is not hard to demonstrate the existence of axially symmetric solutions for arbitrary n, no matter what the value of λ/e^2.

A third example is the theory of a single scalar field in one space dimension governed by a potential $V(\phi)$, the case we studied in Sec. 2.1. If we assume that $V(\phi)$ is equal to zero at its global minima, we can find a function $W(\phi)$ such that

$$V = \frac{1}{2}\left(\frac{dW}{d\phi}\right)^2. \tag{8.26}$$

The energy of a static configuration can then be written as

$$\begin{aligned} E &= \int dx \left[\frac{1}{2}\left(\frac{d\phi}{dx}\right)^2 + V(\phi)\right] \\ &= \int dx \left[\frac{1}{2}\left(\frac{d\phi}{dx} \mp \sqrt{2V(\phi)}\right)^2 \pm \frac{d\phi}{dx}\frac{dW}{d\phi}\right] \\ &= \int dx \left[\frac{1}{2}\left(\frac{d\phi}{dx} \mp \sqrt{2V(\phi)}\right)^2\right] \pm T, \end{aligned} \tag{8.27}$$

where

$$T = W(\phi(\infty)) - W(\phi(-\infty)) \tag{8.28}$$

is a topological quantity. We have seen, in Sec. 2.1, that a solution to $d\phi/dx = \pm\sqrt{2V}$ is a solution to the full field equations. What is new here is the statement that its energy is equal to T.

Although T is topological, it does not always correspond to the topological charge

$$Q_{\text{top}} = \frac{1}{2v}[\phi(\infty) - \phi(-\infty)] \tag{8.29}$$

that was defined in Sec. 2.1. For the ϕ^4 model, with $V = (\lambda/4)(\phi^2 - v^2)^2$,

$$W = \sqrt{\frac{\lambda}{2}}\left(v^2\phi - \frac{1}{3}\phi^3\right). \tag{8.30}$$

In this case, $W(\pm v) = \pm\sqrt{2\lambda}\,v^3/3$, and so T and Q_{top} agree up to a multiplicative constant. This is not so for the sine-Gordon model, where the potential

$$V = \frac{m^4}{\lambda}\left[1 - \cos\left(\frac{\sqrt{\lambda}}{m}\phi\right)\right] \tag{8.31}$$

corresponds to

$$W = \frac{2m^3}{\lambda}\cos\left(\frac{\sqrt{\lambda}}{2m}\phi\right). \tag{8.32}$$

The vacua are at $\phi = 2\pi N m/\sqrt{\lambda}$. The topological charge Q_{top} counts the total kink number; i.e., the difference ΔN between the value of N at $x = \infty$ and that at $x = -\infty$. In contrast, T only counts this modulo 2: it vanishes if ΔN is even, and is equal to $4m^3/\lambda$ if ΔN is odd.

A number of other field theories have been found with similar properties (e.g., [159–162]). All have an energy that can be written as a sum of perfect squares plus a topological term, implying a lower bound that is saturated when a set of first-order equations are satisfied. These properties can be understood as consequences of supersymmetry, as will now be explained.

8.3 Supersymmetry

In four spacetime dimensions the algebra of simple supersymmetry can be written as

$$\{Q_\alpha, \bar{Q}_\beta\} = 2(\gamma^\mu)_{\alpha\beta}P_\mu, \tag{8.33}$$

where the supersymmetry generator Q is a four-component Majorana spinor with $\bar{Q} = Q^\dagger\gamma^0$, P_μ is the four-momentum, and subscripts from the beginning of the Greek alphabet denote Dirac indices. A more complex structure is possible in N-extended supersymmetry, with generators Q_r $(r = 1, \ldots, N)$. Here the most general form of the algebra of the supercharges can be written as

$$\{Q_{r\alpha}, \bar{Q}_{s\beta}\} = 2\delta_{rs}(\gamma^\mu)_{\alpha\beta}P_\mu + 2i\delta_{\alpha\beta}X_{rs} - 2(\gamma^5)_{\alpha\beta}Y_{rs}, \tag{8.34}$$

where $X_{rs} = -X_{sr}$ and $Y_{rs} = -Y_{sr}$ are central charges that commute with all of the supercharges and all of the generators of the Poincaré algebra.

Multiplying the left-hand side of Eq. (8.34) on the right by γ^0 gives the positive definite $4N \times 4N$ matrix $\mathcal{M}_{r\alpha,s\beta} \equiv \{Q_{r\alpha}, Q^\dagger_{s\beta}\}$. The positivity of its eigenvalues implies a lower bound on the mass. For example, taking $N = 2$ and working in the rest frame, with $P^\mu = \delta^{\mu 0} M$, we have

$$\mathcal{M} = 2 \begin{pmatrix} M I_4 & iX\gamma^0 - Y\gamma^5\gamma^0 \\ -iX\gamma^0 + Y\gamma^5\gamma^0 & M I_4 \end{pmatrix}, \tag{8.35}$$

where each entry represents a 4×4 block, with $X \equiv X_{12}$, $Y \equiv Y_{12}$, and I_4 being the 4×4 unit matrix. Requiring that the eigenvalues of $\mathcal{M}$ all be positive gives the inequality

$$M^2 - X^2 - Y^2 \geq 0. \tag{8.36}$$

For arbitrary N, this generalizes to

$$M \geq \sqrt{\frac{X_{rs}X_{rs} + Y_{rs}Y_{rs}}{2}}. \tag{8.37}$$

In order for a state to saturate this lower bound, $\mathcal{M}$ must have a zero eigenvalue, with means that there must be a linear combination of the Q_r and $Q_s^\dagger$ that vanishes when acting on the state. In other words, the state must be invariant under a subset of the supersymmetry transformations.

What does this have to do with the theories discussed in the previous sections? None of these include fermions, and so they clearly are not supersymmetric. However, they all have the property that they can be made supersymmetric by adding a few fields in a relatively simple fashion.

Let us start with the Yang–Mills theory with an adjoint representation scalar field Φ that was considered in the previous section. (Although we originally had the SU(2) theory in mind, the discussion below applies equally well for an arbitrary gauge group G.) To this we add a pseudoscalar field Θ and two Majorana fermion fields χ_r, all transforming under the adjoint representation, and take the Lagrangian to be [163]

$$\mathcal{L} = -\frac{1}{2}\text{tr}\, F_{\mu\nu}^2 + \text{tr}\,(D_\mu\Phi)^2 + \text{tr}\,(D_\mu\Theta)^2 + i\,\text{tr}\,\bar{\chi}_r\gamma^\mu D_\mu\chi_r$$

$$+ie\,\text{tr}\,\epsilon_{rs}\bar{\chi}_r[\chi_s, \Phi] + e\,\text{tr}\,\epsilon_{rs}\bar{\chi}_r\gamma^5[\chi_s, \Theta] - V(\Phi, \Theta), \tag{8.38}$$

where the scalar field potential is

$$V(\Phi, \Theta) = -e^2\text{tr}\,[\Phi, \Theta]^2. \tag{8.39}$$

(This is positive, despite the minus sign, because the commutator is anti-Hermitian.) This potential vanishes whenever Φ and Θ commute, and thus has a continuous family of degenerate minima. We will focus primarily on the vacua for which[2] $\langle\Theta\rangle = 0$. With this choice of vacuum, we can look for solutions of

---

[2] This is actually the most general case in the SU(2) theory, where any mutually commuting Φ and Θ must be proportional to each other. By a transformation corresponding to an element of the R-symmetry, we can then set $\langle\Theta\rangle = 0$. This is not necessarily so for larger gauge groups.

the classical field equations for which Θ vanishes at spatial infinity. It is easy to see that we can consistently set both Θ and the fermion fields χ_r to zero everywhere. The field equations then reduce to the same second-order equations as were obtained by taking the BPS limit of zero coupling. However, there is now no need to invoke a delicately tuned limit of parameters.

This Lagrangian is invariant under $N = 2$ supersymmetry transformations of the form

$$\delta A_\mu = i\bar{\zeta}_r\gamma_\mu\chi_r,$$

$$\delta\Phi = \epsilon_{rs}\bar{\zeta}_r\chi_s,$$

$$\delta\Theta = -i\,\epsilon_{rs}\bar{\zeta}_r\gamma^5\chi_s,$$

$$\delta\chi_r = -\frac{i}{2}\sigma^{\mu\nu}F_{\mu\nu}\zeta_r + i\epsilon_{rs}\gamma^\mu D_\mu(\Phi - i\gamma^5\Theta)\zeta_s - e[\Phi,\Theta]\gamma^5\zeta_r, \qquad (8.40)$$

where $\sigma^{\mu\nu} = (i/2)[\gamma^\mu,\gamma^\nu]$ and the ζ_r $(r = 1,2)$ are two independent Majorana spinor parameters.

The supercharges that generate these transformations can be written as spatial integrals of the time components of supercurrents S_r^μ. Calculating the anticommutators of the S_r^μ and then integrating to obtain the anticommutators of the supercharges shows that the latter are indeed of the form of Eq. (8.34), with the central charges,

$$X_{12} = 2\int d^2S^i\,\mathrm{tr}\,(\Phi E_i + \Theta B_i),$$

$$Y_{12} = 2\int d^2S^i\,\mathrm{tr}\,(\Phi B_i + \Theta E_i), \qquad (8.41)$$

being given by surface integrals at spatial infinity. If Φ, but not Θ, is nonzero as $r \to \infty$, we have

$$X_{12} = \mathcal{Q}_E,$$
$$Y_{12} = \mathcal{Q}_M, \qquad (8.42)$$

where $\mathcal{Q}_M$ and $\mathcal{Q}_E$ were defined in Eq. (8.12). The operator arguments given above then imply the lower bound

$$M \geq \sqrt{\mathcal{Q}_M^2 + \mathcal{Q}_E^2} \qquad (8.43)$$

that we obtained in the previous section.

These arguments also tell us that any state that saturates this bound must be invariant under a portion of the supersymmetry. For magnetically charged states built upon a classical soliton solution, this means that the classical solution should be invariant under the transformations of Eq. (8.40) for at least some choices of the ζ_r. Because the classical solution is purely bosonic, with

$\chi_1 = \chi_2 = 0$, the variations of the bosonic fields vanish automatically. The variations of the fermion fields, on the other hand, are in general nonzero. If $\Theta = 0$ everywhere, the condition for the vanishing of the $\delta\chi_r$ becomes

$$0 = -\frac{i}{2}\sigma^{\mu\nu}F_{\mu\nu}\zeta_r + i\epsilon_{rs}\gamma^\mu D_\mu\Phi\zeta_s. \tag{8.44}$$

This is satisfied for arbitrary choices of the ζ_r only by a vacuum configuration with $F_{\mu\nu} = D_\mu\Phi = 0$. However, if the two supersymmetry parameters are related by

$$\zeta_1 = \gamma^5\gamma^0\zeta_2, \tag{8.45}$$

Eq. (8.44) becomes

$$0 = \left[E_k\gamma^k\gamma^5 - iB_k\gamma^k + i\gamma^\mu D_\mu\Phi\right]\epsilon_{rs}\zeta_s. \tag{8.46}$$

This is satisfied for all ζ_s if

$$\begin{aligned} B_i &= D_i\Phi, \\ E_i &= 0, \\ D_0\Phi &= 0. \end{aligned} \tag{8.47}$$

We have thus recovered the Bogomolny equation for a monopole carrying no electric charge. The corresponding equations for dyons, given in Eq. (8.18), can be obtained by taking

$$\zeta_1 = e^{i\alpha\gamma^5}\gamma^5\gamma^0\zeta_2. \tag{8.48}$$

In both cases, the BPS solutions are invariant under a one-parameter set of transformations, and thus preserve half of the supersymmetry.

Although it may seem that the supersymmetry algebra has simply provided an alternative derivation of the results of the previous section, we actually obtain more. Recall that supermultiplets can be obtained by acting on a base state with the various supercharges. If some subset of the supercharges annihilates the base state, the number of states obtained in this fashion is reduced and we have a "short" supermultiplet. For the Yang–Mills theory with $N = 2$ supersymmetry, multiplets that saturate the bound of Eq. (8.43) are composed of four states, while those with larger masses contain 16. This has implications for the effects of perturbative quantum corrections. *A priori*, one might expect that the classical relation that we have found between the mass and the topological charges of BPS monopoles would be modified by one-loop quantum effects. However, our algebraic results tell us that such a modification would also imply a change in the multiplet size, something that we would not expect to result from a perturbative correction. Hence, we conclude that for these states the relation between the mass and the central charges must be exact [164].

Although this discussion has focused on the BPS monopoles of the Yang–Mills theory, a similar analysis can be applied to the other examples where a soliton mass saturates a lower bound given by the topological charges. In all known cases, the bosonic theory can be simply extended to a supersymmetric one, and the condition that a state be invariant under a subset of the supersymmetry becomes, at the classical level, a set of first-order equations whose solutions necessarily satisfy the full set of field equations. Indeed, such partially supersymmetric states, even when not associated with magnetic monopoles, are now referred to as BPS states.[3]

In the case of the (2+1)-dimensional U(1) theory, we saw that the reduction of the vortex equations to first order was only possible when the vector and scalar masses are equal. This is precisely the case where the theory can be extended to one with $N = 2$ supersymmetry with the vector and scalar particles lying in the same $N = 2$ supermultiplet. For other values of the masses it is still possible to obtain a theory with extended supersymmetry by introducing a sufficient number of additional fields. However, requiring that a fraction of the supersymmetry be preserved gives classical equations that involve the additional fields in a nontrivial manner and so cannot be applied to the original theory.

The case of two spacetime dimensions is special, in that extended supersymmetry is not needed to obtain BPS states. A general supersymmetric scalar theory has a Lagrangian density of the form

$$\mathcal{L} = \frac{1}{2}(\partial_\mu \phi)^2 + \frac{i}{2}\bar{\psi}\gamma^\mu \partial_\mu \psi - \frac{1}{2}[W'(\phi)]^2 - \frac{1}{2}W''(\phi)\bar{\psi}\psi, \tag{8.49}$$

where ψ is a Majorana fermion and the superpotential $W(\phi)$ is an arbitrary function of ϕ. By integrating the time component of the supercurrent and keeping careful track of surface terms, one obtains the supercharges, which satisfy

$$\{Q_\alpha, \bar{Q}_\beta\} = 2(\gamma^\mu)_{\alpha\beta}P_\mu + 2i(\gamma^5)_{\alpha\beta}T, \tag{8.50}$$

where

$$T = W(\phi(\infty)) - W(\phi(-\infty)), \tag{8.51}$$

which we already encountered in Eq. (8.28), is a central charge. If we choose a basis where $\gamma^0 = \sigma_y$, $\gamma^1 = i\sigma_z$, and $\gamma^5 = \gamma^0\gamma^1 = -\sigma_x$, and work in the rest frame, we find that

$$Q_1^2 = M + T,$$
$$Q_2^2 = M - T. \tag{8.52}$$

A BPS state is obtained by requiring $Q_2 = 0$ or $Q_1 = 0$, according to whether T is positive or negative.

---

[3] In all of these cases, the squared quantities enter with positive coefficients. There are other examples, involving gravity, where the action can be written as a topological term plus a sum of squares with coefficients of varying sign. Setting each of the squared quantities to zero yields a stationary point of the action, and thus a classical solution, but these solutions do not necessarily preserve any supersymmetry [165].

In this case, the vanishing of one of the Q_α does not imply a reduction in the size of the supermultiplet, so we cannot use the argument given above to show that perturbative effects cannot change the saturation of the mass bound. For quite some time there was dispute as to whether or not the saturation was maintained, with one-loop calculations using various regularization schemes purporting to show that the kink mass did or did not receive corrections. The issue was eventually resolved and clarified, and it was shown that if a kink obeys $M = T$ at the classical level, this relation is preserved by the leading quantum corrections [166–168]. There is a twist, however. At the quantum level there is an anomaly that contributes to T, so that the central charge becomes [168]

$$T = W(\phi(\infty)) + \frac{W''(\phi(\infty))}{4\pi} - W(\phi(-\infty)) - \frac{W''(\phi(-\infty))}{4\pi}. \tag{8.53}$$

The kink mass does receive perturbative corrections, but these are precisely equal to the corrections to T, provided that the anomaly terms are included.

In all of the above examples the soliton was invariant under the action of one half of the supersymmetry generators. This is not the only possibility. In $N = 4$ supersymmetric Yang–Mills theory there are four spinor parameters entering an arbitrary supersymmetry transformation, as compared to the two in the $N = 2$ theory. Imposing two relations between these leads to the BPS solutions that we have found, which are invariant under one half of the enlarged supersymmetry. However, there is now the possibility of imposing three constraints, leaving only one independent spinor parameter and giving a new class of solutions that preserve only one-fourth of the original supersymmetry [169–171]. Although this turns out not to be possible in the SU(2) theory, such 1/4-BPS solutions do exist with larger gauge groups. They obey a more complex set of first-order Bogomolny-type equations, and the relation between the mass and the electromagnetic charges is more complicated than in the 1/2-BPS case.

8.4 Multisoliton solutions

One would not ordinarily expect to find static solutions describing spatially separated solitons, because at any finite distance there would be nonzero intersoliton forces. A static multisoliton solution would require a remarkable cancellation between the attractive and the repulsive forces. However, the exact proportionality between the energy and the topological charge of BPS solutions suggests that just such a cancellation might take place. Indeed, we have seen that in the BPS monopole the Higgs field has a $1/r$ tail that corresponds to a long-range attractive force that has the potential to exactly cancel the magnetic repulsion. Similarly, the BPS U(1) vortices occur when the vector and scalar masses are equal, so that the exponential tails of the two fields fall at the same rate, again giving the possibility of cancellation between the attractive and the repulsive forces.

We can explore this further by counting the parameters needed to describe an arbitrary solution. In a solution describing noninteracting separated solitons there should be parameters reflecting independent degrees of freedom for each of the component solitons.

One might worry that this parameter counting could only be done after the putative multisoliton solution was actually found. In fact, this is not the case. Let us assume that we are given a BPS solution with some nonzero topological charge. This solution obeys a set of first-order Bogomolny-type equations. To count parameters, we need to know how many zero modes there are about this solution; i.e., how many linearly independent perturbations of the solution preserve the first-order equations. (In doing this counting, the zero modes corresponding to local gauge transformations must be excluded, e.g., by imposing a background gauge condition.) This question can be answered by index theory methods. Because these are most easily introduced in the context of instanton solutions, in Sec. 10.8, I will defer the details to Appendix B, and simply quote the results here.

For the $(2+1)$-dimensional $U(1)$ theory, an index theorem shows that there are precisely $2n$ nongauge zero modes about any BPS solution with vorticity n, assuming that such a solution actually exists [41]. That existence is assured by the arguments of Sec. 3.4 that showed that rotationally invariant solutions exist for arbitrary n. We now see that these are simply particularly symmetric multivortex solutions in which the n component vortices happen to be located at the same point. The zero modes for $n = 1$ correspond to spatial translation of the vortex, so it is natural to assume that for $n > 1$ they correspond to independent translations of n noninteracting vortices and that the $2n$ parameters can be taken to be the two-dimensional positions of the n vortices. Indeed, as was already noted in Sec. 3.4, one can show that static multivortex solutions exist for any choice of the n vortex positions [42].

Similarly, one can show that there are $4n$ zero modes satisfying the background gauge condition about any $SU(2)$ multimonopole solution carrying n units of magnetic charge [83]. Of these, $3n$ are naturally interpreted as specifying the positions of the component monopoles. The remaining n zero modes specify the $U(1)$ phases for these monopoles. To understand these, recall that although the fourth zero mode about a single monopole corresponded simply to a global gauge rotation, a time-dependent excitation of this mode led to a dyonic solution with nonzero electric charge. For the case of n monopoles, one overall phase can be shifted by a global gauge transformation, but the $n - 1$ relative phases are gauge-invariant. Excitation of these n modes corresponds to endowing the component monopoles with independent electric charges. As with the vortices, the index theorem by itself does not establish the existence of multimonopole solutions. However, it can be shown [84] that for any set of n points, none of them too close together, there is an n-monopole BPS solution with Higgs field zeros at the specified points.[4]

Actually obtaining explicit multisoliton solutions is a nontrivial matter. We have seen that the solution describing n vortices superimposed at the origin is rotationally symmetric and can readily be obtained by numerical solution of a pair of ordinary differential equations. By contrast, in the case of monopoles there are no spherically symmetric configurations with $n > 1$, and so no such simple multimonopole solutions. A rather fruitful method for studying solutions is a generalization, due to Nahm [172–175], of the ADHM instanton construction that will be described in Sec. 10.10. This Nahm construction transforms the problem of finding BPS monopoles in three spatial dimensions to that of solving a set of nonlinear differential equations for three matrix functions of a single variable. Once this "Nahm equation" has been solved the spacetime gauge and Higgs fields can be obtained from the solutions of a linear ordinary differential equation.[5]

The Nahm equation can be solved completely for the two-monopole case, with the matrices given in terms of elliptic functions [177]. Some curious features of the resulting monopole solutions are worth noting. First, since a single monopole is spherically symmetric, one might have expected the solutions describing two spatially separated monopoles to be axially symmetric under rotations about the line separating them. This is never the case [178]. To understand this physically, imagine trying to patch together the solutions describing two widely separated monopoles with their centers on the z-axis at $z = \pm R$. In order to make the Higgs fields match up, the monopoles should be written in the string gauge form described in Sec. 5.3. There must then be a Dirac string emanating from each of the monopoles. Let us take these strings to run outward along the z-axis, with one running from $z = R$ to $z = \infty$ and the other from $z = -R$ to $z = -\infty$, so that we do not have to deal with overlapping strings. Considered in isolation, each of the monopoles has an axial symmetry corresponding to a combination of physical rotation about the z-axis and a global phase rotation in the unbroken U(1) that acts on the charged vector W field. Because their strings run in opposite directions, the compensating phase rotations for the two monopoles are in opposite directions. As a result, the combined configuration cannot be exactly axially symmetric. A gauge-invariant measure of the departure from axial symmetry is given by the scalar product

---

[4] There are solutions in which the monopole cores overlap, but in these the connection between zeros of the Higgs field and monopole positions is sometimes lost. As will be described below, there are solutions with more Higgs zeros than monopoles, with the difference accounted for by the presence of antizeros of the field.

[5] A detailed description of the Nahm construction for both SU(2) and larger gauge groups can be found in [155]. Both the Nahm construction and other approaches to multimonopole solutions are discussed in [176].

$\mathbf{W}^*_{(1)} \cdot \mathbf{W}_{(2)}$, where $\mathbf{W}_{(a)}$ denotes the charged vector field of monopole a; this is $O(e^{-evR})$.

If the two monopoles are brought together so that the two zeros of the Higgs field coincide, the solution does become axially symmetric [179–182]. However, in contrast to what one might have expected from merging two spherical objects, the profiles of the energy density and Higgs fields are toroidal in shape.

Some unusual solutions with higher charge and extra symmetry have also been found. There is a solution with tetrahedral symmetry whose energy density contours resemble a tetrahedron with holes in its faces [183, 184]. Remarkably, this is a three-monopole solution. The Higgs field has four zeros, one at each vertex of the tetrahedron, but also a compensating antizero at the center. This demonstrates clearly that the zeros of the Higgs field need not coincide with the monopole positions when several monopoles are brought closely together [185]. There are also solutions corresponding to the other Platonic solids. There is a cubic $n = 4$ solution [183, 184] with a four-fold zero at the origin [186], an octahedral $n = 5$ solution [187] with zeros at the vertices and an antizero at the origin [186], a dodecahedral $n = 7$ solution [187] with a seven-fold zero at the center [186], and an icosahedral $n = 11$ solution [188].

8.5 The moduli space approximation

The multisoliton solutions of a given charge can be viewed as forming a manifold, known as the moduli space. A natural set of coordinates for the moduli space is given by the collective coordinates z_r corresponding to the zero modes about the multisoliton solutions.

This viewpoint leads to a useful method, the moduli space approximation [189], for understanding the dynamics in the low-energy limit. Let us denote the solutions on the moduli space by $\psi^{cl}(\mathbf{x}; z)$, where ψ is to be understood as a multicomponent field comprising all of the fields in the theory. An arbitrary field configuration can then be written as

$$\psi(\mathbf{x}, t) = \psi^{cl}(\mathbf{x}; z(t)) + \delta\psi(\mathbf{x}; z(t), t), \tag{8.54}$$

where $\delta\psi$ is orthogonal to motion on the moduli space. Thus, motion involving the zero modes is described by the time dependence of the collective coordinates, while motion involving the modes with nonzero eigenvalue is described by $\delta\psi$. If the total energy of the system is small compared to the lowest nonzero frequency, excitation of any mode except the zero modes is strongly suppressed and $\delta\psi$ can be ignored. In this approximation, the time dependence of the field comes entirely through the time dependence of the collective coordinates. With the motion thus restricted to the moduli space, the potential energy is constant and just equal to the static multisoliton energy E_{static}. Assuming a kinetic energy of the standard form, the Lagrangian density can be integrated over the D-dimensional space to yield

$$L = -E_{\text{static}} + \frac{1}{2}g_{rs}(z)\dot{z}_r\dot{z}_s + \cdots, \tag{8.55}$$

where the ellipsis represents terms that are ignored in this approximation and

$$g_{rs}(z) = \int d^D x \, \frac{\partial\psi^{\text{cl}}(\mathbf{x}; z)}{\partial z_r} \frac{\partial\psi^{\text{cl}}(\mathbf{x}; z)}{\partial z_s} \tag{8.56}$$

may be viewed as defining a metric on the moduli space.

We have thus reduced a field theory system with an infinite number of degrees of freedom to a system with a finite number of degrees of freedom. Furthermore, the classical trajectories of this reduced system are simply characterized. They are just geodesic motions defined with respect to the metric g_{rs}. To determine these exactly would require knowing the precise form of this metric. However, in some cases we can obtain considerable insight with only partial information about the metric.

As an example, consider the scattering of two BPS vortices. The moduli space is four dimensional. We could choose the z_r to be the x- and y-coordinates of the two vortices. However, because the two vortices are identical, interchanging their positions does not yield a new configuration, so we need to identify the corresponding points on the moduli space. A better choice of the collective coordinates is to have two of them specify the position of the center of mass and two to be the polar coordinates ρ and θ that describe the relative positions of the vortices. We then account for the identity of the two vortices by identifying θ and $\theta + \pi$.

Because the center-of-mass motion separates from the relative motion, the moduli space can be written as the product of a center of mass manifold and a relative space manifold. The former is simply the two-dimensional Euclidean plane, whose geodesics are straight lines describing the constant motion of the center of mass. The nontrivial dynamics is captured in the relative space manifold. Because the forces between the vortices fall exponentially at large distance, this manifold must be locally flat at large ρ. It would asymptotically approach R^2 were it not for the requirement that the points (ρ, θ) and $(\rho, \theta + \pi)$ be identified. With this identification, the asymptotic manifold becomes a portion of a cone. Without interactions, this would continue until the cone came to a point. Including the short-range interactions between the vortices rounds off the tip of the cone and removes this singularity [176].

Now consider two vortices moving directly at each other in the center-of-mass frame. At first thought, it seems almost obvious that the two will either pass right through each other or else be reflected back along their original paths. However, this is incorrect. The initial motion is described by a trajectory moving directly up the cone at some fixed θ_0. From the symmetry of the situation it is clear that this trajectory will continue up to the (rounded) tip of the cone and then straight down the opposite side. Unfolding the cone, we see that this corresponds to a shift from θ_0 to $\theta_0 + \pi/2$. Thus, two vortices that approach head-on scatter at right angles to their initial direction [190].

Two complications arise when we try to apply the moduli space approximation to the scattering of monopoles. First, the presence of the massless electromagnetic field means that there is a continuum of normal frequencies running down to zero. No matter how slowly the monopoles are moving, excitation of some of these modes via electromagnetic radiation is always energetically allowed. This turns out not to be a problem [191, 192]. When the monopoles are separated by distances large compared to their core size, they can be treated as point magnetic charges and standard electromagnetic methods can be used to show that the total radiation falls as the fifth power of the monopole velocities. These methods fail when the monopole cores overlap, but the only modes affected by this overlap have wavelengths comparable to or smaller than the monopole core size and are therefore energetically inaccessible if the monopole velocity is sufficiently low.

The second complication arises from the fact that there is an unbroken gauge symmetry with a corresponding Gauss's law that must be satisfied. If we adopt a (4+1)-dimensional notation where $\Phi = A^4$, but with no dependence on x^4, so that $\partial_4 = 0$, Gauss's law can be written as

$$0 = D_a F^{a0}, \tag{8.57}$$

with a running from 1 to 4. If the time dependence of the fields is only through the collective coordinates, we then have

$$0 = D_a \left[D^a A^0 - \dot{z}_r \frac{\partial (A^{\mathrm{cl}})^a}{\partial z_r} \right]. \tag{8.58}$$

This requires that A_0 be nonzero and proportional to the collective coordinate velocities, and so of the form

$$A_0 = \dot{z}_r \epsilon_r. \tag{8.59}$$

Hence

$$F^{a0} = -\dot{z}_r \delta_r A^a, \tag{8.60}$$

where Gauss's law requires that

$$\delta_r A^a = \frac{\partial (A^{\mathrm{cl}})^a}{\partial z_r} - D_a \epsilon_r \tag{8.61}$$

obey the background gauge condition

$$0 = D_a \delta_r A^a. \tag{8.62}$$

Substituting these results into the field theory Lagrangian and then proceeding as before, we obtain a moduli space Lagrangian of the form in Eq. (8.55), but with the metric now given by

$$g_{rs}(z) = 2 \int d^3x \, \mathrm{tr} \, \delta_r A^a \delta_s A^a. \tag{8.63}$$

Again, the exact form of the moduli space metric can only be found in some special cases. However, the asymptotic form in that part of moduli space describing n well-separated monopoles or dyons can be inferred by noting that in this limit the dynamics should be completely described by the purely Abelian long-range electric, magnetic, and scalar forces between point objects [193, 194]. Examination of this metric reveals that it develops a singularity if any of the intermonopole separations is taken to zero. This is symptomatic of the fact that new interactions, mediated by the massive non-Abelian gauge bosons, come into play at short distances.

The exact metric, including these corrections, for the specific case of two monopoles was found by Atiyah and Hitchin [195–197]. The full moduli space is eight dimensional, but three of the collective coordinates correspond to the center-of-mass position and a fourth to the overall U(1) phase, which is conjugate to the conserved total electric charge. The nontrivial dynamics is confined to the four-dimensional manifold spanned by the relative coordinates. This manifold must satisfy two important constraints. First, it must have an SO(3) isometry that reflects the three-dimensional rotational invariance of the dynamics. The second is more subtle. One consequence of the underlying supersymmetry is that a BPS monopole moduli space must be hyper-Kähler; i.e., it must have a natural quaternionic structure at each point.[6] These two requirements, together with the requirement that the metric have the correct asymptotic behavior at large monopole separation, reduce the problem of finding the metric to the solution of a set of ordinary differential equations for three metric functions that are easily analyzed [198].

With the Atiyah–Hitchin metric in hand, one can address the scattering of two BPS monopoles. In the particular case of head-on collisions, the scattering angle is $\pi/2$, with the monopoles emerging in the plane perpendicular to their initial line of approach [195, 196]. The particular direction in that plane is determined by the departure from axial symmetry in the initial two-monopole configuration.

8.6 BPS monopoles in larger gauge groups

Our discussion of BPS monopoles so far has focused on those arising in the SU(2) gauge theory. The close connection with extended supersymmetry suggests that in generalizing our results to a gauge group G with rank $r > 1$, we should consider only theories with adjoint representation Higgs fields. Such theories can be naturally extended to $N = 2$ ($N = 4$) supersymmetric Yang–Mills theory with two (six) adjoint representation scalar fields that can be viewed as elements of the Lie algebra. The scalar field potential is minimized (and set equal to zero) by any set of commuting scalar field vacuum expectation values.

---

[6] For a more precise definition and a discussion of the relation of the hyper-Kähler structure to the properties of the zero modes, see [155].

Let us assume that G is simple, with gauge coupling e. Let us also assume that only one scalar field, Φ, has a nonzero vacuum expectation value Φ_0. Without any loss of generality, we can take Φ_0 to lie in the Cartan subalgebra, so that it can be written as a linear combination

$$\Phi_0 = \mathbf{h} \cdot \mathbf{H} \tag{8.64}$$

of the commuting generators H_j. The generators of the unbroken gauge group are those which commute with Φ_0. These are all of the H_j, together with any E_α for which $\mathbf{h} \cdot \boldsymbol{\alpha} = 0$. We have two cases to consider. If $\mathbf{h}$ has nonzero inner products with all of the roots of G, then the unbroken gauge group is $\mathrm{U}(1)^r$; I will refer to this case as maximal symmetry breaking. The other possibility is that $\mathbf{h}$ is orthogonal to some of the roots. These roots are then the roots of a rank r' semisimple group K, and the unbroken group is $\mathrm{U}(1)^{r-r'} \times K$.

Let us start with the case of maximal symmetry breaking and consider a solution with total magnetic charge Q_M. At large distance the Higgs field and the magnetic charge must commute, so in a direction where the asymptotic Higgs field is Φ_0 we can require that the magnetic charge also lie in the Cartan subalgebra and write

$$Q_M = \frac{4\pi}{e} \mathbf{g} \cdot \mathbf{H}. \tag{8.65}$$

As was discussed in Sec. 6.1, the generalization of the Dirac quantization condition to larger gauge groups depends on the group representations that appear among the fields in the theory. The conditions on a nonsingular monopole solution are the most restrictive, because they must be consistent with all possible representations that could be added to the theory. The result is that we must require that

$$\mathbf{g} = \sum n_\alpha \boldsymbol{\alpha}^*, \tag{8.66}$$

where

$$\boldsymbol{\alpha}^* = \frac{\boldsymbol{\alpha}}{\alpha^2} \tag{8.67}$$

is a root of the dual lattice.

However, since the simple roots β_a form a basis for the Lie algebra, we can re-express this as

$$\mathbf{g} = \sum_{a=1}^{r} n_a \boldsymbol{\beta}_a^*. \tag{8.68}$$

The integers n_a are the r topological charges corresponding to the r $\mathrm{U}(1)$ factors of the unbroken group.

There are many possible choices for the simple roots. A particularly natural choice is specified by requiring that they all obey

$$\mathbf{h} \cdot \boldsymbol{\beta}_a > 0. \tag{8.69}$$

With this choice, there is a unique set of r fundamental BPS monopole solutions, each carrying one unit of topological charge [199]. If $A_i^{(s)}(\mathbf{r}; v)$ and $\Phi^{(s)}(\mathbf{r}; v)$ are the fields of the SU(2) BPS monopole with Higgs vacuum expectation value v, the ath fundamental monopole solution is given explicitly by

$$A_i(\mathbf{r}) = \sum_{s=1}^{3} A_i^{(s)}(\mathbf{r}; \boldsymbol{\beta}_a \cdot \mathbf{h}) t_s(\boldsymbol{\beta}_a),$$

$$\Phi(\mathbf{r}) = \sum_{s=1}^{3} \Phi^{(s)}(\mathbf{r}; \boldsymbol{\beta}_a \cdot \mathbf{h}) t_s(\boldsymbol{\beta}_a) + [\mathbf{h} - (\mathbf{h} \cdot \boldsymbol{\beta}_a^*)\boldsymbol{\beta}_a] \cdot \mathbf{H}. \qquad (8.70)$$

The $t_s(\boldsymbol{\beta}_a)$ are the generators of the SU(2) subgroup defined by $\boldsymbol{\beta}_a$, and are given by Eq. (A.7).

The topological charges of this monopole are

$$n_b = \delta_{ab}. \qquad (8.71)$$

Its mass can be found by noting that the monopole mass calculation in Sec. 8.2 was independent of the gauge group. With vanishing electric charge, Eqs. (8.12), (8.64), and (8.65) tell us that the mass of the ath fundamental monopole is

$$m_a = \mathcal{Q}_M = \frac{4\pi}{e} \mathbf{h} \cdot \boldsymbol{\beta}_a^*. \qquad (8.72)$$

Extending this calculation to arbitrary magnetic charge, we find that the energy of an arbitrary purely magnetic static solution is

$$M = \sum_{a=1}^{r} n_a m_a. \qquad (8.73)$$

This suggests that any such solution should be viewed as being a multimonopole solution comprising appropriate numbers of fundamental monopoles of the various species. Further support for this interpretation comes from the index theory calculations of Appendix B. A straightforward extension of the calculation for the SU(2) monopole tells us that the number of zero modes about a BPS solution in a larger group is [199]

$$N = 4 \sum_{a=1}^{r} n_a. \qquad (8.74)$$

This is just what should be expected for a collection of fundamental monopoles, each of which has three translational degrees of freedom and one global phase in the U(1) subgroup corresponding to its topological charge.

To make this discussion a bit more concrete, consider an SU(N) gauge theory and choose a basis where Φ_0, represented as an $N \times N$ Hermitian matrix, is diagonal with its eigenvalues decreasing along the diagonal. In a direction where the asymptotic Higgs field is Φ_0, the asymptotic magnetic field is

$$B_i = \frac{4\pi}{e} \frac{\hat{r}_i}{r^2} \,\mathrm{diag}\,(n_1, n_2 - n_1, \ldots, n_{r-1} - n_{r-2}, -n_{r-1}). \qquad (8.75)$$

The fundamental monopole solutions are obtained by inserting the appropriately rescaled SU(2) monopole solution into the 2×2 blocks that lie directly along the diagonal and then adding a suitable diagonal constant to Φ so as to obtain the correct asymptotic value.

Of course, we could also obtain a solution by embedding the SU(2) monopole in the SU(2) subgroup defined by a nonsimple root. For example, using the root $\alpha = \beta_1 + \beta_2$ leads to a solution whose fields have nonzero values in their 11, 13, 31, and 33 matrix elements. This solution is spherically symmetric and at first sight does not appear to be essentially different from the fundamental monopole solutions. However, Eq. (8.74) tells us that it has eight zero modes, rather than four. This indicates that it is actually just a special case of a two-monopole solution. It can be continuously deformed in such a way that the two component monopoles, which are of different species, become spatially separated; this can be verified explicitly by using the Nahm construction to obtain the complete eight-parameter family of solutions [200].

In contrast with the case of two identical SU(2) monopoles, these solutions with two separated monopoles of different species are axially symmetric, reflecting the presence of two independent unbroken U(1) subgroups. We can also study their low-energy interactions using the moduli space approximation. The moduli space metric for this two-monopole case has been found [201–203]. Unlike the SU(2) case, the asymptotic form of the metric is exact.[7] This is a consequence of the presence of two conserved U(1) charges, which eliminates the possibility of interactions via the exchange of massive vector mesons.

Let us now turn to the case of nonmaximal symmetry breaking, where the unbroken gauge group has a non-Abelian factor K. In this case the vector $\mathbf{h}$ defined by the Higgs field vacuum expectation value is orthogonal to some roots, so Eq. (8.69) must be replaced by the weaker requirement that the simple roots obey $\beta_a \cdot \mathbf{h} \geq 0$. This does not uniquely fix the set of simple roots, with the ambiguity in choosing the β_a corresponding to the action of the non-Abelian subgroup K. In addition, Eq. (8.72) seems to indicate that the fundamental monopoles corresponding to the simple roots that are orthogonal to $\mathbf{h}$ must be massless. However, it is easy to see that there can be no massless classical solitons. In fact, substitution of the appropriate β_a into Eq. (8.70) simply yields the vacuum solution. Nevertheless, we will see that there is a sense in which these massless monopoles exist.

To explore this further, consider the case of SU(3). With the symmetry broken to U(1)×U(1), there are two species of massive fundamental monopoles, as indicated in Fig. 8.1a. With the unbroken symmetry being SU(2)×U(1), there are two choices for the simple roots, shown in Fig. 8.1b. The two choices are related by a rotation in the SU(2) subgroup whose roots are β_2 and $-\beta_2 = \beta_2'$.

---

[7] This result can be extended to the case of an arbitrary number of distinct fundamental monopoles [204–206].

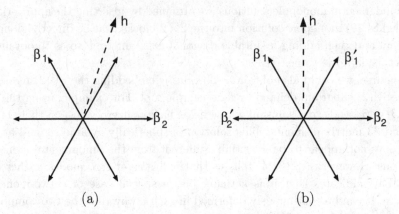

Fig. 8.1. The simple roots for (a) SU(3) maximally broken to U(1)×U(1) and (b) SU(3) broken to U(1)×SU(2). In the latter case there are two alternative pairs of simple roots.

With either choice one fundamental monopole, that corresponding to β_1 or β_1', is massive, while the one corresponding to β_2 or β_2' is massless. Note that β_1 and β_1' lead to gauge-equivalent solutions.

The magnetic charge of these monopoles is given by Eq. (8.65). Because β_1 has nonzero inner products with the roots of the unbroken SU(2), the magnetic charge of the massive monopoles has both Abelian U(1) and non-Abelian SU(2) components; that of the massless monopoles is purely non-Abelian. The case with non-maximal symmetry breaking can be viewed as a limiting case of that with maximal symmetry breaking, corresponding to a rotation of the vector **h** that brings it to the vertical. This suggests that we examine the behavior of the solutions as this limit is approached. Taking this limit has no qualitative effect on the β_1 monopole solution. On the other hand, the core region of the β_2 monopole expands without bound, while at the same time the deviation of the fields from their vacuum values goes to zero. In the limit we are left with simply a vacuum solution.

We can also consider the behavior of a two-monopole solution containing one β_1 monopole and one β_2 monopole. Again, the core of the β_2 monopole grows without bound [207]. In the limit we obtain a solution that is, up to a possible gauge transformation, just a single massive β_1 monopole. This is to be expected, since $\beta_1^* + \beta_2^* = \beta_1'^*$, which is related by an SU(2) rotation to β_1^*.

Now consider the case of two β_1 monopoles and one β_2 monopole. The new feature here is that the total magnetic charge, proportional to $2\beta_1^* + \beta_2^*$, lies entirely within the U(1) subgroup, and so remains purely Abelian in the limit of non-Abelian breaking. Starting with the monopoles at arbitrary locations and reducing the mass of the β_2 monopole, we find that the core of the latter begins to expand, just as in the previous cases. However, when this core reaches the two massive monopoles, its expansion slows down and ends. In the limit, we have a solution containing two massive monopoles enclosed in what might be termed a cloud of non-Abelian field [207]. Inside this cloud the magnetic field is that which would be expected from the two massive monopoles, with both Abelian and non-Abelian components. Outside the cloud only the Abelian component survives.

This distinction between the case where the magnetic field is purely Abelian and that where it is not carries over to the zero modes and the moduli space. As was already noted in Sec. 6.4, when the magnetic field has a non-Abelian component there are non-normalizable zero modes corresponding to the action of the unbroken group on the long-range tail of the magnetic field. Furthermore, the counting of zero modes by the index theory methods described in Appendix B fails because of the poor behavior of the fields at large distance. Neither of these problems is present when the magnetic charge is purely Abelian. In this case, one finds that the number of zero modes is still given by Eq. (8.74), with the sum running over all species of fundamental monopoles, both massive and massless [208, 209]. Thus, even though the massless monopoles do not exist as isolated soliton solutions, their degrees of freedom survive as the collective coordinates describing the properties of the non-Abelian clouds. Furthermore, the moduli space metric behaves smoothly as one goes from the maximally broken case to the limit with nonmaximal breaking [209].

An example that shows this particularly clearly is provided by the case of SO(5) broken to SU(2)×U(1). With maximal symmetry breaking there are two species of massive monopoles, and the metric on the eight-dimensional moduli space of solutions containing one of each type is known exactly. With nonmaximal breaking there is a spherically symmetric solution with a massive monopole surrounded by a non-Abelian cloud whose form is known explicitly [210]. Using the explicit solution and the definition of the moduli space metric, Eq. (8.56), one can calculate the metric for the latter case and verify explicitly that it is the smooth limit of the metric for the former [209].

Examination of solutions with two massive monopoles and a single massless monopole [200, 211–214] reveals an ellipsoidal non-Abelian cloud. However, in solutions with multiple massless monopoles one finds that there is no longer a one-to-one correspondence between clouds and massless monopoles [215].

The moduli space approximation can be used to study the interactions of collections of massive monopoles and non-Abelian clouds [216–219]. Because of the presence of the additional massless fields corresponding to the gauge bosons of the unbroken non-Abelian group, there is no guarantee that the moduli space approximation will be valid. Indeed, application of this approximation leads to a prediction that the cloud radius eventually expands at greater than the speed of light, a clear sign that the approximation has broken down. However, comparison with numerical solution of the full field equations shows that there is a regime in which the moduli space approximation gives reliable predictions for the motion of the massive monopoles and for the exchange of energy between the massive and massless monopoles [220].

Table 8.1 *The particle masses, charges, and spins in the BPS limit of the nonsupersymmetric SU(2) theory.*

	Mass	Q_E	Q_M	Spin
photon	0	0	0	1
ϕ	0	0	0	0
$W^\pm$	ev	$\pm e$	0	1
Monopole	$\dfrac{4\pi v}{e}$	0	$\pm\dfrac{4\pi}{e}$	0

8.7 Montonen–Olive duality

Montonen and Olive [221] pointed out an intriguing feature of the particle spectrum in the BPS limit of the SU(2) theory. Table 8.1 shows the masses and charges for the elementary particles implied by the Lagrangian of Eq. (5.45), together with those of the unit monopole and antimonopole. If one interchanges Q_M and Q_E and at the same time replaces e by $4\pi/e$, the entries for the gauge bosons and the monopoles are interchanged, but the overall spectrum of masses and charges remains the same.

The suggests that there might be a duality relating two equivalent formulations of the quantum field theory, similar to that between the sine-Gordon and massive Thirring models that was described in Sec. 2.7. In this case, however, the dual formulations of the theory would be identical in form except for the interchange of strong and weak coupling. In other words, the duality would actually be a self-duality.

Further support for this conjecture can be found by considering the forces between static objects. We have already seen that the energy of two static monopoles is just twice the monopole mass, regardless of their separation, so there is no force acting between them. Duality would require that the same be true for static W's. This can be tested by examining W-W scattering in the static limit. If the W's have the same sign of the electric charge, then in the limit where their spatial momenta are taken to zero the repulsive $1/r^2$ force corresponding to photon exchange is precisely canceled by the attractive $1/r^2$ force from Higgs exchange, just as required. (If the W's have opposite charges both forces are attractive, and there is a nonvanishing force for any value of the momenta.)

There is one very obvious difficulty. The gauge bosons have spin 1, whereas the supposedly dual quantum state built upon the spherically symmetric monopole solution is spinless. This contradiction is resolved by adding fermions and making the theory supersymmetric, something that we have seen is quite natural in the context of BPS solutions. In particular, recall from Sec. 5.7 that in the presence of a unit SU(2) monopole an adjoint representation Dirac fermion has two zero modes. With one such fermion (as one would have with $N = 2$ supersymmetry), there would be $2^2 = 4$ degenerate monopole states corresponding to the various possible occupation numbers of the zero modes. Two of these states would form a spin-1/2 multiplet, while the other two would be spinless, so there would still be no spin-1 dual to the W. But if we go to $N = 4$ supersymmetry there are two Dirac spinor fields, four fermionic zero modes, and thus $2^4 = 16$ degenerate states. These form one spin-1, four spin-1/2, and five spin-0 multiplets, and thus include not only the spin-1 counterpart of the W, but also the duals to the remaining members of the electrically charged $N = 4$ supermultiplet [222].

These are not the only particles in the theory. We have seen that there are dyons carrying unit magnetic charge and arbitrary integer electric charge. Duality requires that there also be one-particle states with unit electric charge and multiple magnetic charge. The existence of a dyonic state with two units of magnetic charge (in fact a supermultiplet of such states) was demonstrated by Sen, using arguments based on a supersymmetric extension of the moduli space dynamics discussed in Sec. 8.5 [223]. These states can be viewed as threshold bound states composed of two monopoles bound together through their interactions with the fermion fields. Extending this approach to states with higher magnetic charges poses technical difficulties, but there seems little doubt that the required states exist.

Now let us consider the extension of this conjecture to larger gauge groups, beginning with the case of maximal symmetry breaking, where the unbroken gauge group is a product of U(1) factors. In the elementary particle sector there is a massive gauge boson (with an accompanying supermultiplet) carrying electric-type U(1) charge (or charges) corresponding to each root $\boldsymbol{\alpha}$ of the original gauge group. Its mass is

$$M_\alpha = e\mathbf{h} \cdot \boldsymbol{\alpha}. \tag{8.76}$$

If $\boldsymbol{\alpha}$ is a simple root $\boldsymbol{\beta}_a$, then the dual states will be obtained from the fundamental monopole corresponding to this root, which has a mass

$$m_a = \frac{4\pi}{e}\mathbf{h} \cdot \boldsymbol{\beta}_a^* = \frac{4\pi}{e}\mathbf{h} \cdot \frac{\boldsymbol{\beta}_a}{\beta_a^2}. \tag{8.77}$$

If $\boldsymbol{\alpha}$ is not simple, a candidate for generating the dual states might be the solution, with energy

$$m_\alpha = \frac{4\pi}{e}\mathbf{h} \cdot \boldsymbol{\alpha}^* = \frac{4\pi}{e}\mathbf{h} \cdot \frac{\boldsymbol{\alpha}}{\alpha^2}, \tag{8.78}$$

obtained by embedding the SU(2) monopole in the SU(2) subgroup defined by $\boldsymbol{\alpha}$. However, we have seen that the counting of zero modes indicates that this solution is actually a particularly symmetric multimonopole solution, and so cannot give rise to the required one-particle magnetic dual states. Instead, the magnetic dual that we seek must be a threshold bound state of an appropriate collection of fundamental monopoles. The existence of such a bound state was first examined for the case of SU(3) broken to U(1)×U(1). In the notation of Fig. 8.1a, what is required is a bound state of a β_1 monopole and a β_2 monopole. By using the known moduli space metric for this two-monopole sector and applying Sen's method, one can verify the existence of the desired bound state [202, 203]. This calculation can be trivially extended to all other two-monopole cases. The result presumably extends to composite roots that are sums of arbitrary numbers of simple roots. However, this has not been verified, again because of the technical difficulty of the calculation.

Now note that the mass of the electrically charged particles, Eq. (8.77), involves α, while that of the magnetically charged particles, Eq. (8.78), involves α^*. For groups with a single root length μ (the "simply-laced" groups) the roots and their duals differ only by a multiplicative constant, and the only effect is that the Montonen–Olive substitution $e \leftrightarrow 4\pi/e$ must be replaced by $e\mu \leftrightarrow 4\pi/e\mu$. Indeed, the presence of the extra factor is not surprising if one realizes that the normalization of the gauge coupling depends on the convention for the root length.

The situation is more complex if there are two different root lengths. In this case, replacing each root by its dual is equivalent, up to an overall rescaling, to interchanging the long and the short roots. For some Lie algebras—F_4, G_2, and SO(5) = Sp(4)—this gives the original algebra, but with a relabeling of the roots. However, the interchange of long and short roots turns SO($2N+1$) into Sp($2N$), and vice versa.

Thus, the generalization of the Montonen–Olive conjecture is that theories based on simply laced gauge groups are self-dual under the interchange of electric and magnetic charges and weak and strong coupling. If the group is not simply laced, the duality maps the gauge theory onto one based on the dual Lie group.

New issues arise when we consider the case of nonmaximal symmetry breaking, where a non-Abelian subgroup remains unbroken. First, the elementary particle sector now includes some massless electrically charged particles, the "gluons" and their superpartners. The dual states are presumably related to the "massless monopoles". Although these do not exist as isolated classical solutions, they are in some sense manifested through the non-Abelian clouds and through the presence of their degrees of freedom in the moduli space metric. However, the exact relation between these and the gluon states needs to be clarified. Similarly, the precise mechanism by which the required non-Abelian multiplets of monopole states emerge from the classical solutions is yet to be fully worked out.

9

Euclidean solutions

It is easy to understand why static solutions or solutions that oscillate in real time should be relevant in quantum field theory, since these describe objects that are already present in the classical theory. It is less obvious that Euclidean solutions—solutions in spacetimes with imaginary times and hence Euclidean metrics—should be physically interesting. Their importance arises from their connection with a purely quantum mechanical effect, barrier penetration. In this chapter I will focus on developing the Euclidean formalism for systems with a finite number of degrees of freedom (and often just one). The next three chapters will explore the application of these methods to field theories, systems with an infinite number of degrees of freedom.

9.1 Tunneling in one dimension

Consider a quantum mechanical particle in one dimension with the Hamiltonian

$$H = \frac{p^2}{2m} + V(q) \,, \tag{9.1}$$

where the potential energy V has a barrier as shown in Fig. 9.1. If a right-moving wave with an energy E that is less than the value of V at the top of the barrier is incident on the potential energy barrier, most of the wave is reflected, but there is also a small transmitted wave with an amplitude proportional to $e^{-B/2}$, where the WKB approximation gives B as an integral

$$B = 2 \int_{q_1}^{q_2} dq \sqrt{2m[V(q) - E]} \tag{9.2}$$

over the classically forbidden region $q_1 < q < q_2$.

In the region inside the potential energy barrier the total energy is less than the potential energy. If we were to try to interpret the difference as a negative kinetic energy, we would conclude that the velocity was imaginary, just as if it were the

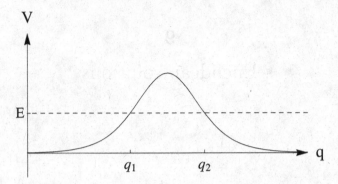

Fig. 9.1. A potential energy barrier in one dimension. The classically forbidden region is the region $q_1 < q < q_2$, where the turning points are defined by $V(q_1) = V(q_2) = E$.

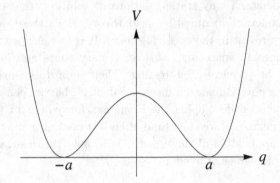

Fig. 9.2. A double-well potential.

derivative of q with respect to an imaginary time. This observation is the fundamental motivation for applying a Euclidean analysis to the problem. Although somewhat trivial in the case with only one degree of freedom, Euclidean methods lead to great simplifications when multiple degrees of freedom are involved.

The field theory applications that will be of primary interest to us are associated with two types of phenomenon that are already seen in quantum mechanical systems with one degree of freedom. The first is the mixing of two or more states that would be degenerate in the absence of tunneling. The simplest example of this occurs in a symmetric double-well potential, such as that shown in Fig. 9.2. If the barrier were infinitely high, and hence impenetrable, there would be a tower of energy eigenstates confined to the left side of the barrier, and a similar tower of states confined to the right side. The two ground states, $|L\rangle$ and $|R\rangle$, would each have an energy E_0. With a finite barrier neither of these is an energy eigenstate. Instead, assuming the barrier to be high relative to E_0, the two lowest eigenstates are given by the symmetric and antisymmetric linear combinations

$$|\pm\rangle = \frac{1}{\sqrt{2}}\left(|L\rangle \pm |R\rangle\right) \qquad (9.3)$$

with energies

$$E_{\pm} = E_0 \mp K\,e^{-B/2} \equiv E_0 \mp \Delta/2\,, \qquad (9.4)$$

where K is a constant whose calculation will be discussed later.

Now consider a particle in a linear combination of these two states. Let us suppose that at $t = 0$ its wavefunction is localized on the left side of the barrier. Taking $|\Psi(0)\rangle = |L\rangle$, we have

$$\begin{aligned}
|\Psi(t)\rangle &= \frac{1}{\sqrt{2}}\,e^{-iE_+ t}\left(|+\rangle + e^{-it\Delta}|-\rangle\right) \\
&= \frac{1}{2}\,e^{-iE_+ t}\left[\left(1 + e^{-it\Delta}\right)|L\rangle + \left(1 - e^{-it\Delta}\right)|R\rangle\right] . \qquad (9.5)
\end{aligned}$$

Thus, the system oscillates back and forth with a frequency Δ.

It is important here that the potential be symmetric, so that the energy levels in the two wells, before tunneling is taken into account, are the same. If the two wells had been different, with their respective energy levels differing by amounts large compared to Δ, then the true energy eigenstates of the full system would be concentrated on one side of the barrier or the other, with only exponentially small contributions from the opposite side. If the particle's wavefunction was initially on one side of the barrier, it would remain concentrated on that side, with only a small probability of the particle ever being found on the opposite side.

The second phenomenon of interest to us is the decay of a metastable state. This can be illustrated by the potentials in Fig. 9.3. The one on the left has a narrow local minimum of V on one side of the barrier and a broad lower minimum on the other side. Without barrier penetration there would be a tower of discrete energy levels on the left side of the barrier, the lowest of which can again be labeled $|L\rangle$, and a much denser spectrum of states on the right side. In contrast with the previous example, where there was a single right-hand state

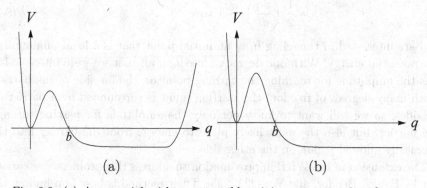

Fig. 9.3. (a) A potential with a metastable minimum at $q = 0$ and a very wide well with a lower energy. (b) The limiting case in which the width of the right-hand well becomes infinite.

close in energy with $|L\rangle$, there are now many such states. With barrier penetration restored the energy eigenstates are again mixtures of left and right states, but with the contribution of $|L\rangle$ to any given eigenstate being small and tending toward zero as the width of the right-hand well increases.

Now consider the limit in which the width of the right-hand well goes to infinity, as shown in Fig. 9.3b. Let us again consider a state where $|\Psi(0)\rangle = |L\rangle$. Rather than finding oscillation, we now find that the magnitude of the overlap $\langle L|\Psi(t)\rangle$ falls exponentially with time, with the exponent itself being proportional to $e^{-B/2}$. Instead of expanding $|L\rangle$ as a linear combination of true energy eigenstates, it is more convenient to view it as a metastable state with a complex energy whose imaginary part is related to the decay width Γ by

$$\text{Im}\, E = -\frac{\Gamma}{2}\,. \tag{9.6}$$

Alternatively, one can view the potential in Fig. 9.3b as being obtained by analytic continuation of a potential with only a single minimum, at $q = 0$. The complex energy of the metastable state in the former potential is then the analytic continuation of the real energy of the stable ground state of the original potential.

9.2 WKB tunneling with many degrees of freedom

The results of the previous section are familiar from elementary quantum mechanics. What we need to do is to generalize these to the case of a system with many degrees of freedom q^1, q^2, ..., q^N. It will be convenient to assemble these into an N-component vector $\mathbf{q}$.

Thus, suppose that we have a system described by the Lagrangian

$$L = \frac{1}{2} \sum_{j=1}^{N} \left(\frac{dq^j}{dt}\right)^2 - V(q^1, q^2, \ldots q^N)$$
$$= \frac{1}{2}\left(\frac{d\mathbf{q}}{dt}\right)^2 - V(\mathbf{q}) \tag{9.7}$$

and are interested in tunneling from an initial point that is a local minimum of the potential energy. With one degree of freedom, all that we were interested in was the amplitude for reaching the turning point on the far side of the barrier. With many degrees of freedom the starting point is surrounded by a barrier on all sides, so we will want to know not only the amplitude for passing through the barrier, but also the most likely place to emerge from the barrier into the classically allowed region on the other side.

The extension of the WKB approximation to address this problem was carried out by Banks, Bender, and Wu [224, 225]. Their method is based on finding the most probable escape path (MPEP). Any path P through the barrier can be specified as a trajectory $\mathbf{q}(s)$, where the parameter s along the path is defined by

$$(ds)^2 = \sum_{j=1}^{N} (dq^j)^2 \equiv (d\mathbf{q})^2\,, \tag{9.8}$$

with the initial condition that $\mathbf{q}(0) = \mathbf{q}_0$. [Note that the final point of the path, $\mathbf{q}(s_f) \equiv \mathbf{q}_f$, is not specified in advance.] Treating this path as a one-dimensional system, we can define a path-dependent barrier penetration integral

$$B[P] = 2 \int_0^{s_f} ds \, \sqrt{2[V(\mathbf{q}(s)) - E]}\,, \tag{9.9}$$

with $E = V(\mathbf{q}_0)$. The MPEP is the path that minimizes $B[P]$. Its end point is the most probable escape point from the barrier, and the leading WKB approximation for the tunneling amplitude is $Ae^{-B/2}$, with B evaluated on the MPEP and the prefactor A still to be determined

The task of minimizing B turns out to be most conveniently carried out in a Lagrangian framework, which also has the advantage of being easily carried over to the field theory context [226]. It is here that the advantages of the Euclidean approach become apparent. In classical mechanics Jacobi's principle tells us that, for a system described by a Lagrangian of the form of Eq. (9.7), a path from $\mathbf{q}_0$ to $\mathbf{q}_f$ that minimizes

$$I = \int_0^{s_f} ds \, \sqrt{2[E - V(\mathbf{q}(s))]} \tag{9.10}$$

gives a solution of the equations of motion whose time evolution is determined by

$$\frac{1}{2}\left(\frac{d\mathbf{q}}{dt}\right)^2 = E - V(\mathbf{q})\,. \tag{9.11}$$

Alternatively, Hamilton's principle tells us that the same solution can be found by looking for a stationary point of the action

$$S = \int_{t_0}^{t_f} dt \, L(\mathbf{q}, \dot{\mathbf{q}})\,, \tag{9.12}$$

where $\mathbf{q}(t_0) = \mathbf{q}_0$ and $\mathbf{q}(t_f) = \mathbf{q}_f$.

With appropriate sign changes, these results translate to the statement that a path that minimizes the $B[P]$ defined in Eq. (9.9) corresponds to a stationary point of the Euclidean action

$$S_E = \int_{\tau_0}^{\tau_f} d\tau \left[\frac{1}{2}\left(\frac{d\mathbf{q}}{d\tau}\right)^2 + V(\mathbf{q})\right] \tag{9.13}$$

and thus is a solution of the Euclidean equation of motion

$$\frac{d^2 q_j}{d\tau^2} = \frac{\partial V}{\partial q_j}\,. \tag{9.14}$$

Let us denote this solution by $\bar{\mathbf{q}}(\tau)$.

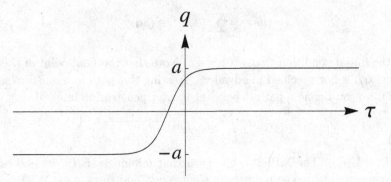

Fig. 9.4. The instanton corresponding to tunneling between the two minima of a potential such as that in Fig. 9.2.

For this solution the motion along the path satisfies

$$\frac{1}{2}\left(\frac{d\bar{\mathbf{q}}}{d\tau}\right)^2 = V(\bar{\mathbf{q}}) - E = V(\bar{\mathbf{q}}) - V(\mathbf{q}_0). \tag{9.15}$$

This, together with Eq. (9.8), implies that

$$\begin{aligned}
S_E[\bar{\mathbf{q}}] &= \int_{\tau_0}^{\tau_f} d\tau\, 2[V(\bar{\mathbf{q}}) - V(\mathbf{q}_0)] + \int_{\tau_0}^{\tau_f} d\tau V(\mathbf{q}_0) \\
&= \int_{\tau_0}^{\tau_f} d\tau\, \sqrt{\left(\frac{d\bar{\mathbf{q}}}{d\tau}\right)^2}\, \sqrt{2[V(\bar{\mathbf{q}}) - V(\mathbf{q}_0)]} + \int_{\tau_0}^{\tau_f} d\tau V(\mathbf{q}_0) \\
&= \int_0^{s_f} ds\, \sqrt{2[V(\bar{\mathbf{q}}) - V(\mathbf{q}_0)]} + \int_{\tau_0}^{\tau_f} d\tau V(\mathbf{q}_0). \tag{9.16}
\end{aligned}$$

This result gives us a relation between the tunneling exponent B and the Euclidean action, provided that we are careful in specifying the limits of the integration.

When the tunneling is between two degenerate minima of V, the approach of $\bar{\mathbf{q}}(\tau)$ to the end points of the trajectory is exponentially slow. In this case, τ runs from $-\infty$ to ∞ and $\bar{\mathbf{q}}(\tau)$ behaves as shown in Fig. 9.4. (Note the similarity with the one-dimensional kink solitons of Chap. 2.) Euclidean time translation invariance tells us that this solution is not unique, with any value allowed for the point where the solution crosses the middle of the barrier. Solutions of this type are referred to as instantons, with the term originating from the "time" at the center of the solution.[1] It is well to keep in mind, however, that τ is not a time,

---

[1] In the original paper on the Yang–Mills case [227], the solutions were referred to as pseudoparticles. Although most researchers in the field soon adopted the term instanton, "pseudoparticle" appears in much of the early literature on the subject because of the reluctance of some journals to accept neologisms.

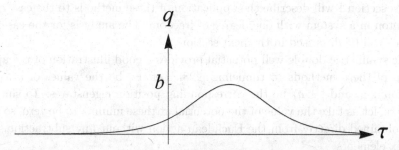

Fig. 9.5. The bounce solution corresponding to decay out of a metastable minimum such as that in Fig. 9.3b. The classical turning point is at $q = b$.

but just a particular parameterization of the favored tunneling path through configuration space.

We see from Eqs. (9.9) and (9.16) that for these instanton solutions

$$\frac{B}{2} = S_E[\bar{\mathbf{q}}] - S_E[\mathbf{q}_0] \qquad \text{(instanton)}, \qquad (9.17)$$

where $S_E[\mathbf{q}_0]$, given by the last integral in Eq. (9.16), is the Euclidean action of the trivial constant solution $\mathbf{q}(\tau) = \mathbf{q}_0$. Note that this relation between B and S_E only holds at their stationary points.

We will also be interested in the case where the tunneling is between a local minimum and a turning point $\mathbf{q}_f$ that is not a minimum of V, such as the decay of a bound state in a potential like that in Fig. 9.3b. In this case, the solution begins at $\tau = -\infty$, but reaches $\mathbf{q}_f$ at a finite value of τ, at which point $d\bar{\mathbf{q}}/d\tau = 0$. Because the Lagrangian is invariant under time reversal, this solution can be continued back to the initial point $\mathbf{q}_i$, which is reached at $\tau = \infty$. This doubles the Euclidean action, so that for the full solution

$$B = S_E[\bar{\mathbf{q}}] - S_E[\mathbf{q}_0] \qquad \text{(bounce)}. \qquad (9.18)$$

For obvious reasons this solution is called a bounce.[2] The bounce corresponding to a potential like that in Fig. 9.3b is shown in Fig. 9.5.

9.3 Path integral approach to tunneling: instantons

As discussed in the previous section, the exponent in the tunneling amplitude is readily obtained by WKB methods. This approach has the advantage of clarifying the physical significance of the Euclidean solution. However, path integral methods turn out to be more convenient for calculating the pre-exponential factor.

---

[2] Terminology varies in the literature, with some authors including bounces within the term instanton.

In this section I will describe the application of these methods to the case of an instanton in a system with one degree of freedom. The analysis for the case of a bounce will be discussed in the next section.

The symmetric double-well potential provides a good illustration of the application of these methods to tunneling [228]. Let $\pm a$ be the values of q at the two minima, and $|\pm a\rangle$ be the corresponding position eigenstates. To simplify matters, let us take the value of the potential at these minima to be zero, so that B is obtained directly from the Euclidean action without any subtraction. The matrix elements

$$\langle a|e^{-HT}|a\rangle = \langle -a|e^{-HT}|-a\rangle \tag{9.19}$$

and

$$\langle a|e^{-HT}|-a\rangle = \langle -a|e^{-HT}|a\rangle \tag{9.20}$$

can be expressed as path integrals

$$\langle \pm a|e^{-HT}|a\rangle = \int [dq(\tau)]e^{-S_E[q]}, \tag{9.21}$$

where the integration is over paths such that $q(-T/2) = a$ and $q(T/2) = \pm a$. (To simplify notation, the subscript E on the action will be omitted for the remainder of this chapter, with all actions understood to be Euclidean.)

The matrix element on the left-hand side can be expanded in terms of energy eigenstates to give

$$\langle \pm a|e^{-HT}|a\rangle = \sum_n e^{-E_n T}\langle \pm a|n\rangle\langle n|a\rangle. \tag{9.22}$$

In the limit of large T this is dominated by the contributions from the states with lowest energy. Thus we can find these energies by evaluating the path integral for large T and picking out the dominant exponentials.

In particular, let us denote the lowest even and odd energy eigenstates by $|+\rangle$ and $|-\rangle$, respectively, and assume that T is large enough that the contributions from all higher states can be ignored. We then have

$$\langle a|e^{-HT}|a\rangle = |\langle a|+\rangle|^2 e^{-E_+ T} + |\langle a|-\rangle|^2 e^{-E_- T} \tag{9.23}$$

and

$$\langle -a|e^{-HT}|a\rangle = \langle -a|+\rangle\langle +|a\rangle e^{-E_+ T} + \langle -a|-\rangle\langle -|a\rangle e^{-E_- T}. \tag{9.24}$$

Clearly $\langle a|\pm\rangle = \pm\langle -a|\pm\rangle$. If these two lowest states are well separated from all the others, we also have $|\langle a|+\rangle| = |\langle a|-\rangle|$. In the limit of large T, we will then have

$$\frac{\langle a|e^{-HT}|a\rangle + \langle -a|e^{-HT}|a\rangle}{\langle a|e^{-HT}|a\rangle - \langle -a|e^{-HT}|a\rangle} = e^{(E_- - E_+)T}. \tag{9.25}$$

We now use path integrals to evaluate the matrix elements in this expression, approximating each of the path integrals by a sum of Gaussian integrals about their stationary points. Given a Euclidean solution $\bar{q}(\tau)$ we write

$$q(\tau) = \bar{q}(\tau) + \sum_n c_n \psi_n(\tau), \qquad (9.26)$$

where $\psi_n(\tau)$ is an eigenmode with eigenvalue λ_n of

$$\frac{\delta^2 S}{\delta q(\tau)\delta q(\tau')}\bigg|_{q=\bar{q}(\tau)} = -\frac{d^2}{d\tau^2} + V''(\bar{q}(\tau)) \equiv S''(\bar{q}). \qquad (9.27)$$

We then change integration variables from $q(\tau)$ to the mode coefficients and write

$$[dq] = \prod_n \frac{dc_n}{\sqrt{2\pi}}. \qquad (9.28)$$

(The factors of 2π are for later convenience. Our results will be insensitive to an overall—and generally divergent—normalization factor for the path integral.)

The contribution to the path integral from this stationary point then becomes

$$I = \int \prod_n \frac{dc_n}{\sqrt{2\pi}} e^{-[S(\bar{q}) + \frac{1}{2}\sum_k \lambda_k c_k^2 + \cdots]}, \qquad (9.29)$$

where the ellipsis denotes terms that are cubic and higher order in the deviation from $\bar{q}$. Treating these as higher-order perturbations then gives

$$\begin{aligned} I &= e^{-S(\bar{q})} \prod \lambda_n^{-1/2} [1 + \cdots] \\ &= e^{-S(\bar{q})} [\det S''(\bar{q})]^{-1/2} [1 + \cdots]. \end{aligned} \qquad (9.30)$$

The ellipsis again denotes higher-order corrections; these will not be shown explicitly in the following equations, but should be understood.

Let us now enumerate the relevant stationary points. For $\langle a|e^{-HT}|a\rangle$ we have the trivial constant solution $q_0(\tau) = a$. With our convention that $V(q)$ vanishes at the minima, $S(q_0) = 0$ and the contribution to the path integral is simply

$$I_0 = [\det S''(q_0)]^{-1/2}. \qquad (9.31)$$

For $\langle -a|e^{-HT}|a\rangle$ we have the instanton solution, modified slightly so that q_1 runs from $-a$ at $\tau = -T/2$ to a at $\tau = T/2$. According to our formula, this should give a contribution

$$e^{-S_1} [\det S''(q_1)]^{-1/2}, \qquad (9.32)$$

where S_1 is the Euclidean action of the instanton. However, there is a problem. There is a zero mode of S'',

$$\psi_0(\tau) = N^{-1/2} \frac{dq_1}{d\tau}, \qquad (9.33)$$

reflecting the broken τ-translation symmetry.[3] Because of the zero eigenvalue, $\det S''$ vanishes and the pre-exponential factor diverges.

This divergence is closely related to the infrared divergences associated with the zero mode that we encountered when calculating the quantum corrections to the kink mass in Sec. 2.3. The solution is again to replace the coefficient of the zero mode by a collective coordinate z that specifies the location of the center of the instanton, and to include only the nonzero modes in the determinant. Instead of a Gaussian integral over the zero-mode coefficient we must integrate over the full range of z from $-T/2$ to $T/2$. Because

$$\psi_0 \, dc_0 = \frac{dq}{d\tau} \, dz \tag{9.35}$$

we have

$$(2\pi)^{-1/2} dc_0 = \left(\frac{N}{2\pi} \right)^{1/2} dz \,. \tag{9.36}$$

The net contribution of the one-instanton stationary points is then

$$I_1 = e^{-S_1} \left(\frac{N}{2\pi} \right)^{1/2} T \left[\det{}' S''(q_1) \right]^{-1/2} \,, \tag{9.37}$$

with the prime on the determinant indicating that only the nonzero modes are to be included. Finally, let us define

$$K = \left(\frac{N}{2\pi} \right)^{1/2} \left[\frac{\det{}' S''(q_1)}{\det S''(q_0)} \right]^{-1/2} \,, \tag{9.38}$$

so that

$$I_1 = e^{-S_1} \left[\det S''(q_0) \right]^{-1/2} KT \,. \tag{9.39}$$

In addition to these, there are approximate stationary points that must also be considered. Although a static configuration with an instanton and an anti-instanton separated by a τ interval much greater than their width is not quite a solution of the Euclidean equations, it is close enough to being a stationary point of the action that it must be included, as must configurations with larger numbers of (necessarily alternating) instantons and anti-instantons. Let us focus on a configuration q_n with a total of n instantons and anti-instantons. Its action is just n times that of a single instanton, $S(q_n) = nS_1$. It has n approximate zero modes corresponding to independent τ-translations of its component instantons.

---

[3] The normalization factor is

$$N = \int_{-T/2}^{T/2} d\tau \left(\frac{dq_1}{d\tau} \right)^2 \,. \tag{9.34}$$

By virial arguments that parallel those leading to Eq. (2.79), one can show that for the example at hand $N = S_1$. Strictly speaking, this is exactly a zero mode only in the limit of infinite T. However, for large T the eigenvalue and the corrections to the mode are exponentially small, and in the end we are only interested in the large T limit.

A collective coordinate z_j must be introduced for each of these, leading to a total Jacobian factor $(N/2\pi)^{n/2}$. Integrating over the instanton locations, which satisfy the constraint $z_1 < z_2 < \cdots < z_n$ gives a factor of

$$\int_{-T/2}^{T/2} dz_1 \int_{z_1}^{T/2} dz_2 \cdots \int_{z_{n-1}}^{T/2} dz_n = \frac{T^n}{n!}. \tag{9.40}$$

The final factor to be included is the functional determinant (with zero modes excluded) of $S''(q_n)$. The key point here is that the effect of an instanton on the spectrum of fluctuations is localized around the position of the instanton; i.e., the ratio of determinants that appears in Eq. (9.38) is independent of T for large T. Hence, the determinant factor for a configuration with n well-separated instantons and anti-instantons can be written as

$$[\det S''(q_0)]^{-1/2} \left[\frac{\det' S''(q_n)}{\det S''(q_0)}\right]^{1/2} = [\det S''(q_0)]^{-1/2} \left[\frac{\det' S''(q_1)}{\det S''(q_0)}\right]^{n/2}. \tag{9.41}$$

Putting all the factors together, we find that the contribution from all configurations with n instantons and anti-instantons is

$$I_n = e^{-nS_1} [\det S''(q_0)]^{-1/2} K^n \frac{T^n}{n!}. \tag{9.42}$$

Adding together the contributions from all of the stationary and approximately stationary points we have

$$\langle a|e^{-HT}|a\rangle = \sum_{\text{even } n} I_n$$

$$= [\det S''(q_0)]^{-1/2} \sum_{\text{even } n} \frac{[e^{-S_1}KT]^n}{n!}$$

$$= [\det S''(q_0)]^{-1/2} \cosh [e^{-S_1}KT] \tag{9.43}$$

and

$$\langle -a|e^{-HT}|a\rangle = \sum_{\text{odd } n} I_n$$

$$= [\det S''(q_0)]^{-1/2} \sum_{\text{odd } n} \frac{[e^{-S_1}KT]^n}{n!}$$

$$= [\det S''(q_0)]^{-1/2} \sinh [e^{-S_1}KT]. \tag{9.44}$$

Recalling Eq. (9.25), we have

$$e^{(E_- - E_+)T} = \exp [2KTe^{-S_1}], \tag{9.45}$$

so that the splitting of the two lowest levels is

$$\Delta = E_- - E_+ = 2Ke^{-S_1}. \tag{9.46}$$

The exponent is the standard WKB result. What has been gained by the path integral calculation is an expression for the pre-exponential factor, Eq. (9.38).

We could also have extracted the absolute energies of the two lowest states from Eqs. (9.43) and (9.44), obtaining

$$E_0 = \frac{1}{2}(E_+ + E_-) = -\lim_{T\to\infty}\frac{1}{2T}\ln\det S''(q_0),\qquad(9.47)$$

which turns out to be equal to $\frac{1}{2}\sqrt{V''(a)}$ [228, 232]. This is just the result one would obtain by neglecting tunneling and approximating each well as a simple harmonic oscillator. The perturbative corrections arising because the wells are anharmonic are much larger than the exponentially small instanton corrections we have calculated. However, these corrections add equally to both the even and the odd states, so the splitting is correctly given by the instanton result.

A note of caution. Our treatment of the n-instanton stationary points assumed that the instantons were well separated. This is known as the dilute-gas approximation. To check its validity we must verify that the dominant contributions to our sums are from the terms where the instanton density n/T is low enough for this assumption to be valid. In an exponential sum $\sum y^n/n!$, the dominant contribution is from the terms with $n \approx y$, so the dominant contribution to Eqs. (9.43) and (9.44) is from the terms with

$$\frac{n}{T} \approx K e^{-S_1}.\qquad(9.48)$$

If $\delta\tau$ is the characteristic width of the instanton, we need that

$$(\delta\tau)K e^{-S_1} \ll 1.\qquad(9.49)$$

We haven't actually evaluated the determinants entering K. However, dimensional arguments suggest that $K\delta\tau$ should be roughly of order unity, so that the dilute-gas approximation will be valid as long as the instanton action is large enough that $e^{-S_1} \ll 1$.

9.4 Path integral approach to tunneling: bounces

Let us now turn to the case of a state that decays by tunneling [229, 230]. Once again, for convenience, the value of the potential energy at the metastable minimum $q = a$ is taken to be zero. To start, let us consider the matrix element

$$\langle a|e^{-HT}|a\rangle = \sum_n e^{-E_n T}|\langle a|n\rangle|^2 = \int [dq(\tau)]e^{-S[q]}.\qquad(9.50)$$

Arguments along the lines of the last section suggest that we can extract the energy E_0 of the lowest state in the well on the left from the large-time behavior of this matrix element. However, as remarked previously, a state with wavefunction concentrated in this well is not an energy eigenstate. Instead, it is a metastable state that can be described as having a complex energy whose imaginary part is related to the decay rate.

Nevertheless, let us proceed as before and evaluate the path integral by summing the contributions from the stationary points and approximate stationary points of the action: the trivial configuration $q_0(\tau) = a$, the bounce solution $q_b(\tau)$, and all possible multibounce solutions [230]. Just like the instanton, the bounce has a zero mode, proportional to its τ-derivative, that must be exchanged for a collective coordinate. There is no constraint that the number of bounces be even or odd, so summing over all numbers of bounces gives an exponential rather than a hyperbolic function, with the result being

$$\int [dq(\tau)]e^{-S[q]} = [\det S''(q_0)]^{-1/2} \exp\left[KT\, e^{-S(q_b)}\right], \qquad (9.51)$$

with

$$K = \left(\frac{N}{2\pi}\right)^{1/2} \left[\frac{\det' S''(q_b)}{\det S''(q_0)}\right]^{-1/2}, \qquad \textbf{(incorrect)} \qquad (9.52)$$

where the normalization constant for the zero mode is $N = S(q_b)$.

Arguing as before, we can extract E_0 from the coefficient of T in the dominant exponential at large T, obtaining

$$E_0 = -\left[\lim_{T\to\infty} \frac{1}{2T} \ln \det S''(q_0)\right] - Ke^{-S(q_b)} = \frac{1}{2}\sqrt{V''(a)} - Ke^{-S(q_b)}. \quad (9.53)$$

As we will see, $S''(q_b)$ has a mode with negative eigenvalue. This makes the determinant factor in K, and thus the entire second term, imaginary. If it were not for this fact, it would be pointless to have calculated the exponentially small bounce contributions, because they are much smaller in magnitude than the (real) perturbative corrections to the contribution from the trivial solution q_0. This complex result for E_0 is not a problem, since we expected to find an imaginary part from which the decay rate could be extracted. What is a problem, however, is that we have missed a factor of $1/2$.

We will come back to this factor shortly, but first let us discuss the negative mode. The existence of such a mode can be inferred from the fact that the zero mode, $\psi_0 = N^{-1/2}dq_b/d\tau$, has a node located at the center of the bounce. Recall from Eq. (9.27) that S'' has the form of a one-dimensional Schrödinger operator. For such a potential the lowest eigenstate has no nodes, the next one node, and so on. Since the zero mode has one node, there must be a lower mode (and only one such mode) that has no nodes.

The origin of this negative mode can be understood by considering the series of configurations shown in Fig. 9.6 [230]. These can be parameterized by c, the extremum value of $q(\tau)$. The curve with $c = b$ is the bounce configuration of Fig. 9.5, and that with $c = 0$ (i.e., the x-axis) is the trivial solution $q(\tau) = 0$. The action of these configurations as a function of c is illustrated in Fig. 9.7.

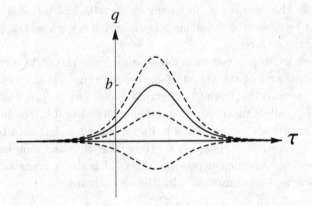

Fig. 9.6. A series of configurations in Euclidean space. The solid line with the maximum value of q equal to b is the bounce solution shown in Fig. 9.5. The dashed curve above this is a configuration that extends into the classically allowed region beyond the barrier, while the dashed curves below it are configurations that remain within the potential barrier.

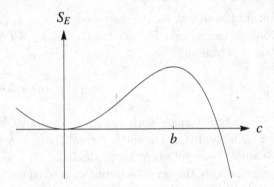

Fig. 9.7. The Euclidean action for a family of configurations such as those shown in Fig. 9.6. The maximum at $c = b$ corresponds to the bounce.

The trivial solution is clearly a local minimum, since any small deviation gives positive contributions to both the potential energy and the τ-derivative term. As c increases from 0 both of these terms, and thus the total action, increase monotonically until the bounce is reached. Since this is the only other stationary point, it must be a local maximum. As c increases further, the configuration spends more and more time in the region with negative potential energy, so the action decreases monotonically.

Thus the bounce, which is a local minimum of the barrier penetration integral $B[P]$, is only a saddle point of the Euclidean action. To understand this, recall that $B[P]$ is defined by an integral over a path from an initial point $\mathbf{q}_0$ to a point $\mathbf{q}_f$ lying on an equipotential surface Σ defined by $V(\mathbf{q}) = V(\mathbf{q}_0)$, and then back to $\mathbf{q}_0$. For the instanton the boundary conditions on S_E restrict it to the same class of paths, so a minimum of $B[P]$ is also a minimum of S_E. For the bounce, on the other hand, one calculates the Euclidean action over a path that starts at $\mathbf{q}_0$, tunnels through the barrier, and then tunnels back through the barrier to return to $\mathbf{q}_0$. Some of these are paths that reach Σ and then turn back; among this subset of paths, S_E is a minimum.

The negative mode arises from paths that do not turn back at Σ. The paths in Fig. 9.6 with $c < b$ never reach Σ, and remain within the potential barrier, while those with $c > b$ go beyond Σ and into the classically allowed region. Thus the negative mode corresponds to paths that are possible paths for the Euclidean action, but not for $B[P]$ [231].

Our path integral is an infinite-dimensional integral, but it is only the integration over the coefficient of the one negative mode that is problematic. We can see how to handle this integration by considering the analogous integral

$$J = \int_{-\infty}^{\infty} dc \, (2\pi)^{-1/2} e^{-S(c)} , \tag{9.54}$$

where $S(c)$ is the function plotted in Fig. 9.7. As $c \to -\infty$ the action increases and the integral converges. On the other hand, as $c \to \infty$ the action tends toward ∞ and the integral diverges. However, we can relate it to a finite integral by analytic continuation. As noted in Sec. 9.1, the potential with the metastable minimum at $q = 0$ can be viewed as the analytic continuation of a potential with a global minimum at $q = 0$. For the latter potential, the minimum corresponds to a stable state with a well-defined real energy, and the integration over c gives a finite integral. In our case, there is no stable state localized around $q = 0$ and the integral over c is ill-defined. However, we can make it well-defined by deforming the contour of integration from the real axis as shown in Fig. 9.8; this gives the integral a small imaginary part. The integration from $-\infty$ to b is clearly real. The imaginary part comes from the remainder of the contour, and in the steepest descent approximation is given by

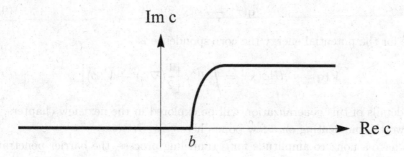

Fig. 9.8. The contour of integration that makes the integral in Eq. (9.54) well-defined.

$$\operatorname{Im} J = \operatorname{Im} \int_b^{b+i\infty} dc\, (2\pi)^{-1/2} e^{-S(b)} e^{-\frac{1}{2}S''(b)(c-b)^2}$$

$$= \frac{1}{2} e^{-S(b)} |S''(b)|^{-1/2} , \qquad (9.55)$$

with the factor of $\frac{1}{2}$ being due to the fact that the contour only goes through half of the Gaussian peak.

With this factor taken into account, the incorrect Eq. (9.52) is replaced by

$$K = \frac{i}{2} \left(\frac{N}{2\pi}\right)^{1/2} \left|\frac{\det' S''(q_b)}{\det S''(q_0)}\right|^{-1/2} . \qquad (9.56)$$

Finally, substituting the value of N, we obtain the decay width

$$\Gamma = -2\operatorname{Im} E_0 = \left(\frac{S(q_b)}{2\pi}\right)^{1/2} \left|\frac{\det' S''(q_b)}{\det S''(q_0)}\right|^{-1/2} e^{-S(q_b)} . \qquad (9.57)$$

As in the previous section, this result was obtained under the assumption that $V(a) = 0$. If $V(a) \neq 0$, we obtain instead

$$\Gamma = \left(\frac{B}{2\pi}\right)^{1/2} \left|\frac{\det' S''(q_b)}{\det S''(q_0)}\right|^{-1/2} e^{-B} , \qquad (9.58)$$

where, as in Eq. (9.18), B is the difference between the bounce action and that of the trivial solution.

9.5 Field theory

The formalism developed in the previous sections is readily carried over to field theory [226, 230], a system with a (continuous) infinity of degrees of freedom. This is largely a matter of translating notation. For a theory with a single scalar field ϕ in D spatial dimensions, the N-dimensional configuration space of the q^j is replaced by the infinite-dimensional space of field configurations $\phi(\mathbf{x})$. For the tunneling path we have the replacement

$$\mathbf{q}(\tau) \longrightarrow \phi(\mathbf{x}, \tau) , \qquad (9.59)$$

while for the potential energy the correspondence is

$$V(\mathbf{q}) \longrightarrow U[\phi(\mathbf{x})] = \int d^D x \left[\frac{1}{2}(\boldsymbol{\nabla}\phi)^2 + V(\phi)\right] . \qquad (9.60)$$

The details of this generalization will be explored in the next few chapters, but it is worth commenting on a few points here.

To have a nonzero amplitude for a tunneling process, the barrier penetration integral must be finite, which in turn requires that the deviation of the fields from their value in the initial configuration must be localized not just in Euclidean time, but also in space. As a result, there are additional zero modes, corresponding to translation of the instanton or bounce in the various spatial directions. A position collective coordinate must be introduced for each such mode. There are also Jacobian factors related to the normalization constants of these modes; in many cases virial theorems relate these to the Euclidean action. We must integrate over the full range of the collective coordinates. This gives a factor of the volume of space, $\mathcal{V}$; we will see in the next few chapters how $\mathcal{V}$ factors out when one looks at physically measurable quantities. The Euclidean solution may also break other symmetries, leading to additional zero modes and collective coordinates, which must also be integrated over.

Although additional zero modes are acceptable, and indeed required, we cannot allow additional negative modes. In our examples with one degree of freedom, it was easy to see that the instanton had no negative modes and that the bounce had precisely one, which was associated with the possibility of extending the path into the classically allowed region. With more degrees of freedom, and in field theory in particular, there is another way in which negative modes can arise. Our method is based in finding the MPEP, the tunneling path that minimizes the barrier penetration integral B. However, finding a stationary point of the Euclidean action only guarantees that we have a stationary point of B. It is quite possible for the Euclidean solution to be a saddle point of B, with some variations of the path lowering the value of B. Such variations would lead to additional negative modes of S''. If these exist, then the Euclidean solution is not an acceptable bounce or instanton.

Finally, the usual short-distance divergences of field theory are inevitably encountered in the calculation of the functional determinants. The worst of these are canceled when the determinant about the instanton or bounce is divided by that about the trivial solution. The remaining divergences disappear by a cancellation between divergences in the functional determinant and those arising from evaluating the divergent counterterms for the Euclidean solution. This closely parallels the treatment of the divergences in the kink mass in Sec. 2.2, with the quotient of determinants being analogous to the difference of vacuum zero-point energies.

10

Yang–Mills instantons

The most important examples of instantons are those, first discovered by Belavin, Polyakov, Schwarz, and Tyupkin [227], that arise in non-Abelian gauge theories. In this chapter I will focus on the properties of these and related solutions, and on their relation to the vacuum structure of the gauge theory. Their physical consequences will be discussed in the next chapter, which will also describe the additional instanton-related effects that arise when fermion fields are added to the theory.

10.1 $A_0 = 0$ gauge

Before delving into the details of these solutions, there are some points that must be addressed at the outset. It was shown in the previous chapter how tunneling amplitudes in field theory are related to solutions of the Euclidean field equations. When we turn to gauge theories, two new issues arise. First, the discussion in Chap. 9 was in the context of a theory with a kinetic energy that was purely quadratic in time derivatives. In general, gauge theory kinetic energies are not of this form. Second, we need to understand how the Euclideanization should affect A_0. If we were to simply replace t by $i\tau$, and leave A_0 unchanged, then the F_{0j} components of the field strength would become complex. This would be avoided if A_0 also gained a factor of i, but that would require some justification.

Both of these issues can be clarified by working in a gauge with $A_0 = 0$. From a canonical point of view, this gauge has some unusual features. With A_0 absent, the Lagrangian density takes the form

$$\mathcal{L} = \text{tr}\, \dot{A}_j^2 - \frac{1}{2}\text{tr}\, F_{ij}^2. \tag{10.1}$$

This describes a theory with three independent dynamical fields, A_1, A_2, and A_3, whose conjugate momenta are the electric components of the field strength, $\Pi^j = F^{j0}$. Varying with respect to the A_j yields the three equations of motion

$$D_\mu F^{\mu j} = 0. \tag{10.2}$$

However, we do not recover the A_0 equation, which is the Gauss's law constraint

$$\mathcal{C}(\mathbf{x}) \equiv D_j F^{j0}(\mathbf{x}) = D_j \Pi^j(\mathbf{x}) = 0. \tag{10.3}$$

It should be kept in mind that $\mathcal{C}$ is an element of the Lie algebra of the gauge group, so this is actually an adjoint multiplet of constraints at each point in space. (Here, and throughout most of this chapter, group indices have been suppressed.)

Although $\mathcal{C}(\mathbf{x})$ does not necessarily vanish, Eq. (10.2) implies that its time derivative does. This vanishing of $\dot{\mathcal{C}}(x)$ can be understood from symmetry considerations. Even after fixing $A_0 = 0$, we still have the freedom to make time-independent gauge transformations. These can be viewed as symmetries of the Lagrangian and, by Noether's theorem, should lead to corresponding conserved charges.

This can be made explicit by considering the infinitesimal gauge transformation corresponding to a gauge function $\Lambda(\mathbf{x})$ that vanishes at spatial infinity. The Noether charge that generates this transformation is

$$
\begin{aligned}
C_\Lambda &= 2 \int d^3x \operatorname{tr} D_j \Lambda(\mathbf{x}) \Pi^j(\mathbf{x}) \\
&= -2 \int d^3x \operatorname{tr} \Lambda(\mathbf{x}) D_j \Pi^j(\mathbf{x}) \\
&= -2 \int d^3x \operatorname{tr} \Lambda(\mathbf{x}) \mathcal{C}(\mathbf{x}).
\end{aligned}
\tag{10.4}
$$

(There is no surface term from the integration by parts in the second line because of the requirement that $\Lambda(\mathbf{x})$ vanish at spatial infinity.) Because the gauge transformations are a symmetry of the Lagrangian, C_Λ is conserved and $\dot{C}_\Lambda = 0$ for arbitrary Λ. This is equivalent to the vanishing of $\dot{\mathcal{C}}(\mathbf{x})$ for all $\mathbf{x}$.

At this point, we have a theory with three independent dynamical fields, A_1, A_2, and A_3. This is too many fields, so it is not quite the theory that we want. However, we also have an infinite number of conserved quantities $\mathcal{C}(\mathbf{x})$. By restricting ourselves to the subspace in which $\mathcal{C}(\mathbf{x}) = 0$ for all $\mathbf{x}$, we arrive at the desired Yang–Mills theory. Classically, this corresponds to imposing Gauss's law as a constraint on the initial conditions. Quantum mechanically, it can be implemented by restricting ourselves to the subspace of the Hilbert space in which all components of the $\mathcal{C}(\mathbf{x})$ vanish.[1]

In this gauge the kinetic energy is purely quadratic, and so the arguments relating tunneling to the Euclidean Lagrangian

$$\mathcal{L}_E = \operatorname{tr} (\partial_4 A_j)^2 + \frac{1}{2} \operatorname{tr} F_{ij}^2 \tag{10.5}$$

---

[1] Because there are nontrivial commutators between the gauge components of $\mathcal{C}$, one cannot in general find a simultaneous eigenstate of all them. However, the commutators are all proportional to components of $\mathcal{C}$, so eigenstates with zero eigenvalue for all components do exist. This is analogous to the situation with angular momentum, for which one can have states with $L_x = L_y = L_z = 0$.

(with $x_4 \equiv \tau$) go through straightforwardly. In analogy with the Lagrangian of Eq. (10.1), this Lagrangian implies three field equations, but not the Euclidean version of Gauss's law,

$$D_j F^{j4} = 0. \tag{10.6}$$

However, the classical instanton solution must satisfy the boundary conditions that its Euclidean time derivatives $\partial_4 A_j$ vanish at the initial and final values of x_4. This, together with the other field equations, implies that the Gauss's law constraint is satisfied at all intermediate values of x_4.

Although we could just work with the Euclidean theory in this form, we can also give the theory a full Euclidean gauge freedom by introducing a field $A_4(x)$ that essentially acts as a Lagrange multiplier to enforce the Euclidean Gauss's law constraint. We define F_{j4} in the obvious way, and take the Euclidean Lagrangian to be

$$\mathcal{L}_E = \frac{1}{2} \operatorname{tr} F_{rs}^2 , \tag{10.7}$$

where r and s run from 1 to 4. Once we have done this, we are free to work in any gauge that we find convenient, with the boundary conditions on the instanton at the initial and final values of x_4 being such that $F_{j4}(\mathbf{x}, x_4^{\text{init}}) = F_{j4}(\mathbf{x}, x_4^{\text{fin}}) = 0$.

In the course of our examination of the Yang–Mills instantons, there will be occasion to refer to both Euclidean and Minkowskian coordinates. I will adopt the convention that Greek indices refer to Minkowskian coordinates that take values from 0 to 3, Roman indices from late in the alphabet $(r, s, \dots)$ refer to Euclidean indices that range from 1 to 4, and Roman indices from the middle of the alphabet $(i, j, \dots)$ are spatial indices running from 1 to 3, and can occur in either a Minkowskian or a Euclidean context.

10.2 Yang–Mills vacua: $A_0 = 0$ gauge

If instantons are associated with a tunneling process, the natural starting point of the discussion is to specify from where and to where the tunneling is taking place. Somewhat surprisingly, the answer to this question—although not the observable physical consequences—depends on the gauge in which one works. Perhaps the simplest picture is the one associated with the $A_0 = 0$ gauge [233, 234]. As described below, in this gauge there are degenerate classical minima that are separated by energy barriers. Quantum mechanical tunneling through these barriers is described by the instanton.

We have already seen in the previous section that $A_0 = 0$ gauge has some unusual features, and the existence of multiple classical vacua is just one more of these. This suggests that it might be helpful to examine the problem in a more "normal" gauge in which there is a unique vacuum. This will be done in the next section.

Classically, the vacuum configurations have $F_{\mu\nu} = 0$. In $A_0 = 0$ gauge this means that

$$A_j = \frac{i}{g} G^{-1} \partial_j G, \tag{10.8}$$

where $G(\mathbf{x})$ is an element of the gauge group, which for the present we take to be SU(2). Let us assume that G tends to a constant value G_∞ as $r \to \infty$; without loss of generality we can take $G_\infty = I$. $G(\mathbf{x})$ is a map from R^3 into the gauge group. Our assumption that $G(r = \infty)$ is a constant means that we can view spatial infinity as a single point, and space as a three-sphere. The map is then one from S^3 into the group. Such maps, and hence the vacuum configurations, can be classified by elements of $\pi_3(SU(2))$ which, we recall from Eq. (4.46), is Z, the additive group of the integers. Thus, the classical vacua fall into disconnected sets, each corresponding to an element of $\pi_3(SU(2))$. The vacua within a given set are related by gauge transformations that can be built up from a series of infinitesimal gauge transformations with gauge functions Λ that vanish at spatial infinity; we will call these "small" gauge transformations. Vacua in different sets are also gauge-equivalent, but the gauge transformations that connect them are "large" gauge transformations that cannot be obtained from a sequence of infinitesimal gauge transformations.

The correspondence between vacua and homotopy classes can be made explicit by defining a quantity

$$N[G] = \frac{1}{24\pi^2} \epsilon^{ijk} \int d^3x \ \mathrm{tr}\, G^{-1} \partial_i G \, G^{-1} \partial_j G \, G^{-1} \partial_k G. \tag{10.9}$$

The normalization here is such that if G is taken to be in the fundamental representation of SU(2), with the generators $T_a = \sigma_a/2$, $N[G]$ is the winding number that counts how many times SU(2), which is topologically a three-sphere, is mapped onto R^3, which we are also viewing as a three-sphere. To see this, consider the neighborhood of a point where G has some value G_0. Near this point, we can write

$$G(x) = G_0 \exp[i\sigma_a \Lambda_a(x)] \approx G_0[I + i\sigma_a \Lambda_a(x)] \tag{10.10}$$

with Λ_a small. The right-hand side of Eq. (10.9) is then the integral of

$$-\frac{i}{24\pi^2} \epsilon^{ijk} (\mathrm{tr}\, \sigma_a \sigma_b \sigma_c) \, \partial_i \Lambda_a \, \partial_j \Lambda_b \, \partial_k \Lambda_c = \frac{1}{12\pi^2} \epsilon^{ijk} \epsilon^{abc} \, \partial_i \Lambda_a \, \partial_j \Lambda_b \, \partial_k \Lambda_c$$

$$= \left(\frac{1}{2\pi^2}\right) \left(\epsilon^{ijk} \partial_i \Lambda_1 \, \partial_j \Lambda_2 \, \partial_k \Lambda_3\right). \tag{10.11}$$

On the last line the second factor is the Jacobian corresponding to the transformation from the spatial coordinates x_j to the group coordinates Λ_a. This converts the integral over space to one over the group. Dividing by the volume of the unit three-sphere, $2\pi^2$, normalizes this integral so that it gives directly the number of times that SU(2) is covered by G as one goes over all of space.[2]

A check of the normalization can be obtained by explicitly calculating the winding number for

$$g_1(x) = \exp[i\hat{r}_a \sigma_a f(r)], \tag{10.12}$$

where $f(r)$ is any monotonic function with $f(0) = -\pi$ and $f(\infty) = 0$. This gives a one-to-one mapping of the physical space into SU(2), with every element of SU(2) occurring at only a single value of $\mathbf{x}$. One verifies that $N[g_1] = 1$, as required.

Now suppose that the gauge transformation can be written as a product $G = G_1 G_2$. Substitution of this into Eq. (10.9) gives

$$N[G_1 G_2] = N[G_1] + N[G_2] + \frac{1}{8\pi^2}\Delta, \tag{10.13}$$

where

$$
\begin{aligned}
\Delta &= \epsilon^{ijk} \int d^3x \operatorname{tr} \left[G_1^{-1}\partial_i G_1 \, \partial_j G_2 \, G_2^{-1}\partial_k G_2 \, G_2^{-1} \right. \\
&\qquad\qquad \left. + \partial_i G_2 \, G_2^{-1} G_1^{-1}\partial_j G_1 \, G_1^{-1}\partial_k G_1 \right] \\
&= -\epsilon^{ijk} \int d^3x \operatorname{tr} \left[G_1^{-1}\partial_i G_1 \, \partial_j G_2 \, \partial_k G_2^{-1} + \partial_i G_2 \, G_2^{-1}\partial_j G_1^{-1} \, \partial_k G_1 \right] \\
&= \epsilon^{ijk} \int d^3x \operatorname{tr} \left[G_1^{-1}\partial_i G_1 \, \partial_j (\partial_k G_2 \, G_2^{-1}) - \partial_i G_2 \, G_2^{-1}\partial_j (G_1^{-1}\partial_k G_1) \right] \\
&= \epsilon^{ijk} \int d^3x \operatorname{tr} \partial_j \left(G_1^{-1}\partial_i G_1 \partial_k G_2 \, G_2^{-1} \right) \\
&= 0. \tag{10.14}
\end{aligned}
$$

Thus, the winding number of a product of two gauge transformations is the sum of the individual winding numbers,

$$N[G_1 G_2] = N[G_1] + N[G_2]. \tag{10.15}$$

Equation (10.11) shows that for an infinitesimal gauge transformation, $G_\Lambda \approx I + i\Lambda(\mathbf{x}) = I + i\sigma_a \Lambda_a$, with $\Lambda(\mathbf{x})$ vanishing at spatial infinity, the leading contribution to $N[G_\Lambda]$ is the integral of

$$\frac{1}{2\pi^2} \epsilon^{ijk} \partial_i \Lambda_1 \, \partial_j \Lambda_2 \, \partial_k \Lambda_3 = \frac{1}{2\pi^2} \epsilon^{ijk} \partial_i (\Lambda_1 \, \partial_j \Lambda_2 \, \partial_k \Lambda_3). \tag{10.16}$$

Since this is the total derivative of a quantity that vanishes at spatial infinity, $N[G_\Lambda] = 0$. It then follows from Eq. (10.15) that any continuous change in G

---

[2] This is quite analogous to the results in lower dimensions that were given in Eqs. (4.38) and (4.39).

leaves $N[G]$ invariant, as expected for a winding number. Furthermore, if G_0 is a constant, then G and $G_0 G$ have the same winding number.

Now let us define a current

$$j_A^\mu = \frac{g^2}{8\pi^2} \epsilon^{\mu\nu\alpha\beta} \operatorname{tr}\left(A_\nu \partial_\alpha A_\beta - \frac{2ig}{3} A_\nu A_\alpha A_\beta\right)$$
$$= \frac{g^2}{16\pi^2} \epsilon^{\mu\nu\alpha\beta} \operatorname{tr}\left(A_\nu F_{\alpha\beta} + \frac{2ig}{3} A_\nu A_\alpha A_\beta\right). \qquad (10.17)$$

Although this current is not gauge independent, its divergence,[3]

$$\partial_\mu j_A^\mu = \frac{g^2}{16\pi^2} \operatorname{tr} F_{\mu\nu} \tilde{F}^{\mu\nu}, \qquad (10.18)$$

is. Here

$$\tilde{F}^{\mu\nu} = \frac{1}{2} \epsilon^{\mu\nu\alpha\beta} F_{\alpha\beta} \qquad (10.19)$$

(or "F dual") is the dual field strength.

The charge associated with this current is

$$Q_A = \int d^3x\, j_A^0 = \frac{g^2}{16\pi^2} \int d^3x\, \epsilon^{ijk} \operatorname{tr}\left(A_i F_{jk} + \frac{2ig}{3} A_i A_j A_k\right). \qquad (10.20)$$

(Note that, because j_A is not divergenceless, there is no reason to expect Q_A to be conserved.) For a vacuum configuration, with $F_{ij} = 0$ and $A_j = (i/g)G^{-1}\partial_j G$, we have

$$Q_A = N[G]. \qquad (10.21)$$

Now let $A_\mu(\mathbf{x}, t)$ be a sequence of finite energy configurations, parameterized by t, that interpolate between two vacuum configurations with winding numbers N_1 and N_2. The integral

$$\int d^4x\, \partial_\mu j_A^\mu = \int d^4x\, [\partial_0 j_A^0 + \partial_i j_A^i] \qquad (10.22)$$

can be converted to a sum of integrals over the bounding surface, which is conveniently viewed as a hypercylinder oriented along the t-axis. The first term on the right-hand side contributes only on the initial and final hypersurfaces of constant t, giving $N_2 - N_1$. The second term gives the integral over the surface at $r = \infty$. In $A_0 = 0$ gauge the only contribution to j_A^i comes from the $A_j F_{0k}$ term, but for finite energy configurations this falls fast enough as $r \to \infty$ that the surface integral vanishes. Hence, using Eq. (10.18), we find that

$$\Delta N = N_2 - N_1 = \frac{g^2}{16\pi^2} \int d^4x\, \operatorname{tr} F_{\mu\nu} \tilde{F}^{\mu\nu}. \qquad (10.23)$$

---

[3] In verifying this equation, one can use the cyclic property of the trace and the fact that $\epsilon^{0123} = -\epsilon^{1230}$ to show that the term quartic in the potential on the right-hand side actually vanishes.

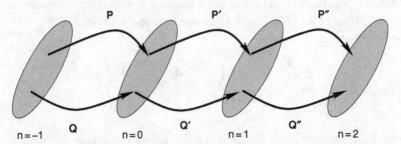

Fig. 10.1. Schematic view of the configuration space in $A_0 = 0$ gauge. The shaded regions correspond to vacuum configurations with various values of the winding number n. These are separated by regions of nonvacuum configurations of nonzero energy. Tunneling paths P, P', and P'' are gauge-equivalent, as are Q, Q', and Q''. Along each of these paths the value of the integral in Eq. (10.24) is determined solely by the difference of the initial and final winding numbers.

A caricature of the vacuum structure in $A_0 = 0$ gauge is shown in Fig. 10.1. Each of the shaded "islands" corresponds to the vacuum configurations of a fixed winding number. These are separated by a "sea" of nonvacuum configurations of higher energy. Each island can be mapped onto the succeeding one by a gauge transformation of unit winding number, such as the $g_1(\mathbf{x})$ of Eq. (10.12). Also shown are a number of paths from one island to the next. These are also gauge equivalent. Paths P' and P'' are obtained by successive applications of g_1 to P, while Q' and Q'' are obtained from Q in a similar manner. Each of these paths connects vacua whose winding numbers differ by one, so integrating along any of these gives

$$\int d^4 x \operatorname{tr} F_{\mu\nu} \tilde{F}^{\mu\nu} = \frac{16\pi^2}{g^2}. \tag{10.24}$$

Note that although Eqs. (10.22)–(10.24) were written with Greek indices, suggesting a Minkowskian interpretation of t as a real time, they are equally valid with a Euclidean interpretation in which x_4 simply parameterizes a sequence of configurations involving only A_j with $j = 1, 2, 3$.

This discussion of vacua has been purely classical up to this point. Let us now turn to the quantum theory. In a theory with a single classical vacuum the wave functional of the quantum ground state is concentrated on field configurations close to the classical vacuum. Here, however, we have many degenerate classical vacua. We are faced with two types of degeneracies. The first, among the states with a given winding number, is handled by imposing the Gauss's law constraint. In general, if a state is an eigenstate of a quantum mechanical charge with zero eigenvalue, then its wavefunction is independent of the conjugate variable; e.g., a one-particle state with zero angular momentum has a wavefunction that is independent of angle. In the present case, Gauss's law restricts us to states for which the C_Λ all vanish, implying that their wave functional is uniform (with respect to an appropriate measure) over all configurations that are related by small gauge transformations. If we only had classical vacua with a single winding number, this would resolve the degeneracy, and there would be a unique quantum vacuum.

However, we have to consider vacuum configurations with all possible winding numbers. Perturbative analysis about these would predict one quantum vacuum state for each winding number. Let us denote these "n-vacua" by $|n\rangle$. If it were not for the possibility of tunneling through the potential energy barriers separating vacuum configurations with different winding numbers, these n-vacua would be eigenstates of the Hamiltonian. However, because there is tunneling, they are not. The situation is similar to that of a particle in a periodic potential, and we can proceed in a similar fashion.

To construct the true eigenstates, let us define an operator T that takes a state $|\Psi\rangle$ with wave functional $\Psi[A_j(\mathbf{x})]$ to a state $T|\Psi\rangle = |\Psi'\rangle$ whose wave functional obeys

$$\Psi'[A_j^g(\mathbf{x})] = \Psi[A_j(\mathbf{x})], \tag{10.25}$$

where A_j^g is the result of acting on A_j with the large gauge transformation g_1 defined in Eq. (10.12). In particular, T takes one n-vacuum to the next, so that

$$T|n\rangle = |n+1\rangle. \tag{10.26}$$

The gauge invariance of the theory ensures that T commutes with the Hamiltonian.

We can then define "θ-vacua,"[4]

$$|\theta\rangle = \sum_{n=-\infty}^{\infty} e^{-in\theta}|n\rangle, \tag{10.27}$$

that are eigenstates of T with eigenvalue $e^{i\theta}$ [233, 234]. (Note that θ is periodic with period 2π.) These θ-vacua are true eigenstates of the Hamiltonian. To verify this, consider the matrix element of the time evolution operator between two θ-vacua,

$$\langle\theta'|e^{-iHt}|\theta\rangle = \sum_{m,n} e^{im\theta'} e^{-in\theta}\langle m|e^{-iHt}|n\rangle$$
$$= \sum_{m,n} e^{im(\theta'-\theta)} e^{i(m-n)\theta}\langle m|e^{-iHt}|n\rangle. \tag{10.28}$$

---

[4] Although an intuitive specification of the n-vacua is fairly clear, a precise definition is a bit more difficult, since their wave functionals have tails extending out to configurations, far from any classical vacuum, for which the winding number does not have even an approximate meaning. One way to proceed is to define the θ-vacua as the lowest energy states with given values of θ, and to then define the n-vacua by inverting the Fourier transform in Eq. (10.27).

Because H is gauge-invariant, the matrix element on the right-hand side can only depend on the difference between the winding numbers, $m - n$. Hence,

$$\langle\theta'|e^{-iHt}|\theta\rangle = \sum_{m,k} e^{im(\theta'-\theta)} e^{ik\theta} \langle k|e^{-iHt}|0\rangle$$

$$= 2\pi\delta(\theta - \theta') \sum_k e^{ik\theta} \langle k|e^{-iHt}|0\rangle. \tag{10.29}$$

Thus, one θ-vacuum cannot evolve into a different one. This argument can clearly be extended to the one- and many-particle states built upon these vacua.[5] If we assume that we live in a state of definite θ, then states with other values of θ are inaccessible to us, and so θ is effectively a constant of nature.

The explicit factor of $e^{ik\theta}$ that appears in the sum in Eq. (10.29) can be absorbed by modifying the Lagrangian. Writing the matrix element of the time evolution operator as a path integral, we have

$$\langle k|e^{-iHt}|0\rangle = \int [dA_\mu] e^{i\int d^4x \mathcal{L}}, \tag{10.30}$$

where the integral is over paths that run from a configuration with zero winding number at $t = -\infty$ to one with winding number k at $t = \infty$. Equation (10.23) tells us that over any such path

$$k = \frac{g^2}{16\pi^2} \int d^4x \operatorname{tr} F_{\mu\nu}\tilde{F}^{\mu\nu}, \tag{10.31}$$

so the factor of $e^{ik\theta}$ is equivalent to adding to the Lagrangian density a term of the form [234, 235]

$$\Delta\mathcal{L} = \frac{\theta g^2}{16\pi^2} \operatorname{tr} F_{\mu\nu}\tilde{F}^{\mu\nu}. \tag{10.32}$$

Recalling Eq. (10.18), we see that this additional term is a total divergence. Classically, such a term would have no effect, since adding a total divergence to the Lagrangian density leaves the Euler–Lagrange equations unchanged. Even quantum mechanically, this has no effect in perturbation theory: If we think of this term as giving rise to a new type of vertex in Feynman graphs, the overall derivative gives a factor of the net momentum entering the vertex which is, of course, zero. This is why it is usually omitted in introductory discussions of Yang–Mills theories, and why the importance, or even the possibility, of such a term was not appreciated until the discovery of the Yang–Mills instanton solutions. This term will be discussed in more detail in Sec. 11.5.

---

[5] Given an arbitrary state $|\Psi\rangle$, a state of definite θ can be defined by $|\Psi,\theta\rangle = \sum_n e^{-in\theta} T^n |\Psi\rangle$. T is clearly invertible, so its negative powers are well defined.

10.3 Yang–Mills vacuum: axial gauge

The discussion in the previous section depended crucially on the fact that we were working in a gauge that admitted multiple classical vacua. One more commonly imagines working in a "physical" gauge where the potential is uniquely determined by the field strength. In such a gauge there is only one classical vacuum. In electrodynamics, Coulomb gauge is a simple example of a gauge that gives a one-to-one mapping between field strength and potential. This is not the case in the non-Abelian theory, where the Gribov ambiguity shows that there are configurations of field strengths that allow several distinct choices of Coulomb gauge potentials, as well as other configurations that cannot be brought into Coulomb gauge at all [85, 236]. These issues are often swept under the rug when working in perturbation theory, but cannot be ignored when considering potentials of order $1/g$, for which the commutator term in the field strength is comparable in size to the derivative terms. As suggested by Eq. (10.8), it is precisely such potentials with which we will be concerned.

A choice that does fix the potentials uniquely is the axial gauge, defined by

$$A_3(x, y, z, t) = 0 \tag{10.33}$$

and the subsidiary conditions

$$\begin{aligned} A_2(x, y, 0, t) &= 0, \\ A_1(x, 0, 0, t) &= 0, \\ A_0(0, 0, 0, t) &= 0. \end{aligned} \tag{10.34}$$

To show that this gauge can always be achieved, let us assume that we are given A_μ everywhere in space, but not necessarily in axial gauge. Our task is to find a gauge transformation $U(x)$ that transforms A_μ to an axial gauge potential A'_μ via

$$A'_\mu = U^{-1} A_\mu U + \frac{i}{g} U^{-1} \partial_\mu U. \tag{10.35}$$

Requiring that $A'_3 = 0$ gives

$$\partial_3 U(x, y, z, t) = ig A_3(x, y, z, t) U(x, y, z, t), \tag{10.36}$$

whose solution is

$$U(x, y, z, t) = P \exp\left[ig \int_0^z dz' \, A_x(x, y, z', t) \right] U(x, y, 0, t), \tag{10.37}$$

where P denotes path ordering. Requiring that A_2 vanish on the hyperplane $z = 0$ then gives

$$\partial_2 U(x, y, 0, t) = -ig A_2(x, y, 0, t) U(x, y, 0, t), \tag{10.38}$$

whose solution can again be written in terms of a path-ordered exponential of an integral along the y-direction on the hyperplane $z = 0$. By proceeding in a similar fashion, the remaining two subsidiary conditions can also be achieved.

If a potential is already in axial gauge, then Eq. (10.33) allows only z-independent gauge transformations. The three subsidiary conditions then rule out, successively, y-dependent, x-dependent, and t-dependent gauge transformations, leaving only the freedom to make global gauge transformations.

In fact, the axial gauge potential is given in terms of the field strength by a rather simple expression. Because $F_{3\mu} = \partial_3 A_\mu$,

$$A_\mu(x, y, z, t) = \int_0^z dz' \, F_{3\mu}(x, y, z', t) + C_\mu(x, y, t). \qquad (10.39)$$

The subsidiary conditions determine C_μ uniquely, giving

$$A_\mu(x, y, z, t) = \int_0^z dz' \, F_{3\mu}(x, y, z', t) + \int_0^y dy' \, F_{2\mu}(x, y', 0, t)$$
$$+ \int_0^x dx' \, F_{1\mu}(x', 0, 0, t) + \int_0^t dt' \, F_{0\mu}(0, 0, 0, t'). \qquad (10.40)$$

In particular, in the classical vacuum with $F_{\mu\nu} = 0$ the gauge potential vanishes everywhere.

Physical results should not depend on the choice of gauge. Let us see how we can recover the results of the previous section [235]. With the multiple vacua of the $A_0 = 0$ gauge, there were families of configurations, such as those lying along the path P of Fig. 10.1b, that connected one vacuum to another. Transforming these configurations into axial gauge gives a path $\tilde{P}$ that starts and ends at the same point. In contrast with the $A_0 = 0$ gauge instantons that describe tunneling through the energy barrier separating distinct vacua, axial gauge instantons correspond to a somewhat peculiar process in which one tunnels into an energy barrier merely to tunnel back out and return to one's starting point.

Because $\operatorname{tr} F_{\mu\nu}\tilde{F}^{\mu\nu}$ is gauge-invariant, its integrals over the paths P and $\tilde{P}$ must be equal. However, when these are converted to surface integrals of the gauge-variant j_A^μ the contributions arise from different parts of the surface. We saw previously that for the $A_0 = 0$ gauge path P the only contributions were from the initial and final surfaces, giving the integral in terms of a difference of winding numbers. In axial gauge, where the vacuum winding number clearly vanishes, the surface integral from path $\tilde{P}$ comes entirely from the surfaces at $z = \pm\infty$.

A particularly striking consequence of the multiple classical vacua in $A_0 = 0$ gauge was the existence of multiple quantum vacua labeled by a continuous periodic parameter θ. This naturally led us to the inclusion of the new, θ-dependent, term in the Lagrangian that was given in Eq. (10.32). The axial gauge analysis doesn't lead so compellingly to this term, but once we realize that it can appear, there is no reason not to include it [235]. There is the difference that θ is a fixed coupling constant in axial gauge, but a continuous variable in $A_0 = 0$ gauge. However, as was already noted, the fact that transitions between states of different θ are forbidden means that θ can be viewed as a constant of nature even in the latter gauge.

An analogy that may be instructive is given by the single-particle Lagrangian

$$L = \frac{1}{2} B \dot{\alpha}^2 - K(1 - \cos \alpha). \tag{10.41}$$

If α is viewed as ranging over all real numbers, this describes a particle of mass B moving in a periodic potential with minima analogous to the $A_0 = 0$ gauge n-vacua. As is well known, the states of this system fall into bands, with states within a band labeled by a periodic variable analogous to the gauge theory θ. Alternatively, α can be taken to be periodic with period 2π, and the system interpreted as a pendulum with moment of inertia B swinging in a constant gravitational field. There is now just a single classical ground state. Tunneling between adjacent minima of the periodic potential corresponds to a process in which the pendulum makes a full 2π rotation even though its energy is too low for this to be classically possible. The analogue of the gauge theory $\theta F \tilde{F}$ term is the total time derivative term

$$\Delta L = \frac{\theta}{2\pi} \dot{\alpha}. \tag{10.42}$$

We will see, in Sec. 11.5, that such a term can have measurable effects.

10.4 Some topology

In the course of our discussion of classical vacua in $A_0 = 0$ gauge we found that the values of the volume integral of $\operatorname{tr} F_{\mu\nu} \tilde{F}^{\mu\nu}$ were quantized. One might guess that this has a topological interpretation. This is indeed the case. We have already encountered homotopy classes in the course of studying topological solitons. What we are concerned with now is a different type of topological object, one associated with a gauge field on a manifold.

Although we are actually interested in fields on four-dimensional Euclidean space, R^4, it is perhaps best to begin by discussing fields on a four-sphere, S^4. Let us therefore suppose that we are given a non-Abelian gauge field A_p on S^4. This field is not assumed to satisfy any particular field equations. We could require that it be everywhere nonsingular, but our experience with the Dirac monopole suggests that this might exclude some physically interesting cases. Instead, let us allow singularities, but only those that can be removed by a gauge transformation. (This would allow singularities similar to those along a Dirac string, but not a singularity such as that at the center of a Dirac monopole.) If this is the case, we can cover the four-sphere by multiple gauge patches. In each patch the corresponding gauge potential is nonsingular, while in the overlap between two patches the corresponding potentials are related by a nonsingular gauge transformation. Thus, in the overlap of Regions I and II there is a nonsingular gauge function U such that

$$A_p^{II} = U A_p^{I} U^{-1} - \frac{i}{g} (\partial_p U) U^{-1}. \qquad (10.43)$$

This is quite analogous to the two-patch treatment of the Dirac monopole field that was introduced in Sec. 5.1. With this in mind, consider the case of an Abelian gauge field on a two-sphere. (This could be thought of as a two-sphere enclosing one or more monopoles in ordinary three-dimensional space, although that is not necessary.) The field strength F_{pq} is assumed to be nonsingular everywhere on the sphere. The potentials A_p are nonsingular on gauge patches, and (because the theory is Abelian) satisfy equations of the form

$$A_p^{II} - A_p^{I} = -\partial_p \Lambda \qquad (10.44)$$

on the overlaps between regions.

Now let us define

$$I_1 = \frac{1}{4\pi} \int d^2 x \, \epsilon^{pq} F_{pq}, \qquad (10.45)$$

where the x^p are coordinates on the two-sphere. (There will be the usual spherical coordinate singularities unless one uses several coordinate charts, but these have no effect on our results.)

This quantity is manifestly gauge-invariant. It has two other important properties. First, it is a topological invariant, in the sense that it is unchanged under a smooth deformation

$$A_p^{I} \to A_p^{I} + \delta A_p^{I}$$
$$A_p^{II} \to A_p^{II} + \delta A_p^{II} \qquad (10.46)$$

of the gauge potentials. To see this, note first that Eq. (10.44) implies that δA_p is continuous from one patch to the next, with

$$\delta A_p^{II} = \delta A_p^{I}. \qquad (10.47)$$

Next, substituting Eq. (10.46) into Eq. (10.45) gives

$$\delta I_1 = \frac{1}{4\pi} \int d^2 x \, \epsilon^{pq} \delta F_{pq} = \frac{1}{2\pi} \int d^2 x \, \partial_p (\epsilon^{pq} \delta A_q). \qquad (10.48)$$

The integrand in the last expression is a total derivative of a quantity that varies continuously from patch to patch. Integrating this over the entire manifold gives a surface integral over the boundary of the manifold. Since the two-sphere has no boundary, the integral vanishes and I_1 is invariant, as promised.

The second property relates I_1 to a homotopy class of the gauge group. To obtain this result, we write

$$I_1 = \frac{1}{2\pi} \int d^2x \, \partial_p(\epsilon^{pq} A_q) \equiv \int d^2x \, \partial_p j^p. \tag{10.49}$$

As in Eq. (10.48), we have the integral of a total divergence. In contrast with the previous case, it is the divergence of a current that varies from patch to patch, so more care is need in evaluating the integral. Let us assume, for simplicity, that there are only two coordinate patches, and let Σ be a circle in the overlap region. The integral can then be evaluated by using A_p^{I} in the portion of Region I bounded by Σ, and A_p^{II} in the complementary portion of Region II, also bounded by Σ. In both cases the integral can be expressed as a line integral over Σ (traversed in opposite directions for the two cases), and we obtain

$$I_1 = \frac{1}{2\pi} \int_\Sigma dl_p \, \epsilon^{pq}(A_q^{\mathrm{I}} - A_q^{\mathrm{II}}) = \frac{1}{2\pi} \int_\Sigma dl_p \, \epsilon^{pq} \partial_q \Lambda. \tag{10.50}$$

The last expression is just the winding number that counts how many times the U(1) is traversed as one goes around Σ. It thus relates I_1 to an element of $\pi_1(U(1))$.

The integrand in Eq. (10.45) for I_1 is known as the first Chern form, and the integral itself as the first Chern number. To see the generalization to a higher dimension, let us now return to our non-Abelian theory on a four-sphere and define the gauge-invariant second Chern number[6]

$$I_2 = \frac{g^2}{32\pi^2} \int d^4x \, \epsilon^{pqrs} \operatorname{tr} F_{pq} F_{rs}. \tag{10.51}$$

We first want to show that I_2 is a topological invariant that is unchanged under deformation of the gauge potentials. In contrast with the Abelian case, δA is not continuous across patches, but instead transforms covariantly, with

$$\delta A_p^{\mathrm{II}} = U \delta A_p^{\mathrm{I}} U^{-1}. \tag{10.52}$$

The change in I_2 resulting from this deformation is

$$\delta I_2 = \frac{g^2}{16\pi^2} \int d^4x \, \epsilon^{pqrs} \operatorname{tr} F_{pq} \, \delta F_{rs} = \frac{g^2}{8\pi^2} \int d^4x \, \epsilon^{pqrs} \operatorname{tr} F_{pq} D_r \delta A_s, \tag{10.53}$$

where the covariant derivative in the last expression is taken with respect to the unperturbed potential. However,

$$\epsilon^{pqrs} \operatorname{tr} F_{pq} D_r \delta A_s = \epsilon^{pqrs} \partial_r (\operatorname{tr} F_{pq} \delta A_s) - \epsilon^{pqrs} \operatorname{tr} (D_r F_{pq}) \delta A_s. \tag{10.54}$$

---

[6] Notice that this expression, like that for I_1 above, does not involve the metric. This can be understood by noting that the gauge potential and field strength are one- and two-forms, respectively, and that the integrand here is a four-form, whose integral over a four-manifold is defined independently of the metric.

The second term on the right-hand side vanishes by the Bianchi identity, and we have

$$\delta I_2 = \frac{g^2}{8\pi^2} \int d^4x \, \partial_r (\epsilon^{pqrs} \operatorname{tr} F_{pq} \delta A_s). \tag{10.55}$$

The quantity in parentheses on the right-hand side is a gauge-invariant quantity. Integrating the divergence of this quantity over the four-sphere, which has no boundary, gives zero. Hence, I_2 is invariant under smooth deformations of the gauge potential, as was claimed.

Next, we want to relate I_2 to the topology of the gauge group. From the discussion in Sec. 10.2 we know that

$$\frac{g^2}{32\pi^2} \epsilon^{pqrs} \operatorname{tr} F_{pq} F_{rs} = \frac{g^2}{16\pi^2} \partial_p \left[\epsilon^{pqrs} \operatorname{tr} \left(A_q F_{rs} + \frac{2ig}{3} A_q A_r A_s \right) \right]$$
$$\equiv \partial_p j_A^p. \tag{10.56}$$

Thus I_2 is itself the integral of a total divergence over the four-sphere. As with I_1, the current involved is gauge-variant, so the integration over the manifold reduces to a sum of surface integrals over the boundaries separating the gauge patches. In analogy with our previous calculation, we can divide the four-sphere into two parts separated by a three-sphere hypersurface Σ that lies in the overlap of the two gauge patches. The integral then becomes

$$I_2 = \int_\Sigma dS_p (j_A^{\mathrm{I}p} - j_A^{\mathrm{II}p}). \tag{10.57}$$

Using Eq. (10.43), we find that

$$j_A^{\mathrm{II}p} - j_A^{\mathrm{I}p} = -\frac{1}{24\pi^2} \epsilon^{pqrs} \operatorname{tr} U^{-1} \partial_q U \, U^{-1} \partial_r U \, U^{-1} \partial_s U$$
$$+ \frac{ig}{8\pi^2} \epsilon^{pqrs} \partial_q (\operatorname{tr} U^{-1} \partial_r U \, A_s). \tag{10.58}$$

In the overlap region, and on Σ in particular, all the fields are nonsingular, so when this expression is substituted into Eq. (10.57) the last term leads to an integral of a total divergence over the closed manifold Σ, which vanishes. This leaves us with

$$I_2 = \frac{1}{24\pi^2} \epsilon^{pqrs} \int_\Sigma dS_p \operatorname{tr} U^{-1} \partial_q U \, U^{-1} \partial_r U \, U^{-1} \partial_s U. \tag{10.59}$$

If the gauge group is SU(2), comparison with Eq. (10.9) shows that I_2 is just the winding number that counts the number of times that SU(2) (which is topologically a three-sphere) is covered as one moves over Σ (which is also a three-sphere). It thus corresponds to an element of $\pi_3(G)$. I_2 is also quantized for larger simple groups. In particular, for $G = \mathrm{SU}(N)$, with the matrices in the fundamental representation, the normalization used here makes $I_2 = 1$ for the unit instanton.

One can also work with a single gauge patch, at the cost of introducing a singularity in the gauge field. This can be viewed as taking the limit of the multipatch formalism in which all but one of the patches is shrunk to a point and patch I covers the entire sphere, with the boundaries between integration regions becoming infinitesimal hypersurfaces enclosing the singularities. The volume integrals over the shrunken regions clearly vanish, while that over region I can be converted into a sum of surface integrals over the hypersurfaces enclosing the singularities.

Our real interest is in a gauge field on four-dimensional Euclidean space, R^4, not on a four-sphere. However, if the field strength on R^4 falls rapidly enough at spatial infinity, we can view the field on R^4 as being on S^4. This can be made more precise by projecting the four-sphere (covered by five-dimensional coordinates r_a with $\sum r_a^2 = 1$) onto R^4 (with coordinates x_p). The mapping is given by

$$r_p = \frac{2x_p}{1+x^2}, \quad p = 1, 2, 3, 4,$$

$$r_5 = \frac{1-x^2}{1+x^2}, \tag{10.60}$$

so the "south pole", $(0, 0, 0, 0, -1)$, is mapped to spatial infinity on R^4. The R^4 potential A_p is related to the S^4 potential $\hat{A}_a$ (which satisfies $\sum r_a \hat{A}_a = 0$) by

$$\frac{1+x^2}{2} A_p = \hat{A}_p - x_p \hat{A}_5, \tag{10.61}$$

so that if the S^4 potential is nonsingular at the south pole, then the R^4 potential falls at least as fast as $1/x^2$ at large distance. This mapping between the two spaces applies whether or not the gauge fields obey the field equations. If one of them happens to be a solution, then the conformal invariance of the classical Yang–Mills theory implies that the other is also a solution [237].

We will, in fact, only be concerned with fields whose field strengths fall rapidly enough at large distance that they can be mapped onto S^4. For such fields, I_2 becomes the integer instanton number

$$k = \frac{g^2}{16\pi^2} \int d^4x \, \mathrm{tr} \, F_{pq} \tilde{F}^{pq} = \frac{g^2}{32\pi^2} \int d^4x \, F_{pq}^a \tilde{F}^{apq}. \tag{10.62}$$

10.5 't Hooft symbols

Before embarking on the study of explicit instanton solutions, let us pause to develop some notation that proves to be quite useful. Recall that SO(4) and SU(2)×SU(2) share the same Lie algebra. To explore this further, let us define two sets of matrices e_p and $e_p^\dagger$ by

$$e_p = \begin{cases} i\tau_p, & p = 1, 2, 3, \\ I, & p = 4, \end{cases} \tag{10.63}$$

$$e_p^\dagger = \begin{cases} -i\tau_p, & p = 1, 2, 3, \\ I, & p = 4, \end{cases} \tag{10.64}$$

where τ_1, τ_2, and τ_3 are the usual Pauli matrices.

An SO(4) vector V_p can be represented as a 2×2 matrix

$$V = e_p^\dagger V_p = \begin{pmatrix} V_4 - iV_3 & -V_2 - iV_1 \\ V_2 - iV_1 & V_4 + iV_3 \end{pmatrix} \tag{10.65}$$

with determinant

$$\det V = V_p V_p. \tag{10.66}$$

Note that $(\det V)^{-1/2} V$ is a unitary matrix. Rotations are linear transformations on V_p that preserve its length. These can be implemented via matrix multiplications of the form

$$V \to U_L^{-1} V U_R, \tag{10.67}$$

where U_L and U_R are unitary matrices (to preserve the form of V as a constant times a unitary matrix) with unit determinant (to preserve $\det V$, and hence the length of V_p). They are thus elements of SU(2), and we have

$$\mathrm{SO}(4) = [\mathrm{SU}(2) \times \mathrm{SU}(2)]/Z_2, \tag{10.68}$$

with the division by Z_2 arising because $U_L = U_R = -I$ leaves V unchanged.

The effect of a rotation on the components of V_p can be obtained by writing

$$U_L = e^{-i\boldsymbol{\omega}_L \cdot \boldsymbol{\tau}}, \qquad U_R = e^{-i\boldsymbol{\omega}_R \cdot \boldsymbol{\tau}}. \tag{10.69}$$

For an infinitesimal transformation, the change in V is

$$\delta V = i(\boldsymbol{\omega}_L \cdot \boldsymbol{\tau}) V - iV(\boldsymbol{\omega}_R \cdot \boldsymbol{\tau}). \tag{10.70}$$

Using the identity

$$V_p = \frac{1}{2} \mathrm{tr}\, V e_p, \tag{10.71}$$

we then obtain

$$\delta V_p = \left(\omega_L^j \bar{\eta}_{pq}^j + \omega_R^j \eta_{pq}^j \right) V_q, \tag{10.72}$$

where the 2×2 matrices

$$\eta_{pq} = \eta_{pq}^j \tau_j, \qquad \bar{\eta}_{pq} = \bar{\eta}_{pq}^j \tau_j \tag{10.73}$$

are defined by [238]

$$\begin{aligned} \eta_{pq} &= -i\left(e_p e_q^\dagger - \delta_{pq} I \right), \\ \bar{\eta}_{pq} &= -i\left(e_p^\dagger e_q - \delta_{pq} I \right). \end{aligned} \tag{10.74}$$

They obey $\eta_{pq} = -\eta_{qp}$ and $\bar{\eta}_{pq} = -\bar{\eta}_{qp}$, with the nonzero η_{pq}^j and $\bar{\eta}_{pq}^j$ fixed by

$$\eta_{ij}^k = \bar{\eta}_{ij}^k = \epsilon_{ijk},$$

$$\eta_{i4}^k = -\bar{\eta}_{i4}^k = \delta_{ik}. \tag{10.75}$$

It follows from the above equations that η_{pq} is self-dual,

$$\eta_{pq} = \frac{1}{2}\epsilon_{pqrs}\,\eta_{rs}, \tag{10.76}$$

while $\bar{\eta}_{pq}$ is anti-self-dual, with

$$\bar{\eta}_{pq} = -\frac{1}{2}\epsilon_{pqrs}\,\bar{\eta}_{rs}. \tag{10.77}$$

An antisymmetric rank two SO(4) tensor transforms as the $(1,0)\oplus(0,1)$ representation. The two terms, corresponding to the self-dual and anti-self-dual parts, respectively, are each left unaltered by one of the two SU(2) factors of the group. This correspondence is clarified by comparing Eqs. (10.76) and (10.77) with the roles played by η_{pq} and $\bar{\eta}_{pq}$ in Eq. (10.72).

Two useful identities are

$$\eta_{pq}\eta_{qr} = -2i\eta_{pr} - 3\delta_{pr}I,$$

$$\bar{\eta}_{pq}\bar{\eta}_{qr} = -2i\bar{\eta}_{pr} - 3\delta_{pr}I. \tag{10.78}$$

Further identities can be found in [238].

10.6 The unit instanton

We are now ready to look for instanton solutions. We could do this by directly attacking the second-order Euclidean Yang–Mills equations. However, an easier path is to adopt the method that we used to find the BPS solutions in Sec. 8.2. Because $F_{rs}^2 = \tilde{F}_{rs}^2$, the Euclidean action can be written as

$$
\begin{aligned}
\int d^4x\, \frac{1}{2}\,\mathrm{tr}\, F_{rs}^2 &= \int d^4x\, \frac{1}{4}\,\mathrm{tr}\,(F_{rs}^2 + \tilde{F}_{rs}^2) \\
&= \int d^4x\, \frac{1}{4}\,\mathrm{tr}\,(F_{rs} \mp \tilde{F}_{rs})^2 \pm \int d^4x\, \frac{1}{2}\,\mathrm{tr}\, F_{rs}\tilde{F}_{rs} \\
&= \pm\left(\frac{8\pi^2}{g^2}\right)k + \int d^4x\, \frac{1}{4}\,\mathrm{tr}\,(F_{rs} \mp \tilde{F}_{rs})^2 \\
&\geq \left(\frac{8\pi^2}{g^2}\right)|k|,
\end{aligned}
\tag{10.79}
$$

where k is the instanton number. For fixed k, a minimum of the Euclidean action, and hence a solution to the field equations, is obtained if the field strength is either self-dual, with

$$F_{rs} = \tilde{F}_{rs}, \tag{10.80}$$

or anti-self-dual, with

$$F_{rs} = -\tilde{F}_{rs}, \tag{10.81}$$

according to whether k is positive or negative, respectively. The standard terminology is to call only the self-dual solutions instantons, with the anti-self-dual solutions being termed anti-instantons.

Let us now seek a solution of the self-duality equation (10.80). It is simplest to not impose any particular gauge condition at the outset; once a solution has been found, it can be transformed into whatever gauge is desired. As we have often done before, we start by adopting an ansatz that exploits the symmetries of the theory. Specifically, we require that the solution be invariant under the combination of a global SU(2) gauge transformation and a transformation by one of the SU(2) factors of SO(4). This suggests trying the ansatz

$$A_p = \eta_{pq}\, x_q\, f(x^2). \tag{10.82}$$

The corresponding field strength is

$$F_{pq} = 2\eta_{qp}f + 2(\eta_{qr}x_r x_p - \eta_{pr}x_r x_q)f' - ig\,[\eta_{pr}, \eta_{qs}]\, x_r x_s f^2, \tag{10.83}$$

where $f' \equiv df/d(x^2)$.

Requiring this to be self-dual imposes constraints on $f(x^2)$. Since our ansatz is rotationally invariant, it is sufficient to examine the fields along the positive x_4-axis. Along this axis we have

$$F_{ij} = -2\eta_{ij}f - ig[\eta_{i4}, \eta_{j4}]x^2 f^2,$$

$$F_{k4} = -2\eta_{k4}f - 2\eta_{k4}x^2 f'. \tag{10.84}$$

The self-duality condition $F_{ij} = \epsilon_{ijk}F_{k4}$ then reduces to

$$\frac{df}{d(x^2)} = -gf^2. \tag{10.85}$$

This is solved by

$$f = \frac{1}{g}\frac{1}{x^2 + \lambda^2}, \tag{10.86}$$

where λ^2 is an integration constant that must be positive for the solution to be nonsingular.

We thus have [227]

$$A_p = \frac{1}{g}\frac{\eta_{pq}x_q}{x^2 + \lambda^2} \tag{10.87}$$

or, in terms of SU(2) components,

$$A_p^j = \frac{2}{g}\frac{\eta_{pq}^j x_q}{x^2 + \lambda^2}. \tag{10.88}$$

It can be seen that λ gives a characteristic size for the instanton. The fact that all values of λ are allowed is a consequence of the scale invariance of the classical Yang–Mills theory. The anti-self-dual anti-instanton solution has exactly the same form, except that η_{pq} is replaced by $\bar\eta_{pq}$.

The field strength that follows from the instanton potential is

$$F_{pq} = -\frac{2}{g}\frac{\eta_{pq}\lambda^2}{(x^2+\lambda^2)^2}.$$
(10.89)

Since A_p falls as $1/|x|$ at large distance, one might have expected F_{pq} to fall as $1/x^2$. Its actual $1/x^4$ falloff reflects the fact that, as we will see shortly, the leading large-distance component of A_p is a pure gauge potential.

Because of the spherical symmetry of the solution, it is trivial to evaluate the instanton number by integrating $\operatorname{tr} F_{pq}\tilde F_{pq} = \operatorname{tr} F_{pq}F_{pq}$ over all space. One finds that $k = 1$. Alternatively, one could use Eq. (10.18) and calculate the instanton number as an integral of the normal component of j_A^p over the surface at spatial infinity. Because $F_{pq} \sim 1/x^4$, the contribution comes entirely from the final, cubic, term in Eq. (10.17). Here again, the calculation can be simplified by using spherical symmetry and evaluating the integrand along the positive z-axis.

Now consider the unitary matrix

$$S = \frac{e_p^\dagger x_p}{|x|}.$$
(10.90)

It is easily verified that

$$\frac{i}{g}S^{-1}\partial_p S = \frac{1}{g}\frac{\eta_{pq}x_q}{x^2}.$$
(10.91)

Hence the instanton solution of Eq. (10.87) can be written as

$$A_p = \frac{i}{g}S^{-1}\partial_p S\left(\frac{x^2}{x^2+\lambda^2}\right),$$
(10.92)

which shows that it approaches a pure gauge form at large distance. Let us now gauge-transform this potential to

$$\begin{aligned}
A'_p &= SA_pS^{-1} - \frac{i}{g}\partial_p S\, S^{-1}\\
&= \frac{i}{g}\partial_p S\, S^{-1}\left(\frac{x^2}{x^2+\lambda^2}-1\right)\\
&= \frac{1}{g}\frac{\bar\eta_{pq}x_q\lambda^2}{x^2(x^2+\lambda^2)}.
\end{aligned}$$
(10.93)

[Note that the self-dual η_{rs} of Eq. (10.87) has been replaced by the anti-self-dual $\bar\eta_{rs}$.] In this form the potential falls rapidly enough at spatial infinity that the surface integral of j_A^p at infinity vanishes. The instanton number is still recovered, because the singularity of the potential at $x = 0$ requires that when we evaluate the instanton number as a surface integral we include an infinitesimal surface enclosing the origin. This surface gives a unit contribution to k, as required.

The relation between these two forms of the instanton solution is particularly simple when they are mapped onto the four-sphere. Both correspond to using a single gauge patch, implying that the potential has a singularity somewhere on the sphere. For the solution of Eq. (10.87) that singularity is at the south pole, while for that of Eq. (10.93) it is at the north pole.

10.7 Multi-instanton solutions

The latter form of the one-instanton solution suggests a rather powerful ansatz, first found by 't Hooft [239], that yields multi-instanton solutions. Consider a potential of the form

$$A_p = \bar\eta_{pr} f_r(x), \qquad (10.94)$$

where the $f_r(x)$ are arbitrary functions of x_p, with no particular symmetry assumed. The field strength is

$$F_{pq} = \bar\eta_{qr}\partial_p f_r - \bar\eta_{pr}\partial_q f_r - ig[\bar\eta_{pr}, \bar\eta_{qs}] f_r f_s. \qquad (10.95)$$

Requiring this to be self-dual is equivalent to requiring that its product with the anti-self-dual matrix $\bar\eta_{pq}$ vanish. Using the second identity in Eq. (10.78), we obtain the condition

$$0 = \bar\eta_{pq} F_{pq}$$
$$= -4i\bar\eta_{pq}\partial_p f_q - 6\partial_p f_p + 12g f_p f_p. \qquad (10.96)$$

The coefficient of $\bar\eta_{pq}$ in the last line vanishes if f_p is taken to be the gradient of a scalar function. The remaining terms then cancel if we write

$$f_p = -\frac{1}{2g}\partial_p \ln\rho, \qquad (10.97)$$

where ρ obeys

$$0 = \frac{\Box\rho}{\rho}. \qquad (10.98)$$

This is solved by

$$\rho = 1 + \sum_{a=1}^{k} \frac{\lambda_a^2}{(x - w_a)^2}. \qquad (10.99)$$

For $k = 1$, with $w_1 = 0$, this reproduces the one-instanton solution in the form given by Eq. (10.93). For $k > 1$, the instanton number receives a unit contribution from a surface integral about each of the poles, and none from the surface at infinity. This is therefore a k-instanton solution, with a four-vector position w_a and a scale λ_a for each of the component instantons.[7]

Jackiw, Nohl, and Rebbi [240] pointed out a curious generalization of this solution. Equation (10.98) is still satisfied if we leave out the first term in Eq. (10.99) and write

$$\rho = \sum_{a=1}^{k+1} \frac{\lambda_a^2}{(x - w_a)^2}.$$
(10.101)

Again, surface integrals enclosing each of the poles make unit contributions to the instanton number. However, with the initial 1 omitted, A_p falls as $1/|x|$, and so there is also a contribution from the surface at infinity. With the η_{pq} replaced by an $\bar{\eta}_{pq}$, the latter surface gives a contribution of -1, so the total instanton number is one less than the number of poles.

This appears to give a k-instanton solution that depends on $5k+4$ parameters. (An overall scale can clearly be factored out of ρ without affecting A_p.) This is more than is needed to specify the positions and scales. For $k = 1$ and 2 some of these extra parameters can be changed by gauge transformations, while maintaining the form of the ansatz. For $k = 1$, four of the apparent parameters can be absorbed in this manner, so there is no truly new solution here. [However, this does give the curious result that the positions of the poles in Eq. (10.101) are not special locations, and are not simply related to the center of the instanton.] For $k = 2$, there is one degree of gauge freedom, and thus 13, rather than the apparent 14 parameters. For $k \geq 3$, the poles can only be moved by a gauge transformation if they all lie on a common circle, so the general solution does indeed depend on $5k + 4$ parameters.

In fact, this is still not the most general multi-instanton solution. This can be seen by using index theory methods to determine the number of parameters needed to specify an arbitrary solution with instanton number k.

10.8 Counting parameters with an index theorem

Counting the number of parameters needed to specify an instanton solution is equivalent to counting the number of linearly independent perturbations of the solution that preserve its self-duality, with the proviso that perturbations that are simply gauge transformations should not be counted. As we will see shortly, the perturbations that we are considering satisfy an equation of the form

---

[7] The instanton scale is modified by the presence of the other instantons. For the case of widely separated instantons, the quantity comparable to the λ that appears in the one-instanton solution is actually

$$\lambda_a' = \lambda_a \left[1 + \sum_{b \neq a} \frac{\lambda_b}{(w_b - w_a)^2} \right].$$
(10.100)

$$\mathcal{D}\chi = 0, \tag{10.102}$$

where $\mathcal{D}$ is a matrix differential operator. The kernel of an operator is defined as the subspace that it annihilates, so the number we want is the dimension of the kernel of $\mathcal{D}$. Several approaches were used to determine this number for the SU(2) instantons [241–244]; the discussion in this section will be closest to that of [244].

What we will actually calculate is the index of $\mathcal{D}$,

$$\mathcal{I}(\mathcal{D}) = \dim \ker(\mathcal{D}) - \dim \ker(\mathcal{D}^\dagger), \tag{10.103}$$

which is the difference in the dimensions of the kernels of $\mathcal{D}$ and of its adjoint.[8] As we will see, there is a "vanishing theorem" that shows that $\mathcal{D}^\dagger$ has no zero eigenvalues, and hence no kernel, so that $\mathcal{I}(\mathcal{D})$ is just the number we want.

The motivation for this somewhat roundabout approach requires some explanation. First, note that the zero modes of $\mathcal{D}$ (i.e., its normalizable eigenfunctions with eigenvalue zero) are the same as those of $\mathcal{D}^\dagger\mathcal{D}$. It is obvious that if $\mathcal{D}\chi = 0$ then $\mathcal{D}^\dagger\mathcal{D}\chi = 0$. To show the converse, let us assume that $\mathcal{D}^\dagger\mathcal{D}\chi = 0$ and consider the inner product with $\chi^\dagger$. We have

$$0 = \int d^4x \, \chi^\dagger (\mathcal{D}^\dagger\mathcal{D}\chi)$$
$$= \int d^4x \, (\mathcal{D}\chi)^\dagger (\mathcal{D}\chi), \tag{10.104}$$

from which it follows that $\mathcal{D}\chi = 0$. Second, the nonzero eigenvalues of $\mathcal{D}^\dagger\mathcal{D}$ and $\mathcal{D}\mathcal{D}^\dagger$ are the same, since if

$$\mathcal{D}^\dagger\mathcal{D}\chi = \lambda\chi \tag{10.105}$$

then

$$\mathcal{D}\mathcal{D}^\dagger(\mathcal{D}\chi) = \lambda(\mathcal{D}\chi). \tag{10.106}$$

The first result tells us that counting the zero eigenvalues of $\mathcal{D}^\dagger\mathcal{D}$ is equivalent to counting those of $\mathcal{D}$, and similarly for $\mathcal{D}\mathcal{D}^\dagger$ and $\mathcal{D}^\dagger$. The second gives us important information about the behavior of these eigenvalues under smooth variations of $\mathcal{D}$. Let us suppose that $\mathcal{D}$ depends upon some external field—in our case, $A_p(x)$. If $A_p(x)$ is smoothly varied, then the eigenvalues of $\mathcal{D}^\dagger\mathcal{D}$ will presumably vary continuously. In particular, some zero eigenvalues might become nonzero, or some nonzero eigenvalues might become zero. However, the same must happen to the eigenvalues of $\mathcal{D}\mathcal{D}^\dagger$, with the zero eigenvalues of the two operators appearing or disappearing simultaneously. As a result, the index of $\mathcal{D}$ is not affected by such smooth variations. One might therefore expect—as indeed turns out to be the case—that the index can be expressed in terms of topological invariants of the fields entering $\mathcal{D}$.

---

[8] For further discussion of index calculations, see Appendix B.

These results lead us to consider the quantity

$$\mathcal{I}(\mathcal{D}, M^2) = \text{Tr} \left(\frac{M^2}{\mathcal{D}^\dagger \mathcal{D} + M^2} - \frac{M^2}{\mathcal{D}\mathcal{D}^\dagger + M^2} \right), \tag{10.107}$$

with the trace being understood in a functional sense and M an arbitrary real number. Let us take the limit $M^2 \to 0$, performing the evaluation in a basis where the operators are diagonal. The terms corresponding to nonzero eigenvalues vanish in this limit, while those from the zero eigenvalues each contribute 1 or -1, depending on the term in which they appear. Hence, we conclude that

$$\lim_{M^2 \to 0} \mathcal{I}(\mathcal{D}, M^2) = \mathcal{I}(\mathcal{D}). \tag{10.108}$$

In fact, for any value of M^2 the terms from the nonzero eigenvalues cancel pairwise between the two expressions in Eq. (10.107), while the zero eigenvalues still contribute 1 or -1. Thus, despite appearances, $\mathcal{I}(\mathcal{D}, M^2)$ is actually independent of M^2, which means that we can evaluate it using any value for M^2 that we find convenient; it will turn out to be simplest to evaluate it in the limit $M^2 \to \infty$.

A word of caution is needed here. The discussion above was phrased as if the spectra of the various operators were discrete. This would indeed be the case if we were working with differential operators on a closed manifold, such as S^4. Instead, we are working on infinite four-dimensional Euclidean space, R^4, and these spectra include continuous portions. As a result, the formal arguments above may lead us into error.

One possibility is that the densities of continuum eigenvalues near zero could be sufficiently singular as to give a nonzero contribution to the left-hand side of Eq. (10.108). This portion of the spectrum is sensitive to the large-distance behavior of the instanton field. Because the part of A_p that is not pure gauge falls as $1/x^3$, there is in fact no problem, and there is no anomalous continuum contribution.[9] This is essentially equivalent to the statement that the falloff of the instanton solutions is fast enough that they can be mapped onto solutions on S^4, as discussed at the end of Sec. 10.4.

A second potential problem is that the pairwise cancellation of terms in Eq. (10.107) is ill-defined for continuous spectra, bringing into question the M^2-independence of $\mathcal{I}(\mathcal{D}, M^2)$. However, we will see presently that this M^2-independence can be verified by an explicit calculation.

Let us now proceed to determine $\mathcal{D}$. We assume that we are given a gauge potential $A_p^a(x)$ that gives rise to a self-dual field strength, and consider a perturbation

$$A_p^a \to A_p^a + \delta A_p^a. \tag{10.109}$$

[9] A continuum contribution does arise in other contexts, as discussed in Appendix B.

The change in the field strength is then

$$\delta F_{pq}^a = \partial_p \delta A_q^a - \partial_q \delta A_p^a + g f_{abc} A_p^b \delta A_q^c - g f_{abc} A_q^b \delta A_p^c$$
$$= (D_p \delta A_q)^a - (D_q \delta A_p)^a, \qquad (10.110)$$

where the covariant derivative of δA_p is defined by

$$(D_p \delta A_q)^a = \partial_p \delta A_q^a - ig A_p^b (T^b)_{ac} \delta A_q^c, \qquad (10.111)$$

with the $(T^b)_{ac} = i f_{abc}$ being the generators of the group in the adjoint representation. If δF_{pq} is self-dual, then contraction with the anti-self-dual matrix $\bar{\eta}_{pq}$ gives

$$0 = \bar{\eta}_{pq} D_p \delta A_q. \qquad (10.112)$$

Many of the solutions of this equation are simply gauge transformations of the original solution and can be written in the form

$$\delta_G A_p = D_p \Lambda. \qquad (10.113)$$

To exclude these physically uninteresting modes, we require that δA_p be orthogonal to all such modes, so that

$$0 = \int d^4x \, \delta_G A_p \, \delta A_p$$
$$= \int d^4x \, D_p \Lambda \, \delta A_p$$
$$= -\int d^4x \, \Lambda \, D_p \delta A_p. \qquad (10.114)$$

Requiring that this hold for all $\Lambda(x)$ gives the background gauge condition

$$0 = D_p \delta A_p. \qquad (10.115)$$

Note, however, that the integration by parts in the last line of Eq. (10.114) assumes that Λ falls sufficiently rapidly at large distance. Consequently, although we have removed modes arising from local gauge transformations, those corresponding to global gauge transformations still remain and have to be handled separately.

By recalling Eq. (10.74), we can combine Eqs. (10.112) and (10.115) into the single equation

$$0 = e_p^\dagger e_q D_p \delta A_q. \qquad (10.116)$$

If we define

$$\Psi = -i e_q \delta A_q, \qquad (10.117)$$

Eq. (10.116) can be recast as the Dirac-type equation

$$0 = -e_p^\dagger D_p \Psi. \qquad (10.118)$$

Thus, our immediate task is to determine the index of the operator

$$\mathcal{D} \equiv -e_p^\dagger D_p = -e_p^\dagger(\partial_p - igA_p^bT^b).$$
(10.119)

For the application to counting instanton parameters, the generator T^b in this equation must be in the adjoint representation. However, it will also be of interest to know the index of $\mathcal{D}$ for arbitrary representations, so we will leave the representation unspecified until the end of the calculation. For the moment, we just note that the generators can be chosen to satisfy

$$\operatorname{tr} T^aT^b = \delta_{ab}T(R),$$
(10.120)

where $T(R)$ depends on the representation. For SU(2), $T(R)$ is most easily calculated by evaluating $\operatorname{tr} T_3^2$, which in the representation with "isospin" t gives

$$T(R) = \sum_{t_z=-t}^{t} t_z^2 = \frac{1}{3}t(t+1)(2t+1).$$
(10.121)

In particular, $T = 2$ for the adjoint representation and $T = 1/2$ for the fundamental doublet representation.

There is an important point to note concerning the relation between Ψ and δA_p. If $\Psi(x)$ is a solution of Eq. (10.118), then so is $i\Psi(x)$. As Dirac solutions these are viewed as being linearly dependent, and so count as a single solution when determining the index. However, when δA_p is extracted from Ψ via

$$\delta A_p = \frac{i}{2}\operatorname{tr} e_p^\dagger\Psi,$$
(10.122)

Ψ and $i\Psi$ produce linearly independent sets of δA_p. Hence, the index $\mathcal{I}_b$ for the operator in the bosonic Eq. (10.116) and the index $\mathcal{I}_f$ of the fermionic operator in Eq. (10.118) are related by

$$\mathcal{I}_b = 2\mathcal{I}_f.$$
(10.123)

Hence, the bosonic index must be an even number. In fact, a stronger result holds. If Ψ is a solution, then not only is $i\Psi$ a solution but so is ΨU, where U is any 2×2 unitary matrix. The result is that the number of bosonic zero modes must be a multiple of four.

Let us now proceed to our calculation. A set of Euclidean gamma matrices obeying

$$\{\gamma_p, \gamma_q\} = 2\delta_{pq}$$
(10.124)

is given by

$$\gamma_p = \begin{pmatrix} 0 & -ie_p \\ ie_p^\dagger & 0 \end{pmatrix},$$
(10.125)

with

$$\gamma_5 = \gamma_1\gamma_2\gamma_3\gamma_4 = \begin{pmatrix} -I & 0 \\ 0 & I \end{pmatrix}.$$
(10.126)

Using these, we can write

$$i\gamma \cdot D \equiv i\gamma_p D_p = \begin{pmatrix} 0 & \mathcal{D}^\dagger \\ \mathcal{D} & 0 \end{pmatrix} \tag{10.127}$$

and

$$(i\gamma \cdot D)^2 = \begin{pmatrix} \mathcal{D}^\dagger \mathcal{D} & 0 \\ 0 & \mathcal{D}\mathcal{D}^\dagger \end{pmatrix}, \tag{10.128}$$

while Eq. (10.107) for the index gives

$$\mathcal{I}_f = \mathrm{Tr} \left[-\gamma_5 \frac{M^2}{-(\gamma \cdot D)^2 + M^2} \right]. \tag{10.129}$$

It must be remembered that the trace in this expression encompasses traces over Dirac indices and group indices, as well as a functional trace.

This is a good point to take a moment to verify that $\mathcal{I}$ really is independent of M^2. Differentiating the right-hand side of Eq. (10.129) gives

$$\frac{\partial \mathcal{I}_f}{\partial M^2} = \mathrm{Tr} \left\{ \gamma_5 \frac{(\gamma \cdot D)^2}{[-(\gamma \cdot D)^2 + M^2]^2} \right\}. \tag{10.130}$$

Next, we bring a factor of $\gamma \cdot D$ from the numerator to the left of the γ_5, giving a minus sign from the anticommutator, and then use the cyclic property of the trace to bring it back to its original position. This shows that the right-hand side of the equation is equal to minus itself, and hence that

$$\frac{\partial \mathcal{I}_f}{\partial M^2} = 0. \tag{10.131}$$

Let us now return to the evaluation of $\mathcal{I}_f$. By separating symmetric and antisymmetric terms, we can write

$$\begin{aligned}
(\gamma \cdot D)^2 &= \frac{1}{4}\{\gamma_p, \gamma_q\}\{D_p, D_q\} + \frac{1}{4}[\gamma_p, \gamma_q][D_p, D_q] \\
&= D^2 - \frac{ig}{2}\gamma_p \gamma_q F_{pq} \\
&\equiv D^2 + \Delta.
\end{aligned} \tag{10.132}$$

We can then expand the right-hand side of Eq. (10.129) in a power series, obtaining

$$\mathcal{I}_f = -\mathrm{Tr} \left\{ \gamma_5 M^2 \left[\frac{1}{(-D^2 + M^2)} + \frac{1}{(-D^2 + M^2)}\Delta\frac{1}{(-D^2 + M^2)} \right. \right.$$

$$\left. \left. + \frac{1}{(-D^2 + M^2)}\Delta\frac{1}{(-D^2 + M^2)}\Delta\frac{1}{(-D^2 + M^2)} + \cdots \right] \right\}. \tag{10.133}$$

Let us consider the terms in this expansion one by one. The first term is proportional to the trace of γ_5, and therefore vanishes. The second term contains the trace of $\gamma_5 \gamma_p \gamma_q$, and so also vanishes. The terms represented by the ellipsis vanish when we take the limit $M^2 \to \infty$. (This can be seen by dimensional analysis, but in any case will be clear after the third term has been evaluated.) Hence, in the $M^2 \to \infty$ limit we only need to consider the third term, and can write

$$\mathcal{I}_f = \lim_{M^2 \to \infty} \left(\frac{g}{2}\right)^2 (\operatorname{tr} \gamma_5 \gamma_p \gamma_q \gamma_r \gamma_s)(\operatorname{tr} T^a T^b)$$

$$\times \int d^4 x \left\langle x \left| \frac{M^2}{(-D^2 + M^2)} F_{pq}^a \frac{1}{(-D^2 + M^2)} F_{rs}^b \frac{1}{(-D^2 + M^2)} \right| x \right\rangle. \quad (10.134)$$

The traces over Dirac and group indices give

$$(\operatorname{tr} \gamma_5 \gamma_p \gamma_q \gamma_r \gamma_s)(\operatorname{tr} T^a T^b) = 4\epsilon_{pqrs} \, \delta_{ab} \, T(R). \quad (10.135)$$

This leaves us with the functional trace over the Green's function in the second line. We can move the F_{pq} past the factor of $(-D^2 + M^2)^{-1}$, because the commutator gives a term, containing the derivative of the field strength, that is suppressed by an additional factor of M^2. For similar reasons, in the large M^2 limit we can replace D^2 by ∂^2. Making these changes, and using Eq. (10.135), we have

$$\mathcal{I}_f = \lim_{M^2 \to \infty} 2g^2 T(R) \int d^4 x \left\langle x \left| \frac{M^2}{(-\partial^2 + M^2)^3} \right| x \right\rangle F_{pq}^a \tilde{F}_{pq}^a. \quad (10.136)$$

The first factor in the integral is actually independent of x, and given by

$$\left\langle x \left| \frac{M^2}{(-\partial^2 + M^2)^3} \right| x \right\rangle = \int \frac{d^4 p}{(2\pi)^4} \frac{M^2}{(p^2 + M^2)^3}$$

$$= \frac{1}{32\pi^2}. \quad (10.137)$$

Using Eq. (10.62) to relate the integral of $F_{pq}^a \tilde{F}_{pq}^a$ to the instanton number, we obtain

$$\mathcal{I}_f = 2T(R)k. \quad (10.138)$$

For counting instanton parameters, the generators are in the $t = 1$ adjoint representation, for which $T(R) = 2$. We get an additional factor of two in going from the fermionic to the bosonic operator, giving

$$\mathcal{I}_b = 8k. \quad (10.139)$$

We are not quite done, however. First, the index only gives us the difference between the number of zero eigenvalues of $\mathcal{D}$ and the corresponding number for $\mathcal{D}^\dagger$, whereas what we actually want is the former quantity. A calculation essentially identical to that which yielded Eq. (10.132) shows that

$$\mathcal{D}\mathcal{D}^\dagger = -D^2 - \frac{g}{2}\bar{\eta}_{pq} F_{pq} = -D^2, \quad (10.140)$$

where the last equality follows because $\bar{\eta}_{pq}$ is anti-self-dual while F_{pq} is self-dual. Because $-D^2$ is a positive definite operator, $\mathcal{D}^\dagger$ has no zero eigenvalues, and so $\mathcal{D}$ has $8k$.

Second, we need to worry about the global gauge modes that were not excluded by the background gauge condition. With a gauge group SU(2), of dimension three, there are three such modes, and so the number of physical parameters entering a solution with instanton number k is

$$n = 8k - 3. \tag{10.141}$$

This result suggests that any such solution should be interpreted as being composed of k unit instantons, with no essentially new objects appearing at higher values of k. For each of the component instantons there are one size and four position parameters, just as in the 't Hooft solution. In addition, each has three parameters that specify its SU(2) orientation. Finally, three parameters must be subtracted, because only the relative SU(2) orientations are physically meaningful. Although this simple picture implicitly assumes that the separation between instantons is large compared to their sizes, the parameter counting is correct for all self-dual solutions.

As a final point, recall that there is a correspondence between the instanton solutions on Euclidean R^4 and those on the four-sphere, S^4. Counting of parameters by index methods should therefore yield the same answer for both cases. On S^4, where one does not need to worry about a surface term in the analogue of Eq. (10.114), the background gauge condition eliminates all gauge modes, and so there are no global gauge modes to be subtracted at the end. However, when the methods of this section are generalized to the sphere, the metric enters the differential operators. This leads to an extra contribution of -3, so the result of Eq. (10.141) is recovered [241]. This result on the four-sphere is a manifestation of the Atiyah–Singer index theorem [245], which states that the index on a compact manifold can be expressed solely in terms of topological invariants.

10.9 Larger gauge groups

The results of the preceding section can be generalized to a larger gauge group G, which I will assume to be a compact connected simple group. Because $\Pi_3(G) = Z$ for any such group, the topological arguments for the existence of instanton solutions again apply. In fact, it is quite easy to explicitly exhibit some such solutions.

For any choice of G there are two or more inequivalent SU(2) subgroups. Using one of these subgroups to embed an SU(2) solution into G clearly gives an instanton solution in the larger group. To be explicit, let T^a be the generators of one such subgroup, normalized to give the standard commutation relations, and satisfying

$$\mathrm{tr}\, T^a T^b = \frac{c}{2} \delta^{ab}. \tag{10.142}$$

A solution is then given by

$$A_p = A_p^a T^a ,$$ (10.143)

where the A_p^a are the components of the SU(2) solution. If the embedding solution had instanton number $k_{SU(2)}$, the new solution has instanton number

$$k_G = \frac{g^2}{16\pi^2} \int d^4x \, \mathrm{tr} \, F_{pq} \tilde{F}_{pq}$$

$$= (\mathrm{tr} \, T^a T^b) \frac{g^2}{16\pi^2} \int d^4x F_{pq}^a \tilde{F}_{pq}^b$$

$$= c \, k_{SU(2)}.$$ (10.144)

For example, if $G = SU(3)$, viewed as a group of 3×3 matrices, there is an obvious SU(2) subgroup corresponding to the upper left 2×2 block. This has generators $\lambda_1/2$, $\lambda_2/2$, and $\lambda_3/2$, with $c = 1$. Embedding via this subgroup leaves the instanton number unchanged. There is also an SU(2) [or, more properly, an SO(3) subgroup] with generators λ_2, λ_5, and λ_7. For these, $c = 4$, so embedding via this subgroup quadruples the topological charge. In particular, embedding the unit SU(2) instanton gives a spherically symmetric instanton with $k = 4$. At first sight, one might think that there were now two different species of fundamental instantons, one with $k = 1$ and one with $k = 4$. This turns out not to be the case. The index theory results given below show that this $k = 4$ solution has precisely the number of zero modes that would be expected for a four-instanton solution, and so should be viewed as just a particularly symmetric configuration of overlapping unit instantons. For SU(3), and for all larger G, there is only one fundamental instanton, with $k = 1$.

For any G, the raising and lowering operators corresponding to any one of the roots can be used to define an SU(2) subgroup. [Explicit formulas for the generators are given in Eq. (A.7).] If the roots of G are all of equal length, embedding via any such subgroup yields a unit instanton; when there are two root lengths, a long root must be used. For the classical groups these subgroups are easily visualized. If $G = SU(N)$, it is a subgroup defined by any 2×2 block of the $N \times N$ matrices. For SO(N), with $N \geq 5$, the relevant subgroup is obtained by taking an $SO(4) = SU(2) \times SU(2)$ subgroup of SO(N) in the obvious way and then taking one of the SU(2) factors. For Sp(2N), viewed as a group of $N \times N$ quaternionic matrices, it is the $SU(2) = Sp(2)$ corresponding to any diagonal element.

Extending the index calculation of Sec. 10.8 to an arbitrary gauge group is relatively straightforward [254]. The steps leading up to Eq. (10.138) do not depend on the gauge group. Taking that result, and choosing R to be the adjoint representation of G, we have[10]

$$\mathcal{I}_b = 4kT(\mathrm{adjoint}(G)).$$ (10.145)

[10] The normalization here is such that the long roots of the group have unit length. In this case $T(\mathrm{adjoint}(G))$ is equal to the dual Coxeter number of the group.

We might therefore expect that after subtracting the global gauge zero modes the number of parameters would be

$$n = 4kT(\text{adjoint}(G)) - \dim(G).\tag{10.146}$$

While this is correct for large values of k, it clearly can't be right for small k. For SU(N), we have $T(\text{adjoint}(G)) = N$ and $\dim(G) = N^2 - 1$, so we would have $n = 4Nk - N^2 + 1$, which for low values of k and $N \geq 5$ predicts a negative number of parameters. The explanation is simple. Consider the one-instanton solution in SU(N). This can always be brought into the upper left 2×2 block by a gauge transformation. Clearly, the U($N - 2$) corresponding to the lower right $(N - 2) \times (N - 2)$ block leaves the instanton invariant, so the total number of global gauge modes is not equal to the dimension of SU(N), but rather

$$\dim[\text{SU}(N)] - \dim[\text{U}(N - 2)] = 4N - 5.\tag{10.147}$$

Subtracting this from $\mathcal{I}_b = 4Nk = 4N$ gives five parameters, which are just the position and scale parameters inherited from the SU(2) solution. A generic two-instanton solution cannot be put into a 2×2 block, but it can be rotated into a 4×4 one and so will be left invariant by a U($N - 4$) subgroup. It is only when $k \geq N/2$ that all of the elements of G act nontrivially on a generic solution. Extending the above analysis to arbitrary $k < N/2$, one finds that for SU(N) the number of parameters is

$$n = \begin{cases} 4Nk - (N^2 - 1), & k \geq N/2, \\ 4k^2 + 1, & k < N/2. \end{cases}\tag{10.148}$$

The $4N$ that appears as the coefficient of k is just the sum of four position, one scale, and $4N - 5$ group orientation parameters for each component unit instanton.

The situation is similar for the other simple groups. For Sp($2N$), with $N \geq 2$,

$$n = \begin{cases} 4(N + 1)k - N(2N + 1), & k \geq N, \\ 2k^2 + 3k, & k < N, \end{cases}\tag{10.149}$$

while for SO(N), with $N \geq 5$,

$$n = \begin{cases} 4(N - 2)k - \frac{N(N-1)}{2}, & k \geq (N - 1)/4, \\ 8k^2 - 6k, & 2 \leq k < (N - 1)/4, \\ 5, & k = 1. \end{cases}\tag{10.150}$$

[Recall that SO(3) = Sp(2) = SU(2), and that SO(4) is not simple.]

For the exceptional groups there are five parameters when $k = 1$, just as with the classical groups. For larger values of k we have

$$G_2: \qquad n = 16k - 14, \qquad k \geq 2,$$

$$F_4: \qquad n = 36k - 52, \qquad k \geq 2,$$

$$E_6: \qquad n = \begin{cases} 20, & k = 2, \\ 48k - 78, & k \geq 3, \end{cases}$$

$$E_7: \qquad n = \begin{cases} 20, & k = 2, \\ 72k - 133, & k \geq 3, \end{cases}$$

$$E_8: \qquad n = \begin{cases} 20, & k = 2, \\ 120k - 248, & k \geq 3. \end{cases} \qquad (10.151)$$

10.10 The Atiyah–Drinfeld–Hitchin–Manin construction

A very powerful construction that transforms the problem of solving the self-duality differential equation (10.80) to one of solving a set of algebraic matrix equations is due to Atiyah, Drinfeld, Hitchin, and Manin [246, 247]. In this section I will present this ADHM construction for the $SU(2)$ theory, leaving the extension to larger groups to the next section.

The construction is most easily expressed in terms of quaternions. Recall that quaternions are generalizations of complex numbers that involve three imaginary units i, j, and k that anticommute and obey

$$i^2 = j^2 = k^2 = -1, \qquad ijk = -1. \qquad (10.152)$$

An arbitrary quaternion can be written as

$$q = a + bi + cj + dk \qquad (10.153)$$

where a, b, c, and d are real numbers. If only a is nonzero, q is said to be real; if $a = 0$, then q is pure imaginary. The conjugate of a quaternion generalizes complex conjugation, so that the conjugate of the above quaternion is defined to be

$$q^* = a - bi - cj - dk. \qquad (10.154)$$

Finally, a unit quaternion is one obeying $q^*q = 1$.

For our purposes here it will be convenient to define

$$\hat{e}_1 = -i, \quad \hat{e}_2 = -j, \quad \hat{e}_3 = -k, \quad \hat{e}_4 = 1 \qquad (10.155)$$

and to expand an arbitrary quaternion as

$$q = q_r \hat{e}_r. \qquad (10.156)$$

These conventions have been chosen so that the combination

$$\chi_{rs} = \hat{e}_r \hat{e}_s^* - \hat{e}_s \hat{e}_r^* \qquad (10.157)$$

is self-dual. [An easy way to see this is to note that the $\hat{e}_r$ can be represented by the matrices e_r, and then use Eq. (10.74).]

With these preliminaries completed, let us proceed with the construction. We begin with a $(k+1) \times k$ quaternionic matrix of the form

$$M = B - Cx, \qquad (10.158)$$

where B and C are constant matrices and x is the quaternionic representation of a point in Euclidean space via

$$x = x_r \hat{e}_r. \qquad (10.159)$$

We require that the $k \times k$ matrix

$$R = M^\dagger M \qquad (10.160)$$

(where the dagger denotes the combination of the matrix transpose and quaternionic conjugation) be real and invertible.

The assertion is that if N is a $(k+1)$-component column vector obeying

$$N^\dagger M = 0 \qquad (10.161)$$

and

$$N^\dagger N = I, \qquad (10.162)$$

then a self-dual SU(2) gauge field with instanton number k is given by the quaternion

$$\mathcal{A}_p = \frac{1}{g} N^\dagger \partial_p N = -\frac{1}{g}(\partial_p N^\dagger)N, \qquad (10.163)$$

with the second equality following from the normalization condition on N.

To see that $\mathcal{A}_p$ defines an SU(2) field, note first that the two expressions for $\mathcal{A}_p$ are minus the conjugates of each other, showing that $\mathcal{A}_p$ is a purely imaginary quaternion and so can be expanded as

$$\mathcal{A}_p = -\sum_{a=1}^{3} \frac{\hat{e}_a}{2} A_p^a. \qquad (10.164)$$

The A_p^a are then the components of the potential in the standard SU(2) notation.

To show that the field is self-dual, we write the field strength

$$\begin{aligned}
\mathcal{F}_{pq} &= -\sum_{a=1}^{3} \frac{\hat{e}_a}{2} F_{pq}^a \\
&= \partial_p \mathcal{A}_q + g \mathcal{A}_p \mathcal{A}_q - (p \leftrightarrow q) \\
&= \frac{1}{g} \left[\partial_p N^\dagger \partial_q N - \partial_p N^\dagger N N^\dagger \partial_q N - (p \leftrightarrow q) \right] \\
&= \frac{1}{g} \left[\partial_p N^\dagger \left(I - N N^\dagger \right) \partial_q N - (p \leftrightarrow q) \right].
\end{aligned} \qquad (10.165)$$

The quantity in parentheses on the last line is a projection operator that projects onto the subspace orthogonal to N, which is the same as the subspace spanned by M. The identity of these two subspaces implies that

$$I - NN^\dagger = M(M^\dagger M)^{-1}M^\dagger = MR^{-1}M^\dagger. \tag{10.166}$$

Substituting this into the last line of Eq. (10.165) and using the orthogonality of N and M gives

$$
\begin{aligned}
\mathcal{F}_{pq} &= \frac{1}{g}\left[\partial_p N^\dagger M R^{-1} M^\dagger \partial_q N - (p \leftrightarrow q)\right] \\
&= \frac{1}{g}\left[N^\dagger \partial_p M\, R^{-1} \partial_q M^\dagger\, N - (p \leftrightarrow q)\right] \\
&= \frac{1}{g}\left[N^\dagger C \hat{e}_p R^{-1} \hat{e}_q^* C^\dagger N - (p \leftrightarrow q)\right].
\end{aligned} \tag{10.167}
$$

It is at this point that we use the requirement that $R = M^\dagger M$ be real. Because all of its components are real, it commutes with all of the $\hat{e}_p$, allowing us to write

$$\mathcal{F}_{pq} = \frac{1}{g} N^\dagger C R^{-1}(e_p \hat{e}_q^* - e_q \hat{e}_p^*)C^\dagger N. \tag{10.168}$$

We saw above that the quantity in parentheses was self-dual, thus showing that the field strength is self-dual, as claimed.

Directly demonstrating that the instanton number is equal to k is a less trivial matter. However, we will see presently that the 't Hooft solution with instanton number k can be obtained in this manner. Because the instanton number takes only integer values, any matrices M that satisfy the reality condition and are continuously connected to the M that yields the 't Hooft solution must also have instanton number k. Assuming that the solutions of the reality condition with a given k are all continuously connected then yields the desired result.

The matrix M is not uniquely determined by the gauge field. The gauge field A_p is unchanged if we make the replacement

$$
\begin{aligned}
M &\to SMT, \\
N &\to SN,
\end{aligned} \tag{10.169}
$$

where S and T are x-independent matrices with S a quaternionic matrix obeying $S^\dagger S = I$ and T any invertible real matrix. We can use this freedom to put the matrix C into a canonical form [248, 249]

$$C = \begin{pmatrix} 0 & 0 & \cdots & 0 \\ 1 & 0 & \cdots & 0 \\ 0 & 1 & \cdots & 0 \\ \cdots & \cdots & \cdots & \cdots \\ 0 & 0 & \cdots & 1 \end{pmatrix} \tag{10.170}$$

in which the first row vanishes and the remainder form a $k \times k$ unit matrix. This does not eliminate all the freedom, but still allows transformations with the $(k+1) \times (k+1)$ matrix S in the block diagonal form

$$S = \begin{pmatrix} u & 0 \\ 0 & T^{-1} \end{pmatrix}, \tag{10.171}$$

where u is a unit quaternion and T is a $k \times k$ orthogonal matrix.

With C in this canonical form, it is convenient to decompose B as

$$B = \begin{pmatrix} v \\ b \end{pmatrix} \tag{10.172}$$

where v is a k-component row vector of quaternions and b is a $k \times k$ quaternionic matrix.

The reality condition is that

$$\begin{aligned} M^\dagger M &= B^\dagger B - x^* C^\dagger B - B^\dagger C x + x^* C^\dagger C x \\ &= \left(v^\dagger v + b^\dagger b \right) - \left(x^* b + b^\dagger x \right) + x^* x I \end{aligned} \tag{10.173}$$

be real. The last term is manifestly real. The terms linear in x are of the form

$$x^* b_{mn} + (b^\dagger)_{mn} x = x^* b_{mn} + (b_{nm})^* x \tag{10.174}$$

and are real for arbitrary x if and only if b is symmetric. With b symmetric, the reality condition then reduces to the requirement that

$$B^\dagger B = v^\dagger v + b^\dagger b \tag{10.175}$$

be real.

At this point let us stop and count the number of free parameters. The vector v has $4k$ real components, and the symmetric matrix b has $4[k(k+1)/2]$. The diagonal elements of $B^\dagger B$ are automatically real, while the elements in the upper right are the conjugates of those in the lower left. Thus, the vanishing of the imaginary components of $B^\dagger B$ gives $3[k(k-1)/2]$ constraints on the parameters in v and b, leaving $k(k+15)/2$ independent components. Of these, $[k(k-1)/2]+3$ represent the freedom to make $SO(k) \times SU(2)$ gauge transformations while still preserving the canonical form of C. Subtracting these leaves precisely $8k - 3$ physical parameters. This is just the number required for the general instanton number k solution, strongly suggesting that the ADHM construction yields all instanton solutions. An explicit confirmation that it does is given by an inverse construction [250]. Here one starts with a self-dual gauge field with instanton number k. Equation (10.138) shows that the massless Dirac equation for a fermion in the fundamental doublet representation in this background has k zero modes. The vector v and the matrix b can be written in terms of these zero modes, and one can then show that they obey the requisite reality condition, thus completing the demonstration of completeness.

As an explicit example of the ADHM construction, consider the case where the components of v are all real and b is diagonal. We then have

$$B = \begin{pmatrix} \lambda_1 & \lambda_2 & \cdots & \lambda_k \\ w_1 & 0 & \cdots & 0 \\ 0 & w_2 & \cdots & 0 \\ \cdots & \cdots & \cdots & \cdots \\ 0 & 0 & \cdots & w_k \end{pmatrix}, \tag{10.176}$$

with the λ_j being real and the w_j being quaternions that can be interpreted as positions as in Eq. (10.159). This clearly satisfies the reality constraint.

We now need to find a quaternionic vector N that satisfies Eq. (10.161). If we label the first row by 0, this equation can be written as

$$0 = (N^\dagger M)_j = N_0^\dagger \lambda_j + N_j^\dagger (w_j - x). \tag{10.177}$$

Without any loss of generality, we can write the first component of N as

$$N_0 = \frac{u(x)}{\sqrt{\rho(x)}}, \tag{10.178}$$

where u is a unit quaternion satisfying $u^* u = 1$ and ρ is a real normalization factor to be determined shortly. Equation (10.177) then gives

$$N_j = \frac{\lambda_j (x - w_j)}{\sqrt{\rho} |x - w_j|^2} u(x) \tag{10.179}$$

and the normalization condition, Eq. (10.162), becomes

$$\rho = u^* u + \sum_j \frac{\lambda_j^2}{(x - w_j)^2} = 1 + \sum_j \frac{\lambda_j^2}{(x - w_j)^2}. \tag{10.180}$$

Setting $u(x) = 1$ yields the 't Hooft solution of Eqs. (10.94) and (10.97). Making a different choice for $u(x)$ gives a gauge transformation of this solution, since the transformation $N \to NW$ gives

$$A_p = \frac{1}{g} N^\dagger \partial_p N \to W^\dagger A_p W + \frac{1}{g} W^\dagger \partial_p W. \tag{10.181}$$

If we allow the components of v to be arbitrary quaternions, which can be written as $\lambda_j s_j$, with λ_j a positive real number and s_j a unit quaternion, b must have nonzero off-diagonal elements, and solving Eq. (10.175) becomes a rather nontrivial matter. However, if the λ_j are all small compared to the separations $|w_i - w_j|$, one can solve perturbatively for the off-diagonal elements of b [248]. This gives a solution that corresponds to k well-separated instantons. As before, the λ_j and w_j are scale and position parameters for the individual instantons. The $3k$ angles in the s_j are SU(2) orientation angles, of which three are redundant because multiplying all of the s_j by a common unit quaternion has no effect on the Euclidean space gauge fields.

10.11 The ADHM construction for larger gauge groups

The ADHM formalism can be generalized from $SU(2)$ to the classical groups $SO(n)$, $SU(n)$, and $Sp(2n)$ [246–249]. As with $SU(2)$, one begins with a matrix

$$M = B - Cx \tag{10.182}$$

and then finds a matrix N obeying

$$N^\dagger M = 0, \qquad N^\dagger N = I. \tag{10.183}$$

The gauge potential is again given by

$$A_p = \frac{1}{g} N^\dagger \partial_p N. \tag{10.184}$$

In each case we want A_p to be an $n \times n$ matrix. Its elements, and therefore those of N, should be quaternions, complex numbers, or real numbers, for $Sp(2n)$, $SU(n)$, and $SO(n)$, respectively. Satisfying these requirements imposes conditions on the size and properties of M.

As one might guess from the quaternionic character of the ADHM construction, the simplest case is that of $Sp(2n)$. For $SU(2) = Sp(2)$, M is an $(k+1) \times k$ matrix satisfying the requirement that $M^\dagger M$ be real and invertible. The latter requirement ensures that M has rank k. Hence, the k columns of M together with the vector N span the $(k+1)$-dimensional quaternionic vector space. The extension to $Sp(2n)$ is obtained by taking M to be a $(k+n) \times k$ matrix, with $M^\dagger M$ again real and invertible. N is then $(k+n) \times n$, so that A_p is $n \times n$, as required. Taken together, the columns of M and N span the $(k+n)$-dimensional quaternionic vector space.

As with the $SU(2)$ case, it is convenient to put B and C into a canonical form, which in this case is

$$C = \begin{pmatrix} 0_{n \times k} \\ I_k \end{pmatrix} \qquad B = \begin{pmatrix} v \\ b \end{pmatrix}, \tag{10.185}$$

where v is $n \times k$ and $B^\dagger B$ is real and invertible.

An embedding of the $SU(2)$ 't Hooft solutions is obtained by taking b to be diagonal and the columns of v to be real multiples of a fixed quaternionic vector. The nonzero elements of b then give the instanton positions. The magnitudes of the columns of v give the instanton scales, while their common direction as quaternionic vectors gives the gauge orientation parameters that define an $Sp(2)$ subgroup.

Let us now turn to $SO(n)$. To obtain an $SO(n)$ gauge field, we need N to be a real matrix with n columns. Although M is still naturally viewed as a quaternionic matrix, it can also be decomposed into four real matrices M_p, with

$$M = M_p \hat{e}_p. \tag{10.186}$$

One is led to take M to be a $(4k+n) \times k$ quaternionic matrix. Its columns define k quaternionic vectors or, by separating out the coefficients of the $\hat{e}_p$, $4k$ real $(4k+n)$-component vectors. For the construction to work, we need that these, together with the n columns of N, span the $(4k+n)$-dimensional vector space. This requires that we impose additional conditions on M, since its columns can be linearly independent as quaternionic vectors without their components all being linearly independent as real vectors. A hint comes from noting that requiring that a quaternionic vector v be such that $v^\dagger \hat{e}_p v$ is real for all four values of p is equivalent to requiring that the four real vectors derived from v be mutually orthogonal and of equal length. Indeed, the proper condition on M turns out to be that

$$M^\dagger \hat{e}_q M \equiv 4R_q \tag{10.187}$$

be real for all four values of q and that

$$R = R_q \hat{e}_q \tag{10.188}$$

be invertible. These conditions imply that

$$M_p^t M_q = \delta_{pq} R_4 + \eta_{pq}^j R^j. \tag{10.189}$$

The canonical forms for C and B are

$$C = \begin{pmatrix} 0_{n\times k} \\ I_k \\ \hat{e}_1 I_k \\ \hat{e}_2 I_k \\ \hat{e}_3 I_k \end{pmatrix} \qquad B = \begin{pmatrix} v \\ b_0 + b_1 + b_2 + b_3 \\ \hat{e}_1(b_0 - b_1) \\ \hat{e}_2(b_0 - b_2) \\ \hat{e}_3(b_0 - b_3) \end{pmatrix}. \tag{10.190}$$

Here v is an $n \times k$ quaternionic matrix, while the b_r are $k \times k$ quaternionic matrices with b_0 symmetric and the other three antisymmetric.

To obtain the 't Hooft solutions we set $b_1 = b_2 = b_3 = 0$ and take b_0 to be diagonal, with its elements specifying the instanton positions. The columns of v must then be real multiples of a common quaternionic vector u for which the $u^\dagger e_p u$ are all real. As noted above, this implies that the four component vectors of u are mutually orthogonal. These then define an SO(4) subgroup of SO(n), and the construction associates the 't Hooft solution with one of the SU(2) factors of this SO(4).

It is now easy to see how the construction for SU(n) works. We take M to be a $(2k+n) \times k$ quaternionic matrix such that

$$M^\dagger M = R_4, \qquad M^\dagger \hat{e}_1 M = R_1 \tag{10.191}$$

are both real, with $R_4 + \hat{e}_1 R_1$ invertible. N is a $(2k + n) \times n$ complex matrix (i.e., one containing $\hat{e}_1$ and $\hat{e}_4$, but not $\hat{e}_2$ or $\hat{e}_3$). The canonical form is

$$C = \begin{pmatrix} 0_{n \times k} \\ I_k \\ \hat{e}_2 I_k \end{pmatrix} \qquad B = \begin{pmatrix} v \\ b_0 + b_2 \\ \hat{e}_2(b_0 - b_2) \end{pmatrix}, \qquad (10.192)$$

with b_0 and b_2 as in the SO(n) construction. The 't Hooft solutions are obtained by setting $b_2 = 0$, taking b_0 to be diagonal, and choosing the columns of v to be real multiples of a quaternionic vector u with $u^\dagger \hat{e}_1 u$ real.

The general multi-instanton solutions are obtained by allowing the columns of v to be independently chosen quaternionic vectors. For well-separated instantons in Sp($2n$) the diagonal elements of b give the instanton positions and the remaining elements of B can be determined perturbatively, as in the SU(2) case. The input data for this consist of $n + 1$ quaternions (one for the position and n for the components of the column of v) for each instanton. This gives a total of $4(n + 1)$ real parameters per instanton, which agrees with the coefficient of k in Eq. (10.149). The remaining terms in that equation correspond to the freedom to perform linear transformations on the rows of v.

When this approach is extended to the case of well-separated instantons in SO(n) and SU(n), the columns of v must be chosen so that they each obey the appropriate reality conditions. For SO(n), this can be done by defining an n-component vector u whose first four elements are 1, $\hat{e}_1$, $\hat{e}_2$, and $\hat{e}_3$, with the remaining $n - 4$ elements vanishing. An acceptable column for v is then obtained by acting on u with an arbitrary SO(n) matrix and then multiplying by a real scale parameter. The procedure for SU(n) is similar, except that one starts with a vector u whose first components are 1 and $\hat{e}_2$, with the remainder vanishing. One can check that these prescriptions again give the correct number of parameters.

There is an equivalent, although superficially somewhat different, formulation of the ADHM construction for SU(n). We start by regrouping the eight real matrices that make up the quaternionic matrices b_0 and b_2 into four Hermitian complex matrices Z_p. To do this, we write

$$b_0 = \hat{e}_p X_p$$

$$b_2 = \hat{e}_1 \hat{e}_p Y_p \qquad (10.193)$$

and define

$$Z_p = X_p + iY_p. \qquad (10.194)$$

Similarly, we trade the quaternionic $n \times k$ matrix v for a complex $n \times 2k$ matrix

$$\omega_a^r = \begin{pmatrix} \omega_{1a}^r \\ \omega_{2a}^r \end{pmatrix}, \qquad (10.195)$$

where r runs from 1 to k and a from 1 to n. These matrices can then be combined to give the $(n + 2k) \times 2k$ complex matrix

$$\Delta = \begin{pmatrix} \omega^t \\ (Z_p - x_p I_k) \otimes e_p \end{pmatrix}. \tag{10.196}$$

We then proceed as before; i.e., we look for an $(n + 2k) \times n$ complex matrix N satisfying

$$\Delta^\dagger N = 0$$
$$N^\dagger N = I \tag{10.197}$$

and again define the gauge field to be

$$A_p = \frac{1}{g} N^\dagger \partial_p N. \tag{10.198}$$

In this language, the self-duality conditions that $B^\dagger B$ and $B^\dagger \hat{e}_1 B$ be real become the triplet of constraints

$$0 = i[Z_p, Z_q]\bar{\eta}^j_{pq} - \sum_{a=1}^{n} \omega^\dagger_{a\alpha}(\sigma^i)_{\alpha\beta}\omega_{a\beta}. \tag{10.199}$$

[For clarity, I have suppressed the SU(k) indices on Z_p and ω.] Interestingly, this latter formulation of the construction with the self-duality condition as given in Eq. (10.199) can be naturally obtained as the condition for minimizing the energy of a configuration of k D0-branes in the presence of n D4-branes [251, 252].

10.12 One-loop corrections

The classical instanton solution gives the dominant exponential factor associated with tunneling. The leading contribution to the pre-exponential factor, corresponding to the quantity K in Eq. (9.39), arises from the fluctuations about the instanton configuration. It was first obtained by 't Hooft [238] in the context of an SU(2) gauge theory; the extension to larger groups is straightforward. The full details of his calculation are too lengthy to reproduce here. Instead, I will give an overview of the calculation, explaining how the various contributions arise. An alternative approach, using the O(5) invariance that follows by projecting the solution onto a four-sphere and using the conformal invariance of the theory, is given in [253].

The path integral derivation of this "one-loop" correction about an instanton was described in Sec. 9.3. That discussion was phrased in terms of a scalar field. When extending this to the Yang–Mills case, it is necessary to impose a gauge condition. This is most conveniently done by adding to the Lagrangian a gauge-fixing term of the form $-\frac{1}{2}(D_p A_p^{\mathrm{qu}})^2$, where A_p^{qu} denotes the fluctuation about the instanton solution and the covariant derivative is defined using only the background instanton potential. This then determines the form of the ghost Lagrangian.

To start, let us consider the contribution due to a single instanton with fixed position and scale size λ, written in the form given in Eq. (10.87). The relevant parts of the action can then be written schematically as

$$S = S_{\rm cl} - \frac{1}{2} A^{\rm qu} \mathcal{M}^A A^{\rm qu} - \bar{c} \mathcal{M}^{\rm gh} c + \cdots. \tag{10.200}$$

Here the terms for the gauge and ghost fields are shown explicitly, with integration over the spatial coordinates and traces over the implied indices understood, while the contributions from any matter fields that might be present are represented by an ellipsis. Ignoring these matter fields for the present, we wish to calculate

$$W(\lambda) = e^{-S_{\rm cl}} \left[\frac{\det' \mathcal{M}^A}{\det \mathcal{M}_0^A} \right]^{-1/2} \left[\frac{\det \mathcal{M}^{\rm gh}}{\det \mathcal{M}_0^{\rm gh}} \right] \prod_a J_a. \tag{10.201}$$

The prime indicates that the zero modes are to be omitted in the calculation of the determinant of $\mathcal{M}_A$. Instead, they are handled by introducing collective coordinates, which entails a Jacobian factor J_a for each zero mode. A subscript 0 denotes the corresponding determinant calculated about the vacuum configuration $A_p = 0$.

The determinants can be understood formally as products of eigenvalues. One way to realize this calculationally is to put the system in a box of finite size, so that the spectra of the various operators are discrete. The resulting infinite products are divergent, even after the division by the vacuum determinants, and so a regularization scheme is needed. Dimensional regularization cannot be used, because the background instanton solution is inherently four-dimensional. An alternative is to use a Pauli–Villars-type regulator, which leads to regularized determinant ratios of the form[11]

$$\frac{\det \mathcal{M}}{\det \mathcal{M}_0} = \prod_{j=1}^{\infty} \frac{\omega_j \left(\omega_{j0} + M^2 \right)}{\omega_{j0} \left(\omega_j + M^2 \right)}, \tag{10.202}$$

where the ω_j are the eigenvalues of $\mathcal{M}$ and M is the regulator mass.

Because of the spherical symmetry of the instanton, the modes can be chosen to be eigenstates of the spin and orbital $SO(4) = SU(2) \times SU(2)$ angular momenta and the $SU(2)$ gauge symmetry. The determination of the eigenvalues is then reduced to the solution of a radial Schrödinger-type equation that can be converted to a hypergeometric equation whose eigenvalues can be found. After

---

[11] I am oversimplifying somewhat. For the full details of the regularization, see [238].

taking the size of the spatial box to infinity, the net result for the gauge and ghost field determinants is

$$\left[\frac{\det' \mathcal{M}^A}{\det \mathcal{M}_0^A}\right]^{-1/2} \left[\frac{\det \mathcal{M}^{\text{gh}}}{\det \mathcal{M}_0^{\text{gh}}}\right] = C_1 (M\lambda)^{-2/3}, \tag{10.203}$$

where C_1 is a numerical constant. Note that this ratio of determinants is dimensionless, despite the omission of the zero modes in the numerator of the first factor, because of the way the regularized determinants have been defined in Eq. (10.202).

We must next consider the zero modes. Each of these corresponds to the variation of a collective coordinate z_a and so is of the form

$$a_p^a = \mathcal{N}_a^{-1/2} \left(\frac{\partial A_p^{\text{inst}}}{\partial z_a} + D_p \Lambda^a\right), \tag{10.204}$$

where $\mathcal{N}_a$ is a normalization constant and Λ^a is the gauge transformation that brings the mode into background gauge. The Jacobian factor in Eq. (10.201) is related to the normalization constant by

$$J_a = \sqrt{\frac{\mathcal{N}_a}{2\pi}} M. \tag{10.205}$$

The factor of M arises from the regulator mode associated with the zero mode, while the $\sqrt{2\pi}$ appears because the integration over the collective coordinate replaces a Gaussian integral over a mode coefficient.

There are four translational zero modes. The derivatives with respect to instanton position are, up to a sign, the same as the spatial derivatives of the instanton solution. These modes therefore fall as $1/x^2$ at large distance and would lead to non-normalizable modes were it not for the transformation to background gauge. In background gauge, these modes are proportional to the field strength, fall as $1/x^4$, and are normalizable.

A fifth zero mode is associated with the scale parameter. It is given in background gauge simply by the derivative of the instanton with respect to λ, with no gauge transformation needed

Finally, there are three gauge zero modes that are not eliminated by the gauge-fixing term in the Lagrangian. These are of the form $a_p = D_p \Lambda(x)$, where the gauge function Λ is nonvanishing at spatial infinity. They can be viewed as corresponding to global gauge rotations.

Before normalization, each of these eight modes inherits a factor of $1/g$ from the instanton solution, leading to a factor of $1/g^2$ in the corresponding $\mathcal{N}_a$. The spatial integral of the square of each mode should equal unity, so the a_p^a must have dimensions of $[\text{mass}]^2$. The $\mathcal{N}_a$ for the translation and scale modes are therefore dimensionless, while those for the three global gauge modes have dimensions of

$[\text{mass}]^{-2} \sim [\text{length}]^2$. Since λ is the only scale in the problem, these last three $\mathcal{N}_a$ must be proportional to λ^2. Hence, we see that

$$\prod_a J_a = C_2\, g^{-8} M^8 \lambda^3 , \tag{10.206}$$

where C_2 is another numerical constant.

Substituting our results into Eq. (10.201) gives

$$W(\lambda) = C_1 C_2\, g^{-8}\, \lambda^{-5} \exp\left[-\frac{8\pi^2}{g^2} + \frac{22}{3} \ln(M\lambda) \right]. \tag{10.207}$$

Let us examine the exponent in this expression. Up to now, we have been simply writing the gauge coupling g without taking into account that it runs with energy and so is not well-defined until a scale is specified. Because we have not explicitly included any counterterms in our Lagrangian, the g we are using must be the bare coupling, or rather, since we have introduced a regulator to make everything finite, $g(M)$. The one-loop beta function for SU(2) gives the running of the coupling as

$$\frac{8\pi^2}{g^2(\mu)} = \frac{8\pi^2}{g^2(M)} - \frac{22}{3} \ln\left(\frac{M}{\mu} \right), \tag{10.208}$$

so the effect of the regulator contribution in Eq. (10.207) is to specify that the gauge coupling appearing in the exponent should be evaluated at an energy $1/\lambda$ or, equivalently, at a distance scale λ. This is not a surprise, but it is nice to see how it naturally comes about.[12] It also clear that a calculation to the next order would ensure that the factors of g in the prefactor would also be evaluated at a scale λ.

We now need to integrate over the gauge orientation and scale collective coordinates. After performing the integral over the SU(2) gauge orientation parameters χ^a and evaluating the numerical constants, we find that the weighting factor for instantons of size λ is

$$\int d^3\chi\, W(\lambda) = 2^{14} \pi^6 e^A \frac{1}{\lambda^5} \left[\frac{1}{g(\lambda)} \right]^8 \exp\left[-\frac{8\pi^2}{g(\lambda)^2} \right], \tag{10.209}$$

where $A = 6.9984$ is a constant for which an analytic expression is given in [238]. (Note that product of the numerical prefactors is 1.7×10^{10}.)

We now have to consider the integral over λ. At short distance and high energy the gauge coupling tends to zero, the instanton action tends to infinity, and W, which contains the exponential of a large negative number, tends to zero fast enough that this integral converges at the short-distance end. For large instantons

---

[12] With a gauge group larger than SU(2), the contributions to Eq. (10.203) from the additional global gauge modes give precisely the extra factors of $M\lambda$ that are needed to give the correct beta function, and thus the proper running coupling, for the larger group. The extra modes also give additional factors of g^{-1} in the prefactor.

we find the opposite behavior. With increasing λ the gauge coupling grows and the instanton action decreases, so that our semiclassical expansion eventually becomes unreliable. Unless one is considering a process that puts a natural upper cutoff on λ, the integration over instanton size diverges at large λ.

Furthermore, the dilute-gas approximation, which underlies this calculation, breaks down for large instantons. This approximation assumes that the density of instantons is small relative to the instanton size, which requires that the integral

$$\int d\lambda\, \lambda^4 W(\lambda) \ll 1. \tag{10.210}$$

With $W(\lambda)$ given by Eq. (10.209), this integral diverges at large λ, and the required inequality is clearly violated.

In some cases, such as the electroweak instantons that will be discussed in Sec. 11.4, other considerations restrict the instanton size to a narrow range. The integration over λ is then finite, and for weak coupling the instanton effects are small, just as expected for a tunneling process. For many QCD applications, on the other hand, there is no such cutoff on the instanton size and large instantons, with a strong effective coupling, are important. The net result is that instanton effects can be expected to be large, although they can no longer be reliably calculated with precision. Nevertheless, important qualitative conclusions can be drawn.

11

Instantons, fermions, and physical consequences

The most important physical consequences of the Yang–Mills instantons are associated with the presence of fermions in the theory. Because these turn out to be closely related to the axial anomaly, I begin with a brief review of that topic.

11.1 Anomalies

It sometimes happens that a transformation that is a symmetry of a classical field theory ceases to be a symmetry when the theory is quantized. Perhaps the best-known example of such an anomaly, and the one of relevance for us here, is that associated with the chiral symmetry of a theory with massless quarks. Classical analysis predicts a number of conserved vector and axial vector currents. However, it can happen that after quantization some of the axial currents are not conserved, but instead have anomalous divergences [255–257].

The simplest example of this axial anomaly occurs in a theory with massless fermion fields ψ_r, where the subscript $r = 1, 2, \ldots, N_f$ is a "flavor" index, and a Lagrangian density

$$\mathcal{L} = \bar{\psi}_r \left(i\gamma^\mu \partial_\mu + g\gamma^\mu A_\mu^a T^a \right) \psi_r + \cdots, \tag{11.1}$$

where the ellipsis denotes terms that do not contain the fermion fields. Here I will take the A_μ^a to be SU(N_c) gauge fields, with the T^a the corresponding generators. The "color" gauge indices on the fermions and the corresponding indices on the T^a have been suppressed. For simplicity, let us assume that all flavors of fermions transform under the fundamental representation of the gauge group.

Classically, this Lagrangian is invariant under both the U(1)×U(1) chiral transformations

$$\psi_r \to e^{i\alpha_0} \psi_r,$$
$$\psi_r \to e^{i\beta_0 \gamma^5} \psi_r \tag{11.2}$$

and the SU(N_f)×SU(N_f) chiral transformations

$$\psi_r \to \left(e^{i\alpha^a T^a}\right)_{rs} \psi_s \,,$$

$$\psi_r \to \left(e^{i\beta^a T^a \gamma^5}\right)_{rs} \psi_s \,, \tag{11.3}$$

where the T^a are generators of SU(N_f).

These symmetries imply conserved currents

$$j^\mu = \bar{\psi}_r \gamma^\mu \psi_r \,,$$

$$j_5^\mu = \bar{\psi}_r \gamma^\mu \gamma^5 \psi_r \tag{11.4}$$

and

$$j_a^\mu = \bar{\psi}_r T_{rs}^a \gamma^\mu \psi_s \,,$$

$$j_{5a}^\mu = \bar{\psi}_r T_{rs}^a \gamma^\mu \gamma^5 \psi_s \,, \tag{11.5}$$

with corresponding conserved charges. In particular, if we define right- and left-handed fields

$$\psi_{Rr} = \frac{1+\gamma^5}{2} \psi_r \,,$$

$$\psi_{Lr} = \frac{1-\gamma^5}{2} \psi_r \,, \tag{11.6}$$

the charges corresponding to j^μ and j_5^μ are

$$Q = \int d^3x\, \bar{\psi}_r^\dagger \psi_r = n_R + n_L \,,$$

$$Q_5 = \int d^3x\, \bar{\psi}_r^\dagger \gamma^5 \psi_r = n_R - n_L \,, \tag{11.7}$$

where n_R is the number of right-handed particles minus the number of left-handed antiparticles, and similarly for n_L. Thus, Q is the total particle number and Q_5 the net chirality.

The problematic axial current is j_5^μ. If it were divergenceless, as predicted by the classical analysis, we would have the Ward identity

$$0 = k^\mu T_{\mu\alpha\beta}(k, q_1, q_2) \,, \tag{11.8}$$

where

$$T_{\mu\alpha\beta}(k, q_1, q_2) = i \int d^4x_1\, d^4x_2 \langle 0|T[j_5^\mu(0) j_a^\alpha(x_1) j_a^\beta(x_2)]|0\rangle e^{iq_1 \cdot x_1 + iq_2 \cdot x_2} \,. \tag{11.9}$$

Here T denotes time-ordering, j_a^μ is the gauge current that couples to A_μ^a, and $k = q_1 + q_2$. The leading contribution to the matrix element comes from the two

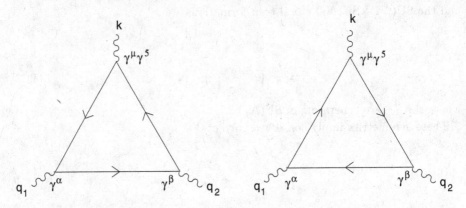

Fig. 11.1. The graphs that give the leading contribution to the matrix element in Eq. (11.9). The incoming momenta and the gamma matrix factors at each vertex are indicated.

triangle graphs shown in Fig. 11.1. Explicit calculation of these gives the nonzero result[1]

$$k^\mu T_{\mu\alpha\beta} = -\frac{N_f}{2\pi^2}\epsilon_{\alpha\beta\rho\sigma}q_1^\rho q_2^\sigma . \tag{11.10}$$

The failure of Eq. (11.8) can be understood by recalling that properly defining the quantum field theory requires specifying a regulator scheme. Any method for regulating the triangle graphs that respects the gauge symmetry, as is required for renormalizability, violates the chiral symmetry. Dimensional regularization has the problem that the γ^5 in the chiral transformation is an explicitly four-dimensional quantity that cannot be naturally continued to $4 + \epsilon$ dimensions. Pauli–Villars regulation is gauge invariant and four-dimensional, but the massive Pauli–Villars regulator field explicitly breaks the chiral symmetry. This field gives a contribution to $\partial_\mu j_5^\mu$ that is proportional to the regulator mass M. This explicit factor of M multiplies a regulator graph proportional to $1/M$ to give a finite contribution in the $M \to \infty$ limit.

A useful way to interpret Eq. (11.10) is to include factors of A_α^a on the external gauge lines of the triangle graphs. We can then view $T_{\mu\alpha\beta}$ as a contribution to the expectation value of j_5^μ in the presence of a background gauge field. Equation (11.10) gives the divergence of this current in the background field as

$$\partial_\mu j_5^\mu = \frac{N_f g^2}{16\pi^2} \epsilon^{\mu\nu\rho\sigma} \mathrm{tr}\,(\partial_\mu A_\nu - \partial_\nu A_\mu)(\partial_\rho A_\sigma - \partial_\sigma A_\rho) . \tag{11.11}$$

---

[1] This assumes a regularization that keeps the two vector currents divergenceless.

Including the effects of the analogous square and pentagon diagrams gives the additional terms needed to obtain the gauge-invariant result[2]

$$\partial_\mu j_5^\mu = \frac{N_f g^2}{8\pi^2} \operatorname{tr} F_{\mu\nu} \tilde{F}^{\mu\nu}. \tag{11.12}$$

11.2 Spectral flow and fermion zero modes

Notice that the anomalous divergence of the axial current, Eq. (11.12), is, up to a multiplicative constant, the same as the current j_A^μ that was defined in Eq. (10.17). This suggests that we combine them to form a divergenceless current

$$\mathcal{J}_5^\mu = j_5^\mu - 2N_f j_A^\mu. \tag{11.13}$$

Like j_A^μ, this current is gauge-variant, and so its analysis depends on the gauge in which we work. Measurable physical consequences must, of course, be the same in all gauges.

Let us start by working in $A_0 = 0$ gauge, where the instanton corresponds to tunneling between two vacua of different winding number. The charge associated with $\mathcal{J}_5^\mu$ is

$$\mathcal{Q}_5 = n_R - n_L - 2N_f n, \tag{11.14}$$

where n is the winding number of the gauge field. The divergenceless of $\mathcal{J}_5^\mu$ implies that $\mathcal{Q}_5$ is conserved. Hence, any change in winding number must be accompanied by a change in fermion chirality, with

$$\Delta(n_R - n_L) = 2N_f \Delta n. \tag{11.15}$$

To see how this comes about, let us consider the spectrum of the Dirac Hamiltonian in the presence of a background field A_μ. It is simplest to view this from the Dirac sea viewpoint, with both positive and negative energies in the spectrum, and antiparticles being unoccupied negative-energy states. For our massless fermions the Hamiltonian is

$$H = -i\alpha^j D_j, \tag{11.16}$$

where $\alpha^j = \gamma^0 \gamma^j$. In a basis with

$$\gamma^j = \begin{pmatrix} 0 & i\sigma^j \\ i\sigma^j & 0 \end{pmatrix}, \quad \gamma^0 = \begin{pmatrix} 0 & -iI \\ iI & 0 \end{pmatrix}, \quad \gamma^5 = \begin{pmatrix} I & 0 \\ 0 & -I \end{pmatrix}, \tag{11.17}$$

we have

$$\alpha^j = \begin{pmatrix} \sigma^j & 0 \\ 0 & -\sigma^j \end{pmatrix}. \tag{11.18}$$

---

[2] The anomaly can also be calculated from a careful examination of the behavior of the path integral measure under chiral transformations [258].

The four-component fermions naturally split into a pair of two-component Weyl fermions. For the upper two components, corresponding to right-handed particles, the Hamiltonian becomes

$$H_R = -i\sigma^j D_j \,, \tag{11.19}$$

while for the lower two, left-handed, components we have

$$H_L = i\sigma^j D_j \,. \tag{11.20}$$

Because $H_L = -H_R$, each negative eigenvalue of one Hamiltonian corresponds to a positive eigenvalue of the other.

Let us start with the background field being a vacuum configuration with winding number n, and then consider the series of configurations along the tunneling path defined by an instanton. At intermediate stages along the way, A_μ is not in a vacuum state, there are nonzero field strengths, and the fermion spectrum is certainly different from the initial spectrum. Nevertheless, since the final configuration is also a vacuum configuration, albeit one with winding number $n + 1$, the spectrum, which is gauge invariant, must be the same at the end as it was at the beginning.

This does not mean, however, that individual states must end up where they started. As shown schematically in Fig. 11.2, an individual energy level can have a net movement up or down the spectrum as the gauge field flows along the tunneling path. The only requirement is that its place be taken up by some other level. The relationship between the two Hamiltonians requires that for every level of H_R that moves up, an energy level of H_L with the opposite sign must move down, and vice versa.

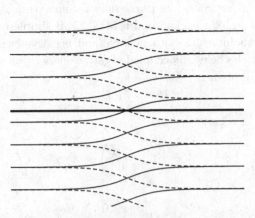

Fig. 11.2. Schematic illustration of the flow of fermion energy levels as the gauge field is varied from a vacuum configuration with winding number n (on the far left) to one with winding number $n + 1$ (on the far right). The right-handed levels are indicated by solid lines and the left-handed ones by the dashed lines. The heavy horizontal line represents the zero of energy.

In particular, as indicated in the figure, some negative-energy states can become positive-energy states and some positive-energy states can become negative-energy states. Let us suppose that the fermions were originally in a vacuum state, with all negative-energy levels filled and all positive-energy levels empty. In the adiabatic approximation the occupation of the individual levels would not change, so the movement of a level from negative to positive energy would lead to the creation of a positive-energy particle, while a flow from positive to negative-energy would give a negative-energy hole, corresponding to a positive-energy antiparticle of the opposite chirality. If the adiabatic approximation is not applicable, particles may move between levels. However, because the original Hamiltonian only couples fermion fields of the same chirality, this movement must be between levels of the same chirality, and so won't affect the total chirality.

Thus, the relation between the changes in chirality and winding number in Eq. (11.15) can be understood if in the presence of a gauge field $A_\mu(x)$ with instanton number k there is a net flow upward of k right-handed levels from negative to positive energy, and an equal flow of left-handed levels from positive to negative energy. We can show that this is the case by means of an index theorem.

In order to follow the flow of the energy levels, let us write $x_4 = \tau$ and define

$$H_R(\tau) = -i\sigma^j \left[\partial_j - igA_j^a(\mathbf{x}, \tau)T^a\right] \tag{11.21}$$

and label its instantaneous eigenvalues and two-component eigenfunctions as $\omega_n(\tau)$ and $\chi_n(\mathbf{x}; \tau)$, respectively. Now consider the equation

$$0 = \mathcal{D}\chi = \left[-\frac{\partial}{\partial\tau} - H_R(\tau)\right]\chi. \tag{11.22}$$

For A_μ sufficiently slowly varying as a function of τ, this equation is solved by

$$\chi = e^{-\int_0^\tau d\tau'\, \omega_n(\tau')} \chi_n(\mathbf{x}; \tau). \tag{11.23}$$

If $\omega_n(\tau)$ has the same sign at $\tau = -\infty$ and $\tau = \infty$, this solution diverges as τ goes to either one limit or the other. On the other hand, if $\omega(-\infty) < 0$ and $\omega(\infty) > 0$, the solution goes to zero in both directions, and is a normalizable zero mode of $\mathcal{D}$. If instead $\omega(\tau)$ goes from positive to negative, a similar construction gives a normalizable zero mode of

$$\mathcal{D}^\dagger = \frac{\partial}{\partial\tau} - H_R(\tau). \tag{11.24}$$

Thus, if k_+ levels of H_R move from negative energy to positive energy and k_- modes move from positive to negative, $\mathcal{D}$ will have k_+ zero modes, $\mathcal{D}^\dagger$ will have k_- zero modes, and the index of $\mathcal{D}$ will be[3]

$$\mathcal{I}(\mathcal{D}) = k_+ - k_- . \tag{11.25}$$

In fact, we have already calculated this index. The operator $\mathcal{D}$ defined above is the same as the one defined in Eq. (10.119), written in $A_0 = 0$ gauge. In Sec. 10.8 we showed that the index of $\mathcal{D}$ was given by Eq. (10.138). For fermions in the fundamental representation of SU(N), we have $T(R) = 1/2$, and hence

$$\mathcal{I}(\mathcal{D}) = k . \tag{11.26}$$

Thus, along the flow defined by a Euclidean gauge field with instanton number k there are k energy levels of H_R that move from negative to positive energy. Because $H_L = -H_R$, there are also k levels of H_L that move from positive to negative energy. This is repeated for each of the N_f flavors, so the net increase in chirality is $2N_f k$, just as required by Eq. (11.15).

The index calculation in Sec. 10.8 did not use the fact that the background field was self-dual, so our result applies even if $A_\mu(x)$ is not a solution of the Euclidean field equations. On the other hand, the vanishing theorem that showed that $\mathcal{D}^\dagger$ had no zero modes did use the self-duality of the gauge field, so for general background fields Eq. (11.26) only gives the difference between the numbers of zero modes of $\mathcal{D}$ and $\mathcal{D}^\dagger$, and thus a lower bound on the number of zero modes.

This analysis of the spectral flow was done in the $A_0 = 0$ gauge, where the instanton corresponds to a tunneling path between vacua of different winding number. In a gauge with a unique vacuum, the tunneling path represented by the instanton is gauge-transformed to one that begins and ends at the same point. This gauge transformation cannot change the physical consequences of the instanton. Hence, although the gauge field eventually returns to its initial value, the flow of the fermion eigenstates as the gauge field background evolves does not return them to their initial position. Instead, they are shifted, with some moving from negative to positive energy, or vice versa, just as in $A_0 = 0$ gauge. This leads to a net change in the fermion chirality, even though the winding number is unchanged. Thus, Q_5 is not conserved. This is not in contradiction with the vanishing of $\partial_\mu \mathcal{J}_5^\mu$. The standard demonstration that a divergenceless current implies a conserved charge proceeds by integrating over the region between two spacelike surfaces and converting the integral to a sum of surface integrals. Conservation of the charge follows if the surface integrals at spatial infinity vanish, as is usually the case. However, although the gauge-variant current j_A^μ vanishes

---

[3] One might worry about the fact that Eq. (11.23) only solves Eq. (11.22) in the limit of slowly varying A_μ. However, the index is a topological invariant, and so the result for more rapidly varying fields must be the same as for a slowly varying field with the same instanton number.

at large distance in $A_0 = 0$ gauge, it is nonvanishing and gives a finite surface integral in other gauges, with the result that conservation of Q_5 is violated by instanton effects.

Let us find the explicit form of the fermion zero mode in the background of a single SU(2) instanton centered at the origin. Rather than working in $A_0 = 0$ gauge, it is simpler to work in a gauge where the instanton is given by Eq. (10.87). In terms of the four e_p defined in Eq. (10.63), the zero-mode equation takes the form

$$0 = (e_p^\dagger)_{\alpha\beta} [\delta_{ab}\partial_p - ig(A_p)_{ab}] \Phi_{b\beta}. \qquad (11.27)$$

Here Greek subscripts denote spinor indices 1 or 2 while Latin subscripts from the beginning of the alphabet denote SU(2) indices, which also take values 1 or 2. The distinction between the two types of indices can be ignored if we view $\Phi_{b\beta}$ as a 2×2 matrix and rewrite the zero-mode equation as the matrix equation

$$0 = i [\partial_p - igA_p] \Phi (e_p^\dagger)^t. \qquad (11.28)$$

Now recall that the Pauli matrices satisfy

$$\sigma_j^t = (i\sigma_2)\sigma_j(i\sigma_2), \qquad j = 1, 2, 3, \qquad (11.29)$$

which implies that

$$(e_p^\dagger)^t = e_p^* = -(i\sigma_2)e_p(i\sigma_2), \qquad p = 1, 2, 3, 4, \qquad (11.30)$$

so that our zero-mode equation becomes

$$0 = i [\partial_p - igA_p] \Phi(i\sigma_2) e_p. \qquad (11.31)$$

The instanton is invariant under the combination of a rotation and an SU(2) global gauge transformation. Since there is only one fermion zero mode, it must be a singlet under such a transformation. Because $\epsilon_{b\beta} = (i\sigma_2)_{b\beta}$ is an SU(2) invariant tensor, a natural ansatz is

$$\Phi_{b\beta} = i\epsilon_{b\beta} h(x^2). \qquad (11.32)$$

Substituting this into Eq. (11.31) leads to

$$0 = [2x_p h'(x^2) - igA_p h(x^2)] e_p, \qquad (11.33)$$

where the prime indicates differentiation with respect to x^2. Using the explicit form of the instanton,

$$A_p = \frac{1}{g} \frac{\eta_{pq}x_q}{x^2 + \lambda^2}, \qquad (11.34)$$

and the easily verified identity

$$\eta_{pq}e_p = i \left(\delta_{pq} - e_p e_q^\dagger\right) e_p = 3ie_q, \qquad (11.35)$$

we obtain

$$0 = h' + \frac{3}{2}\frac{h}{x^2 + \lambda^2}, \tag{11.36}$$

whose solution is

$$h = \frac{B}{(x^2 + \lambda^2)^{3/2}}, \tag{11.37}$$

with B an arbitrary integration constant. The normalized fermion zero mode can therefore be written as

$$\Psi = \frac{\sqrt{2}}{\pi}\frac{\lambda}{(x^2 + \lambda^2)^{3/2}}\chi, \tag{11.38}$$

where χ is a fixed isospinor four-component Dirac spinor. In the basis in which we have been working, the lower components of χ vanish and the upper ones are given by the two-component Weyl spinor $\Phi_{a\alpha} = \epsilon_{a\alpha}$.

This zero mode is for a fermion in the doublet representation of SU(2). Because the unit instanton in any larger gauge group is an embedding of the SU(2) instanton, the generalization to other groups is straightforward. In particular, the unit instanton for $SU(N_c)$ is the embedding of the SU(2) instanton in a 2×2 block. The fundamental representation zero mode is obtained by inserting the zero mode of Eq. (11.38) into the corresponding two components of the fermion field, and then setting the remaining $N_c - 2$ components equal to zero.

It is instructive to summarize the chirality-violating processes associated with an instanton by a nonlocal effective Lagrangian density. This must contain the product of $2N_f$ fermion fields, one ψ_L and one $\bar{\psi}_R$ for each flavor. It can be written in the form [238]

$$\mathcal{L}_{\text{eff}} = Ce^{-8\pi^2/g^2(\lambda)}\prod_{s=1}^{N_f}(\bar{\psi}_R^s\omega)(\bar{\omega}\psi_L^s), \tag{11.39}$$

where an integration over the positions of the fermion fields is understood, ω is a fixed Dirac spinor transforming under the fundamental representation of the gauge group, and the constant C is obtained from the one-loop corrections to the instanton. The nontrivial contribution from $\mathcal{L}_{\text{eff}}$ comes from the terms in the fermion fields corresponding to the zero mode; i.e., the product of the zero mode and the corresponding creation or annihilation operator. The term shown here is not Hermitian; one must add its Hermitian conjugate, which gives the effects of an anti-instanton. One must also, of course, integrate over all instanton positions and scales.[4]

---

[4] This effective Lagrangian contains ω, whose form depends on the gauge orientation of the instanton. However, physical results must be gauge-independent. A gauge-invariant effective Lagrangian can be obtained by integrating over the gauge orientations of ω [238, 259].

11.3 QCD and the U(1) problem

Even before QCD was discovered, it was realized that the eight light pseudoscalar mesons could be understood as approximate Goldstone bosons arising from the spontaneous breaking of an approximate SU(3)×SU(3) chiral symmetry, with the especially low masses of the pions indicating that a chiral SU(2)×SU(2) was even closer to being an exact symmetry of the Lagrangian.

The form of the QCD Lagrangian clarifies the origin of these symmetries. The part containing the quark fields q_r (with the subscript labeling quark flavor) is

$$\mathcal{L} = \sum_r \left[\bar{q}_r \left(i\gamma^\mu \partial_\mu + g\gamma^\mu A^a_\mu T^a \right) q_r + m_r \bar{q}_r q_r \right] . \tag{11.40}$$

The quark "masses" m_r are neither the positions of poles in Green's functions nor effective constituent masses in hadrons, but simply parameters in the Lagrangian. Current-algebra arguments lead to the conclusion that the up and down quark masses are only a few MeV, while the strange quark mass is roughly 100 MeV. All three are small compared to a typical QCD scale (e.g., constituent up and down quark masses of about 300 MeV) suggesting that this Lagrangian can be viewed as an approximation to one with either two or three massless quarks. The SU(2)×SU(2) and SU(3)×SU(3) chiral symmetries would then correspond to the transformations in Eq. (11.3).

However, a Lagrangian of this form is also invariant under the transformations of Eq. (11.2), making the symmetry either U(2)×U(2) or U(3)×U(3) and predicting either a fourth or a ninth approximate Goldstone boson. In the former case, with two quarks considered to be light, the only plausible candidate for the extra Goldstone boson is the η, but its mass of 548 MeV seems to be far too high compared to that of the pions. (Indeed, one can show that the mass of the fourth Goldstone boson cannot be more than $\sqrt{3}$ times that of the other three [260].) With three quarks considered to be light, the Goldstone bosons from the breaking of SU(3)×SU(3) are the three pions (135 and 140 MeV), the four kaons (494 and 498 MeV), and the η. Again, the only candidate for the ninth Goldstone boson, the η' at 958 MeV, is much too massive. The absence of a satisfactory candidate for the ninth (or fourth) Goldstone boson was known as the U(1) problem.

Note that this U(1) problem only arises after the form of the Lagrangian is determined. It is quite possible to write down theories (the sigma model is an example) that are invariant under SU(N_f)×SU(N_f) symmetry but do not have a U(N_f)×U(N_f) symmetry.

The effective Lagrangian of Eq. (11.39), generalized to an SU(3) gauge group, provides a resolution of the U(1) problem. If we define

$$\mathcal{M}_{rs} = (\bar{q}_{Rr}\omega)(\bar{\omega}q_{Ls}), \tag{11.41}$$

the anticommutivity and Grassmann nature of the fermion fields implies that

$$\prod_{s=1}^{N_f}(\bar{q}_{Rs}\omega)(\bar{\omega}q_{Ls}) = \frac{1}{(N_f)!}\det\mathcal{M}.\tag{11.42}$$

The transformations in Eqs. (11.2) and (11.3) can all be written in the form

$$q_L \rightarrow U_L q_L, \qquad q_R \rightarrow U_R q_R,\tag{11.43}$$

where U_L and U_R are $N_f \times N_f$ unitary matrices. Under such transformations

$$\mathcal{M} \rightarrow U_R^\dagger \mathcal{M} U_L.\tag{11.44}$$

The determinant of $\mathcal{M}$, and hence $\mathcal{L}_{\text{eff}}$, is unchanged if U_L and U_R are both $SU(N_f)$ matrices, with unit determinant, or if $U_L = U_R$. However, $\mathcal{L}_{\text{eff}}$ is not invariant under U(1) transformations with $U_L = U_R^\dagger$, which are precisely those transformations corresponding to the anomalous current j_5^μ. Thus, the U(1) that appeared to be a spontaneously broken symmetry in the massless quark limit is not a symmetry at all once instanton effects are included, and so there is no longer any prediction of an extra Goldstone boson.

It is important to recognize that even though $\mathcal{L}_{\text{eff}}$ arises from instanton effects, it is not a small correction to the Lagrangian. We saw in Sec. 10.12 that the integration over instanton scales diverges and the dilute-gas approximation breaks down for large instantons. Although this prevents us from obtaining reliable quantitative results for the instanton effects, we can expect them to be large and comparable to other strong interaction effects.

11.4 Baryon number violation by electroweak processes

We have seen that instanton effects in QCD can be large, but cannot be reliably calculated because of the divergence of the integration over instanton size. By contrast, in the SU(2)×U(1) electroweak theory there are instanton effects that are calculable although, at first sight, they seem to be negligibly small. These are associated with the weak isospin SU(2) factor, with the U(1) playing no role.

The essential new factor here is the presence of the Higgs doublet, which must approach its nonzero vacuum value $\langle\phi\rangle$ as $x^2 \rightarrow \infty$ but vanish at the center of the instanton. The Higgs vacuum expectation value breaks the classical scale invariance. Its effects favor smaller instanton size, so that the fields deviate from the vacuum over a smaller region, thus reducing the classical action. On the other hand, we have seen that the renormalization of the gauge coupling by the one-loop quantum effects favors large instantons. The net result is that the integration over instanton size peaks around $\lambda \sim 1/\langle\phi\rangle$.

The left-handed fermions of the standard model fall into SU(2) doublets. For each generation there are three quark doublets (because of the three colors) and one lepton doublet. The former each carry baryon number $B = 1/3$, while the latter have lepton number $L = 1$. With three generations, each instanton leads to violations of baryon and lepton numbers [261]

$$\Delta B = \Delta L = 3\,. \tag{11.45}$$

Although B and L are not separately conserved, their difference $B - L$ is conserved. This corresponds to the fact that the $B - L$ current (unlike the currents of B and L separately) does not have an anomalous divergence in the Standard Model.

This is a truly striking result. Even though all perturbative Standard Model processes conserve baryon number, nonperturbative instanton effects allow baryon number violating processes such as the annihilation of a proton and neutron to yield an antinucleon and three leptons.[5] However, the prospects for experimentally observing such a process are rather dim, since the rate is suppressed by a factor of

$$\left(e^{-8\pi^2/g^2}\right)^2 = e^{-16\pi^2 \sin^2 \theta_W / e^2}. \tag{11.46}$$

Because the size of the instantons responsible for this process is given by the electroweak scale, let us evaluate the quantities on the right-hand side of this equation at the Z mass. This gives 10^{-161}. The observable universe contains about 10^{78} protons and has an age of approximately 10^{10} years, or 10^{40} times a typical strong interaction time scale of 10^{-23} seconds. Thus, the probability that baryon number violation by such a process ever happened in our past light cone would seem to be vanishingly small.

However, matters are not quite so simple. This instanton-mediated process involves tunneling through the potential energy barrier separating two vacua with different winding number. An alternative possibility is to pass over the top of the barrier via a thermal fluctuation. Although unfeasible today, such a process might have been possible at the much higher temperatures that were present in the very early universe.

The crucial quantity here is the height of the barrier that must be traversed. On any path over the barrier there is a high point of maximum energy. If there is a lowest such maximum, it will dominate the rate for thermal fluctuations; its energy is the minimum energy needed to be able to cross the barrier without tunneling. Because this lowest maximum is a stationary point of the potential energy, it must correspond to a static solution (in $A_0 = 0$ gauge) of the field equations. Since it is a saddle point, rather than a local minimum, it is an unstable solution. This solution is known as a sphaleron [262].

---

[5] Note that the spatial location of this process gives a physical interpretation to the instanton position, which did not have a directly observable meaning in the effects considered previously.

The size of the sphaleron is set by the Higgs vacuum expectation value v. This leads to a rough estimate of its energy,

$$E_{\mathrm{sph}} \sim \frac{4\pi v}{g}, \tag{11.47}$$

where g is the SU(2) gauge coupling. [The value of the U(1) coupling g' plays a lesser role, because it is the non-Abelian part of the theory that gives rise to the effect.] Using a spherically symmetric ansatz, Manton and Klinkhamer [263] found a sphaleron solution with an energy that ranged between 7.5 TeV and 13.3 TeV, depending on the Higgs boson mass; current experimental bounds on the Higgs mass put it in the upper part of this range.

A naïve estimate of the rate for sphaleron processes would be $\Gamma \sim \exp(-E_{\mathrm{sph}}/T)$. However at finite temperature it is the free energy that is the relevant quantity, so we expect the somewhat larger rate

$$\Gamma \sim e^{-F_{\mathrm{sph}}/T}, \tag{11.48}$$

where the calculation of F_{sph} must take into account the finite temperature corrections to the effective potential and the fact that these reduce the expectation value of the Higgs field. Even at this level of approximation it is clear that at temperatures near (and above) that of the electroweak phase transition, baryon number violation via electroweak processes can proceed at significant rates. This has the potential of washing out, or at least significantly diluting, any pre-existing baryon asymmetry. For a detailed review of this and related processes, see [264].

11.5 CP violation and the $\theta F\tilde{F}$ term

The possibility of vacuum tunneling led to the realization that a term

$$\Delta\mathcal{L} = \frac{\theta g^2}{16\pi^2}\operatorname{tr} F_{\mu\nu}\tilde{F}^{\mu\nu} = \frac{\theta g^2}{32\pi^2}\epsilon^{\mu\nu\alpha\beta}\operatorname{tr} F_{\mu\nu}F_{\alpha\beta} \tag{11.49}$$

can be added to the Yang–Mills Lagrangian. Because this term is a total divergence, it has no effect classically. Even in the quantum theory, it has no effect on Feynman diagrams, as was already noted below Eq. (10.32).

A clear demonstration that terms such as this can, nevertheless, have an effect in the full quantum theory is provided by the one-particle Lagrangian of Eq. (10.41) with the total time derivative term of Eq. (10.42) included. If we drop the potential energy term, we have

$$L = \frac{1}{2}B\dot{\alpha}^2 + \frac{\theta}{2\pi}\dot{\alpha}. \tag{11.50}$$

If α is taken to be an angle with period 2π, this can be interpreted as the Lagrangian for a rigid rotor. The momentum conjugate to α, $J = B\dot{\alpha} + \theta/2\pi$, is therefore quantized and takes on integer values (in units where $\hbar = 1$). Converting from the Lagrangian to the Hamiltonian and setting $J = n$ gives the energy eigenvalues

$$E_n = \frac{1}{2B} \left(n - \frac{\theta}{2\pi} \right)^2 . \tag{11.51}$$

The shift $\theta \to \theta + 2\pi$ leaves the spectrum invariant, although with a relabeling of states, reflecting the periodicity of the θ parameter.

For $\theta = 0$ all levels but the ground state are degenerate, with $E_n = E_{-n}$. For $\theta = \pi$ all levels are paired, with $E_n = E_{1-n}$. In both cases the degeneracy is a consequence of the invariance of the Hamiltonian under time reversal. For all other values of θ this time-reversal invariance is broken and the degeneracy is absent.

Now suppose that we add a potential energy $K(1 - \cos \alpha)$, with $K \gg 1/B$; the rotor is now perhaps better viewed as a rigid pendulum. There is now an energy barrier against motions in which the pendulum makes a full rotation. For energies less than the height of this barrier the classical pendulum only oscillates back and forth, but the quantum pendulum can also tunnel through the energy barrier and make a full rotation. In the path integral, it is only the latter type of path that is sensitive to θ. Because these paths are associated with tunneling processes, the θ-dependence of the low-energy eigenvalues is exponentially suppressed, as are the T-violating energy splittings.

Let us now return to the Yang–Mills theory, and ask what observable effects the θ term might have. There could be a θ-dependence of the vacuum energy, but since θ is fixed there would be no way of observing this.[6] A more promising possibility is to look for signals of the symmetry-breaking properties of the θ term. The presence of $\epsilon^{\mu\nu\alpha\beta}$ means that this term violates both parity and time reversal; because of CPT invariance, the latter implies CP violation.

In QCD, the apparent divergence of the integral over instanton sizes means that, although these are instanton effects, there is no exponential suppression. Could a QCD θ term, then, provide an explanation for the experimentally observed CP violation that would be an alternative to that based on a phase in the CKM matrix? Just two pieces of data are sufficient to show that it cannot.

The first observations of CP violation were in kaon decays. For example, the ratio of the amplitudes for the CP-violating decay $K_L^0 \to \pi^+\pi^-$ and the CP-conserving decay $K_S^0 \to \pi^+\pi^-$ is [265]

$$|\eta_{+-}| = \left| \frac{A(K_L^0 \to \pi^+\pi^-)}{A(K_S^0 \to \pi^+\pi^-)} \right| = 2.2 \times 10^{-3} . \tag{11.52}$$

On the other hand, a neutron electric dipole moment d_n, which would be T-violating (and therefore CP-violating) has not been observed. A natural scale

---

[6] However, in axion theories the axion field effectively plays the role of a spacetime-dependent θ and has an effect on the energy density that is in principle observable.

for such a moment would be e times 10^{-13} cm, the characteristic length associated with the nucleon. Experimentally, however, [265]

$$d_n < 2.9 \times 10^{-26} \, e\text{-cm} \,, \tag{11.53}$$

thirteen orders of magnitude below the natural scale.

In the standard model, this extra 10 orders of magnitude suppression of the neutron electric dipole moment is attributable to the fact that CP violation is a weak interaction effect. There is no such suppression in the kaon decays, because the decays themselves are already weak interaction processes. On the other hand, if the θ term were the origin of the CP violation, there would be no need to invoke the weak interactions and therefore no reason to expect such a large discrepancy in the magnitudes of the two effects. Any value of θ large enough to explain CP violation in the kaon system would produce a neutron dipole moment far in excess of the experimental bounds. Hence, a θ term cannot be the explanation of the observed CP violation.

Rather than being a solution, the possibility of such a term is actually a serious problem. Because there is no weak interaction suppression, the smallness of the neutron electric dipole moment places a stringent limit on θ. An estimate using current-algebra methods gives [266]

$$d_n \approx 5 \times 10^{-16} \, \theta \, e\text{-cm} \,. \tag{11.54}$$

Comparing this with Eq. (11.53), we see that θ must be less than 10^{-10} or so. Understanding why this parameter in the Lagrangian should take on such an unnaturally small value has become known as the strong CP problem.

This problem is exacerbated by the presence of fermions in the theory. To understand this, consider the mass term for a single fermion field. This is usually written in the form

$$\mathcal{L}_m = -m\bar{\psi}\psi = -m\bar{\psi}_L\psi_R + \text{h.c.} \tag{11.55}$$

with m real and h.c. denoting the Hermitian conjugate. However, by redefining fields we can convert this to a complex mass parameter. With $\psi' = e^{i\alpha\gamma^5}\psi$, Eq. (11.55) becomes

$$\mathcal{L}_m = -me^{-2i\alpha}\bar{\psi}'_L\psi'_R + \text{h.c.}$$
$$\equiv -m'\bar{\psi}'_L\psi'_R + \text{h.c.} \tag{11.56}$$

This redefinition can be viewed as a chiral transformation of the Lagrangian. Classically, this would be a symmetry if the mass term were absent and would imply the conservation of the Noether current j_5^μ. Now recall that in the presence of a symmetry-breaking term the divergence of a Noether current is related to the infinitesimal change in the Lagrangian $\Delta\mathcal{L}$ by

$$\partial_\mu J^\mu_{\text{Noether}} = \Delta\mathcal{L} \,. \tag{11.57}$$

Taking α in Eq. (11.56) to be infinitesimal, we would then conclude that

$$\alpha\partial_\mu j_5^\mu = 2i\alpha\left(m\bar{\psi}_L\psi_R - \text{h.c.}\right) . \tag{11.58}$$

However, we know that this classical result is not the whole story, since even in the absence of a fermion mass j_5^μ has a nonzero anomalous divergence given by Eq. (11.12). This, in turn, tells us that the change in the Lagrangian is not given just by the right-hand side of Eq. (11.58), but rather by

$$\Delta\mathcal{L} = 2i\left(m\bar{\psi}_L\psi_R - \text{h.c.}\right) + \frac{g^2}{8\pi^2}\text{tr}\, F_{\mu\nu}\tilde{F}^{\mu\nu} . \tag{11.59}$$

Thus, the same transformation that rotates the phase of m also adds an additional $F\tilde{F}$ term; i.e., it shifts the value of θ. The invariant quantity is $\bar{\theta} = \theta - \arg m$. If there are several flavors of fermions the mass term becomes

$$\mathcal{L}_m = M_{rs}\bar{\psi}_{rR}\psi_{sL} + \text{h.c.} , \tag{11.60}$$

where the possibility of flavor mixing means that the mass matrix M may have nonzero off-diagonal terms. Generalizing the one-flavor calculation shows that the invariant CP-violating parameter is

$$\bar{\theta} = \theta - \arg\det M . \tag{11.61}$$

Before considering the effects of fermions it was hard to understand why θ should be so small. This becomes even harder to understand once we realize that the effective value of θ has a contribution from a fermion mass matrix that arises from the coupling to a Higgs field whose vacuum expectation value has an arbitrary phase.

One possible solution is for one of the fermions to be massless. The phase of its mass (and, more generally, of the determinant of the mass matrix) would then be ambiguous and θ could be shifted at will. Hence, all values of θ would be equivalent and there would be no θ-dependent physical quantities. It has therefore been suggested that the strong CP problem would be resolved if the up quark were massless. However, current algebra and other evidence suggests that it is not.

Perhaps a more plausible resolution of the puzzle is a Peccei–Quinn [267, 268] type mechanism, in which there is a coupling to a complex scalar field that dynamically sets $\bar{\theta}$ to zero. There is then a spontaneously broken approximate U(1) symmetry whose pseudo-Goldstone boson is the axion [269, 270].

Another consequence of a nonzero θ, first pointed out by Witten, is seen in the electric charges of magnetically charged objects [69]. Recall that the electric charge plays a dual role in theories that include charged fields. On the one hand, it is a quantity that is dynamically coupled to electric fields and that can be measured by examining the Coulomb tail of these fields. On the other hand, it is proportional to the conserved Noether charge that arises from the invariance of the theory under phase rotations of the complex charged fields.

The Noether charge is the generator of the symmetry transformation and is constructed from products of the variation of the fields and the corresponding conjugate momenta. Consider a transformation of the form

$$\delta \mathbf{A}_\mu = \frac{1}{e} \mathbf{D}_\mu (\phi/|\phi|),$$

$$\delta \phi = 0 \tag{11.62}$$

in a theory with gauge coupling e where SU(2) is broken to U(1) by a triplet field ϕ. [As in Chap. 5, boldface vector notation refers to SU(2) group indices.] The corresponding Noether charge is

$$Q_{\text{Noether}} = \int d^3 x \, \delta \mathbf{A}_j \cdot \mathbf{\Pi}^j , \tag{11.63}$$

where $\mathbf{\Pi}^j$ is the momentum conjugate to $\mathbf{A}_j$. If $\theta = 0$,

$$\mathbf{\Pi}_j = \frac{\partial \mathcal{L}}{\partial_0 \mathbf{A}_j} = \mathbf{F}^{j0} , \tag{11.64}$$

so

$$e Q_{\text{Noether}} = \int d^3 x \, \mathbf{D}_j \hat{\phi} \cdot \mathbf{F}^{j0} = \int d^3 x \left[\partial_j \left(\hat{\phi} \cdot \mathbf{F}^{j0} \right) - \hat{\phi} \cdot \mathbf{D}_j \mathbf{F}^{j0} \right] . \tag{11.65}$$

The field equation $\mathbf{D}_j \mathbf{F}^{j0} = \phi \times \mathbf{D}^0 \phi$ shows that the last term in the second integral vanishes, so

$$e Q_{\text{Noether}} = \int d^2 S_j \hat{\phi} \cdot \mathbf{F}^{j0} = Q_E , \tag{11.66}$$

where the integral is over the surface at spatial infinity. In a gauge with constant $\hat{\phi}$, Eqs. (11.65) and (11.66) reduce to the expression for the electric charge in Eq. (5.84). Because Q_{Noether} is an integer, we recover the familiar result $Q_E = ne$.

Adding the θ term of Eq. (10.32) changes the conjugate momenta, so that now

$$\mathbf{\Pi}^j = \frac{\partial \mathcal{L}}{\partial_0 \mathbf{A}_j} = \mathbf{F}^{j0} - \frac{\theta e^2}{16\pi^2} \epsilon^{jkl} \mathbf{F}_{kl} . \tag{11.67}$$

Repeating the above steps using this modified $\mathbf{\Pi}^j$ and using the Bianchi identity $\epsilon^{jkl} \mathbf{D}_j \mathbf{F}_{kl} = 0$ leads to

$$e Q_{\text{Noether}} = \int d^2 S_j \left(\hat{\phi} \cdot \mathbf{F}^{j0} - \frac{\theta e^2}{16\pi^2} \epsilon^{jkl} \hat{\phi} \cdot \mathbf{F}_{kl} \right)$$

$$= Q_E - \frac{e\theta}{2\pi} \left(\frac{e}{4\pi} Q_M \right) . \tag{11.68}$$

The Noether charge, being conjugate to a periodic variable, remains quantized in integer units. Hence, a monopole with magnetic charge $Q_M = 4\pi/e$ must actually be a dyon with electric charge

$$Q_E = ne + \frac{e\theta}{2\pi} \qquad (11.69)$$

for some integer n. The periodicity of θ can be seen here by noting that the replacement $\theta \to \theta + 2\pi$ leaves the spectrum of allowed electric charges invariant. Because all magnetically charged objects have electric charges of this form, these noninteger charges are consistent with the generalized charge quantization condition of Eq. (5.16).

Looked at from a distance, the monopole appears like a point object and the charged fields whose transformation led to the Noether charge are not evident. The anomalous θ-dependent charge can then be understood by noting that we effectively have ordinary electromagnetism supplemented by a term

$$\Delta \mathcal{L}_{\rm em} = -\frac{\theta e^2}{8\pi^2} \mathbf{E} \cdot \mathbf{B}. \qquad (11.70)$$

The Abelian Gauss's law then becomes

$$\boldsymbol{\nabla} \cdot \mathbf{E} - \frac{\theta e^2}{8\pi^2} \boldsymbol{\nabla} \cdot \mathbf{B} = \rho, \qquad (11.71)$$

where ρ represents any purely electric sources.[7] It immediately follows that any magnetic charge must be accompanied by an electric charge at the same point.

It should be stressed that this is not an instanton effect. The connection with instantons (and the reason for including it here rather than in Chap. 5) is merely the historical accident that it was the discovery of the instanton solutions that led to the consideration of the effects of a θ-term.

Finally, there is an apparent puzzle if the theory contains massless fermions. As was noted already, all values of θ would then be equivalent, which would seem to be in conflict with the presence of θ in the dyon charge quantization condition. The resolution can be understood in light of the discussion of fermions and monopoles in Sec. 5.7. We saw there that the charged fermions create a condensate cloud about the monopole core, with the radius of the cloud inversely proportional to the fermion mass. As the fermion mass goes to zero, the charge density of the fermion–dyon system moves out to spatial infinity, and so the charge as measured at any finite radius goes to zero, regardless of the value of θ [93, 118].

---

[7] The presence of the θ-term here might seem to contradict the statement that adding a total divergence to the Lagrangian density should not affect the equations of motion. The explanation lies in the singularities of the Abelian gauge potential when a monopole is present.

12

Vacuum decay

Consider a scalar field theory with a Lagrangian density of the form

$$\mathcal{L} = \frac{1}{2}(\partial_\mu \phi)^2 - V(\phi),\tag{12.1}$$

where the scalar field potential has two unequal minima, as in Fig. 12.1. The lower of the two, at $\phi = \phi_{\text{tv}}$, corresponds to the vacuum state of the theory and is called the true vacuum. The higher minimum, at $\phi = \phi_{\text{fv}}$, is called the false vacuum. The values of the potential at these minima will be denoted V_{tv} and V_{fv}, respectively. The false vacuum would be stable classically, but quantum mechanically it is only metastable and can decay via tunneling processes that we will study in this chapter.

These processes can be of importance in a cosmological context. For example, if the universe underwent any first-order phase transitions, below the critical temperature the metastable high-temperature phase would play the role of the false vacuum. If there is a string theory landscape, vacuum tunneling plays a crucial role in the evolution of the multiverse. Finally, we cannot exclude the possibility that our universe today is not in its ground state, but instead only in a very long-lived metastable state.

Although thermal and gravitational effects can often be significant, I will start with the case of vacuum decay at zero temperature in flat spacetime, and then later consider the effects of nonzero temperature and of gravity.

12.1 Bounces in a scalar field theory

Although Fig. 12.1 looks very much like Fig. 9.2, there is a crucial difference. The latter figure is a plot of potential energy, and the barrier is a finite one. The former is a plot of $V(\phi)$, which is an energy density. The actual potential energy is

$$U[\phi(\mathbf{x})] = \int d^3x \left[\frac{1}{2}(\boldsymbol{\nabla}\phi)^2 + V(\phi)\right].\tag{12.2}$$

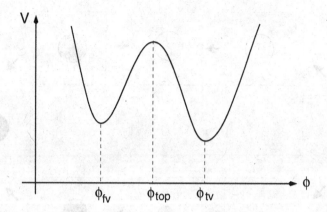

Fig. 12.1. A typical potential with a false vacuum.

To go from the false vacuum through a series of spatially homogeneous configurations would require traversing an infinite potential energy barrier. The tunneling amplitude for this vanishes. Instead, the false vacuum decays by a tunneling process that takes a spatially homogeneous state to one with a region of approximate true vacuum—a bubble—embedded in a false vacuum background. Because the bubble can be nucleated anywhere, the decay rate is proportional to the volume of space, and thus formally infinite. The finite physically measurable quantity that we need is the bubble nucleation rate per unit volume, $\Gamma/\mathcal{V}$.

One can envision many paths through the space of field configurations that connect the pure false vacuum to a configuration with a bubble. Two of these are illustrated in Fig. 12.2. Each path specifies a series of field configurations that define a slice through the potential energy barrier. A plot of $U[\phi(\mathbf{x})]$ along the path would be similar to the one-dimensional potential energy barrier shown in Fig. 9.3. The end point of the path, corresponding to the field configuration at the time that the bubble nucleates, has the same potential energy as the initial, pure false vacuum, configuration; quantum tunneling conserves energy.

As described in Chap. 9, the tunneling amplitude is dominated by the path that minimizes the barrier penetration integral B. This path can be found by finding the bounce solution to the Euclidean equation of motion [226], which in the present case is the field equation

$$0 = \frac{d^2\phi}{d\tau^2} + \boldsymbol{\nabla}^2\phi - \frac{dV}{d\phi} \tag{12.3}$$

that follows from the Euclidean action[1]

---

[1] Because almost all actions in this chapter will be Euclidean, I will generally omit an explicit subscript E on the action.

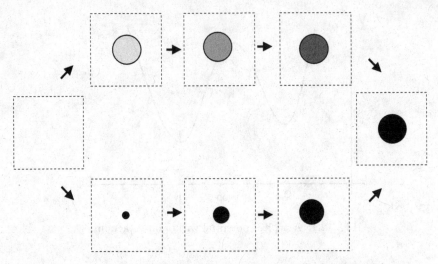

Fig. 12.2. Two of the infinite number of paths through the potential energy barrier that connect the pure false vacuum, on the left, with a configuration of the same energy, on the right, that contains a true vacuum bubble in a false vacuum background. In the upper path the bubbles are all of the same size, with the field in the interior progressing through the barrier in $V(\phi)$. In the lower path the bubble interior is always in the true vacuum, while the bubble radius increases from zero to that of the nucleated bubble.

$$S = \int d\tau \, d^3x \left[\frac{1}{2} \left(\frac{\partial \phi}{\partial \tau} \right)^2 + \frac{1}{2} (\boldsymbol{\nabla} \phi)^2 + V(\phi) \right]. \tag{12.4}$$

I will denote the bounce solution by $\phi_b(\mathbf{x}, t)$. The nucleation rate is then of the form

$$\frac{\Gamma}{\mathcal{V}} = A e^{-B} \tag{12.5}$$

where

$$B = S(\phi_b) - S(\phi_{\text{fv}}). \tag{12.6}$$

Here $S(\phi_b)$ is the Euclidean action of the bounce solution and

$$S(\phi_{\text{fv}}) = \int d\tau \, d^3x \, V_{\text{fv}} \tag{12.7}$$

that of the homogeneous false vacuum. Although both are infinite, their difference is finite.

Solving Eq. (12.3) is equivalent to finding a static solution in four spatial dimensions. It might appear that this is forbidden by Derrick's theorem. However, the proof of Derrick's theorem assumes that ϕ approaches the absolute minimum of V at spatial infinity, which is not the case for the bounce. For a theory in D spatial dimensions we can define

$$I_K = \frac{1}{2} \int d\tau \, d^D x \, (\partial_a \phi)^2 \tag{12.8}$$

and

$$I_V = \int d\tau \, d^D x \, [V(\phi) - V_{\text{fv}}] . \tag{12.9}$$

The manipulations that led to Derrick's theorem then give

$$I_V = \frac{1 - D}{1 + D} I_K . \tag{12.10}$$

This tells us that I_V must be negative if $D > 1$. In contrast with the case of the soliton, this is not a problem because $V(\phi)$ is less than V_{fv} near the center of the bounce. Similarly, we can have $I_V = 0$, as required for $D = 1$, and yet have a nontrivial scalar field configuration. It follows from this virial identity that

$$B = S(\phi_b) - S(\phi_{\text{fv}}) = I_K + I_V = \left(\frac{2}{1 + D} \right) I_K . \tag{12.11}$$

The boundary conditions for Eq. (12.3) follow from the fact that the bounce represents a path through configuration space that begins with the initial configuration (the homogeneous false vacuum) at τ_{init}, reaches a turning point on the opposite side of the potential energy barrier (a configuration with a bubble) at $\bar{\tau}$ and then returns[2] in a τ-reversed fashion to the initial configuration at τ_{fin}. This is illustrated by Fig. 12.3. Note that $\partial\phi/\partial\tau = 0$ everywhere on the hypersurfaces $\tau = \tau_{\text{init}}$ and $\tau = \bar{\tau}$ because these are turning points at which the kinetic energy vanishes.

Examination of the field equations reveals that τ_{init} and τ_{fin} must be $-\infty$ and ∞, respectively. We thus have

$$\phi(\mathbf{x}, \pm\infty) = \phi_{\text{fv}} . \tag{12.12}$$

We must also require that the configurations along the tunneling path all have finite potential energy, measured relative to the false vacuum. This gives the requirement that

$$\phi(|\mathbf{x}| = \infty, \tau) = \phi_{\text{fv}} . \tag{12.13}$$

Both the field equation and the boundary conditions display an O(4) invariance, despite the fact that τ and $\mathbf{x}$ have rather different physical meanings. This suggests setting $\bar{\tau} = 0$ and using an O(4)-invariant ansatz when seeking the bounce. Indeed, one might guess that the bounce with the lowest action (and therefore the dominant one) would have such a symmetry; this has been rigorously demonstrated, subject to mild assumptions about the form of $V(\phi)$, for the case of a theory with a single scalar field [271]. Thus, let us define $s = \sqrt{\tau^2 + \mathbf{x}^2}$

---

[2] Recall, from the discussion in Sec. 9.2, that the return part of the path just gives the factor of two needed to go from the tunneling amplitude to the tunneling probability.

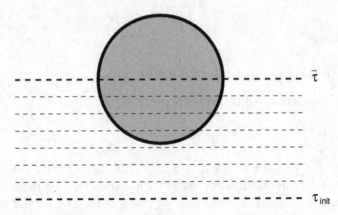

Fig. 12.3. Schematic view of a bounce solution. The field is approximately at its true vacuum value in the shaded region. The heavy circle surrounding this region is a wall region where the field crosses the barrier in $V(\phi)$. The dashed lines represent a sequence of spatial hypersurfaces of constant τ that describe a path through the configuration space. The lower heavy dashed line corresponds to the initial configuration, at τ_{init}, and the upper one, through the center of the true vacuum region, the configuration after emergence from the potential energy barrier. Although not indicated in the figure, there is an exponentially small deviation from the false vacuum that slowly increases as the configurations progress from τ_{init} to the first appearance of the wall region. The upper half of the bounce is a mirror image of the lower half.

and seek a bounce solution that depends only on s. The Euclidean field equation then becomes

$$\phi'' + \frac{3}{s}\phi' = \frac{dV}{d\phi}\,, \tag{12.14}$$

where primes denote differentiation with respect to s. The boundary conditions on the bounce reduce to

$$\phi(\infty) = \phi_{\text{fv}}\,, \tag{12.15}$$

while requiring that the solution be nonsingular at the origin gives

$$\phi'(0) = 0\,. \tag{12.16}$$

The existence of a solution to these equations can be demonstrated by an "overshoot–undershoot" argument due to Coleman [226]. The idea, illustrated in Fig. 12.4, is to reinterpret Eq. (12.14) as the equation for a particle with position ϕ moving in an "upside-down" potential energy $-V(\phi)$, with s denoting the time. The ϕ' term is interpreted as a damping force that decreases with time. The boundary conditions require that the particle start at rest at some initial point ϕ_0 on the true vacuum side of the potential well, and that it come to rest at ϕ_{fv} as $s \to \infty$. What we need to do is to show that a suitable ϕ_0 exists.

First, suppose that we take ϕ_0 to be a point on the true vacuum side such that $-V(\phi_0) \leq -V_{\text{fv}}$. The particle's energy decreases with time, because of the

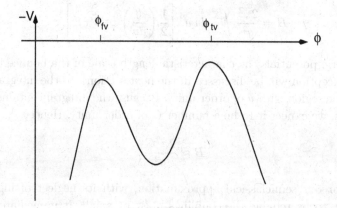

Fig. 12.4. The upside-down potential used in the overshoot–undershoot argument. The field begins, at $s = 0$, near the true vacuum peak and must finally come to rest at ϕ_{fv}.

damping term, and so the particle will never have sufficient energy to reach ϕ_{fv}; i.e., it undershoots. Now suppose instead that ϕ_0 is taken to differ only infinitesimally from ϕ_{tv}. It will then start to move very slowly, so that by the time it has moved appreciably from its initial position the damping force will have almost died away and the particle's energy will be essentially conserved. It will then have nonzero kinetic energy when it reaches ϕ_{fv} and so will continue on, never to return; it overshoots. By continuity, there must be a range of ϕ_0 near the top of the potential that lead to overshoots and another range, lower down, giving undershoots. The boundary between these two ranges gives the desired ϕ_0 that determines the bounce. (While this argument demonstrates the existence of a bounce, it does not prove the uniqueness of the bounce. It is not hard to find potentials that have several ranges of overshoot and undershoot, and thus several distinct bounce solutions.)

Assuming O(4) symmetry, the tunneling exponent from the bounce can be written as

$$B = 2\pi^2 \int_0^\infty ds\, s^3 \left[\frac{1}{2} (\phi')^2 + V(\phi) - V_{\mathrm{fv}} \right], \tag{12.17}$$

where the factor of $2\pi^2$ is from the four-dimensional angular integration. We can estimate the magnitude of B by scaling arguments. Let us assume that the theory has a single characteristic mass scale m and that the scalar field potential can be written as

$$V(\phi) = g^2 v^4 \tilde{V}(\phi/v) + V_{\mathrm{fv}}, \tag{12.18}$$

where $v = m/g$ and the dimensionless parameters in $\tilde{V}$ are all of order unity. (In particular, this implies that $|\phi_{\mathrm{tv}} - \phi_{\mathrm{fv}}|$ is of order v.) We can then define rescaled variables $u = ms$ and $f = \phi/v$ and write

$$B = \frac{2\pi^2}{g^2} \int_0^\infty du\, u^3 \left[\frac{1}{2} \left(\frac{df}{du} \right)^2 + \tilde{V}(f) \right].$$ (12.19)

For generic potentials the characteristic length scale of the bounce is of order m^{-1} (an exception will be discussed in the next section), so the integral is dominated by the region with u of order unity. Because the integral contains no small parameters, we expect it to be a number C of order unity. Hence,

$$B \approx \frac{2\pi^2 C}{g^2}.$$ (12.20)

In order for our semiclassical approximation, with its neglect of higher-order corrections, to be reliable the couplings must be small. It immediately follows that $e^{-B} \ll 1$, which guarantees the validity of the dilute-gas approximation that was used in the derivations of Chap. 9.

As a concrete example, let us consider tunneling with a quartic field potential $V(\phi)$ that has two unequal minima. If an overall additive constant is chosen so that V vanishes at the false vacuum, then the potential depends on four constants. Two of these can be fixed by rescaling and shifting the field so that the false and true vacua are at $\phi = -a$ and $\phi = a$, respectively. An overall dimensionless constant can be factored out, as described above. The actual shape of the potential is determined by the remaining constant. If we take this to be a quantity b lying in the range $0 < b < 1$, and write

$$V(\phi) = \frac{g^2}{4} \left[(\phi^2 - a^2)^2 + \frac{4b}{3}(a\phi^3 - 3a^3\phi - 2a^4) \right],$$ (12.21)

then the top of the barrier separating the false and true vacua is at $\phi = -ba$. The shape of this potential for several values of b is shown in Fig. 12.5.

Numerically solving the bounce equation for several values of b gives the solutions shown in Fig. 12.6. Notice that for $b \gtrsim 0.6$ the field at the center of the bounce, $s = 0$, is visibly different from its true vacuum value. The actions for these bounces are shown in Fig. 12.7, where $g^2 B$ is plotted as a function of b. It can be seen that the quantity C in Eq. (12.20), although formally of order unity, can be quite large.

An expression for the pre-exponential factor in the decay rate for a single degree of freedom was obtained in Sec. 9.4. This is easily generalized to the field theory case [230]. Because the bounce solution is not spatially homogeneous, there are three zero modes, corresponding to spatial translation of the bounce, in addition to the zero mode corresponding to translation in Euclidean time.[3]

---

[3] Further zero modes can appear if the theory has an internal symmetry that is broken by the bounce [272]. However, as discussed in Sec. 9.5, there must be only a single negative mode because a solution with multiple negative modes gives only a saddle point, rather than a minimum, of the tunneling exponent B.

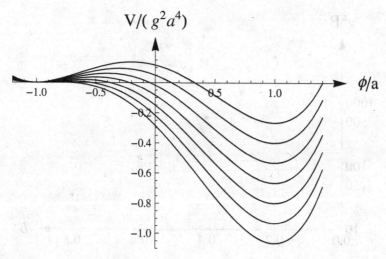

Fig. 12.5. The potential of Eq. (12.21) for various values of b. From top to bottom, the curves correspond to $b = 0.2, 0.3, 0.4, 0.5, 0.6, 0.7, 0.8$.

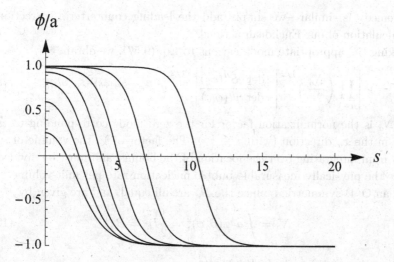

Fig. 12.6. The bounce solutions for the potential of Eq. (12.21). Reading from left to right, these correspond to $b = 0.8, 0.7, 0.6, 0.5, 0.4, 0.3, 0.2$.

These new zero modes require the introduction of three collective coordinates, which can be chosen to be the location of the center of the bounce.

A second factor to be taken into account is that in a field theory the determinant factors have divergences that remain even after their ratio is taken. These are completely analogous to the divergence that we encountered in the course of calculating the first quantum corrections to the energy of a soliton in Sec. 2.2.

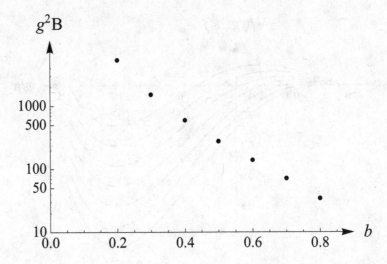

Fig. 12.7. The quantity $g^2 B$ as a function of b for the potential of Eq. (12.21).

The remedy is similar—we simply add the leading counterterm corrections to the calculation of our Euclidean actions.

Making the appropriate modifications to Eq. (9.57), we obtain

$$\Gamma = \mathcal{V} \left[\prod_{a=1}^{4} \left(\frac{N_a}{2\pi} \right)^{1/2} \right] \left| \frac{\det' S''(\phi_b)}{\det S''(\phi_{\text{fv}})} \right|^{-1/2} e^{-B} e^{-[S_{\text{ct}}(\phi_b) - S_{\text{ct}}(\phi_{\text{fv}})]}. \qquad (12.22)$$

Here N_a is the normalization factor for the zero mode corresponding to translation in the x_a direction (with $\tau \equiv x_4$). The factor of $\mathcal{V}$, the volume of space, comes from integrating over the location of the bounce and must be divided out to give the physically measurable bubble nucleation rate per unit volume.[4]

For an O(4)-symmetric bounce the N_a are all equal, and are given by

$$N_a = \int d^4 x (\partial_a \phi)^2 = \frac{1}{2} I_K = B, \qquad (12.23)$$

so

$$\frac{\Gamma}{\mathcal{V}} = \left(\frac{B}{2\pi} \right)^2 \left| \frac{\det' S''(\phi_b)}{\det S''(\phi_{\text{fv}})} \right|^{-1/2} e^{-[S_{\text{ct}}(\phi_b) - S_{\text{ct}}(\phi_{\text{fv}})]} e^{-B}. \qquad (12.24)$$

Evaluation of the determinants in the pre-exponential factor is a nontrivial matter. A rough estimate of their ratio can be obtained by dimensional arguments. Because of the omission of the zero modes from the determinant about

---

[4] The generalization of this to a theory with many fields is fairly straightforward. However, extra care is required in situations where the vacuum structure is not evident at tree level, but only after one-loop corrections to the effective potential are taken into account [125]. In such cases one must be careful to avoid double-counting of one-loop terms in both the effective action and in the determinant factors. For a discussion of this, see Ref. [273].

the bounce, the ratio of determinant factors has the dimensions of $(\text{mass})^4$, just as needed to give the correct dimensions for $\Gamma/\mathcal{V}$. For a generic potential the only relevant mass is m. A rescaling such as that described above Eq. (12.19) shows that the determinants give no additional powers of g. Hence, suppressing numerical factors in the prefactor, we have

$$\frac{\Gamma}{\mathcal{V}} \approx \left(\frac{m}{g}\right)^4 e^{-B} \approx v^4 e^{-B}, \tag{12.25}$$

with the factor of $1/g^4$ coming from the B^2 in the prefactor.

12.2 The thin-wall approximation

An interesting and instructive case to consider is the one where

$$\epsilon \equiv V_{\text{fv}} - V_{\text{tv}} \tag{12.26}$$

is small compared to the height of the barrier in $V(\phi)$ [226]. If we analyze this in terms of the analogous problem of a particle moving in the upside-down potential, we see that the particle, starting near ϕ_{tv}, can only lose a very small amount of energy to damping forces if it is to avoid an undershoot. This can only happen if the particle stays very close to ϕ_{tv} until some large time $s \approx R$ when the damping force has become negligible. It can then move quickly down the valley and back up the other side to come to rest at ϕ_{fv}.

A potential of this form can be constructed by starting with a potential $V_1(\phi)$ that has two degenerate minima, at $\phi = \phi_{\text{fv}}$ and $\phi = \phi_{\text{tv}}$, and then defining

$$V(\phi) = V_1(\phi) + \epsilon \left(\frac{\phi_{\text{tv}} - \phi}{\phi_{\text{tv}} - \phi_{\text{fv}}}\right). \tag{12.27}$$

(Note that the actual minima of V deviate from ϕ_{fv} and ϕ_{tv} by amounts of order ϵ, which I will ignore.) To a first approximation, near $s = R$ we can neglect both the $3\phi'/s$ term in the field equation and the ϵ-dependent term in $V(\phi)$. We then have

$$\phi'' = \frac{dV_1}{d\phi}. \tag{12.28}$$

This is just the equation for a one-dimensional soliton. Applying the methods of Sec. 2.1 [see Eq. (2.21)] gives a solution $\phi_1(x)$ defined by

$$x = \int_{\phi_{\text{top}}}^{\phi_1} \frac{d\phi}{\sqrt{2[V_1(\phi) - V_1(\phi_{\text{fv}})]}}. \tag{12.29}$$

Away from $x = 0$, ϕ_1 approaches the vacuum values exponentially fast. Ignoring terms of order ϵ, we find that the soliton's one-dimensional energy is

$$\sigma = \int dx \left\{ \frac{1}{2}(\phi_1')^2 + [V(\phi) - V_{\mathrm{fv}}] \right\}$$

$$= \int_{\phi_{\mathrm{fv}}}^{\phi_{\mathrm{tv}}} d\phi \sqrt{2[V(\phi) - V_{\mathrm{fv}}]}. \tag{12.30}$$

Our approximate solution for the bounce is then

$$\phi(s) = \begin{cases} \phi_{\mathrm{tv}}, & s \ll R, \\ \phi_1(s - R), & s \approx R, \\ \phi_{\mathrm{fv}}, & s \gg R, \end{cases} \tag{12.31}$$

if $\phi_{\mathrm{tv}} > \phi_{\mathrm{fv}}$; if instead $\phi_{\mathrm{tv}} < \phi_{\mathrm{fv}}$, then the field in the wall region is $\phi_1(R - s)$. This is essentially a four-dimensional bubble of true vacuum surrounded by a thin wall. Its Euclidean action, relative to that of the false vacuum, is

$$S[R] = 2\pi^2 R^3 \sigma - \frac{1}{2}\pi^2 R^4 \epsilon. \tag{12.32}$$

The first term is the contribution from the wall and shows that σ can be viewed as a surface tension, while the second is from the true vacuum interior.

The bounce radius R is determined by requiring that this be a stationary point of the Euclidean action, which in turn requires that

$$0 = \frac{dS}{dR} = 6\pi^2 R^2 \sigma - 2\pi^2 R^3 \epsilon. \tag{12.33}$$

This is solved by

$$R_b = \frac{3\sigma}{\epsilon}, \tag{12.34}$$

which gives

$$B = S = \frac{27\pi^2 \sigma^4}{2\epsilon^3}. \tag{12.35}$$

Differentiating Eq. (12.33) yields

$$\left. \frac{d^2S}{dR^2} \right|_{R_b} = -\frac{18\pi^2 \sigma^2}{\epsilon} < 0. \tag{12.36}$$

This shows that R_b maximizes the action among the family of configurations given by Eq. (12.31), and that the expected negative mode about the bounce corresponds to variation of R. More explicitly, differentiating Eq. (12.14) with respect to s gives

$$\phi''' + \frac{3}{s}\phi'' - \frac{3}{s^2}\phi' = \frac{d^2V}{d\phi^2}\phi'. \tag{12.37}$$

or, equivalently,

$$S''(\phi_b)\phi' = \left[-\frac{d^2}{ds^2} - \frac{3}{s}\frac{d}{ds} + \frac{d^2V}{d\phi^2} \right]\phi'$$

$$= -\frac{3}{s^2}\phi'. \tag{12.38}$$

In the thin-wall approximation, where ϕ' is non-negligible only for $s \approx R$, we can replace s by R in the last line. Hence, in this approximation ϕ' is an eigenfunction of S'' with a negative eigenvalue $-3/R^2$.

Despite the attractive simplicity of the formulas obtained by the thin-wall approximation, one must keep in mind that they are not generally applicable. In the regime of nearly degenerate minima where the approximation is valid the bounce action is generally quite large, as can be seen, for example, from the data shown in Fig. 12.7.

12.3 Evolution of the bubble after nucleation

The bounce plays a dual role. Not only does its action determine the bubble nucleation rate, but a spatial slice through its center gives the most probable emergence point from the potential energy barrier; i.e., the most likely initial configuration for the classical evolution of the bubble after nucleation. In fact, the bounce solution also gives us much of the future development of ϕ directly, without the need for further calculation.

Let us denote the Euclidean space bounce solution by $\phi_E(\mathbf{x}, \tau)$ and the Minkowski spacetime field by $\phi_M(\mathbf{x}, t)$. These satisfy

$$0 = \frac{d^2\phi_E}{d\tau^2} + \nabla^2\phi_E - \frac{dV}{d\phi} \tag{12.39}$$

and

$$0 = -\frac{d^2\phi_M}{dt^2} + \nabla^2\phi_M - \frac{dV}{d\phi}, \tag{12.40}$$

as well as the initial value conditions

$$0 = \left.\frac{d\phi_E(\mathbf{x}, \tau)}{d\tau}\right|_{\tau=0},$$
$$0 = \left.\frac{d\phi_M(\mathbf{x}, \tau)}{dt}\right|_{t=0}. \tag{12.41}$$

Examining these, we see that starting with a Euclidean solution and making the replacement $\tau \to it$ gives a solution of the Minkowskian equations. If the bounce has the O(4)-symmetric form $\phi_E = f(\sqrt{\mathbf{x}^2 + \tau^2})$, then the Minkowskian solution is given by the O(3,1) symmetric $\phi_M = f(\sqrt{\mathbf{x}^2 - t^2})$ with the same function f. In the region outside the light cone of the origin, $r = \sqrt{\mathbf{x}^2} > t$, this gives ϕ_M directly. The region $r < t$ inside the light cone has no Euclidean counterpart. In principle, ϕ_M could be obtained here by analytic continuation, but in practice one writes $\phi_M = \phi_M(u)$, with $u = \sqrt{t^2 - r^2}$, and solves

$$\phi_M''(u) + \frac{3}{u}\phi_M' = -\frac{dV}{d\phi}. \tag{12.42}$$

In the thin-wall approximation the bubble at nucleation has a thin wall of radius $R_0 = 3\sigma/\epsilon$ that separates an exactly true vacuum interior from an exactly false vacuum exterior.[5] Because the Minkowskian solution depends only on $r^2 - t^2$, the bubble wall at future times must be at

$$R(t) = \sqrt{R_0^2 + t^2},$$ (12.43)

which means that it is expanding at a velocity

$$v = \frac{dR}{dt} = \frac{t}{R} = \frac{\sqrt{R^2 - R_0^2}}{R}$$ (12.44)

that approaches the speed of light.

The energy density in this expanding bubble wall can be obtained by Lorentz-boosting the surface tension of the wall at rest, giving

$$E_{\text{wall}} = 4\pi R^2 \frac{\sigma}{\sqrt{1 - v^2}} = 4\pi\sigma \frac{R^3}{R_0} = \frac{4\pi}{3} R^3 \epsilon.$$ (12.45)

This is precisely equal to the latent heat released by the conversion of false vacuum to true. All of the released energy has gone into the kinetic energy of the walls, with none of it exciting the field in the bubble interior.

This is somewhat surprising, and certainly contrasts with our experience with nonrelativistic bubbles expanding in a medium. One might wonder whether this result is an artifact of the thin-wall approximation. To see that it is not, let us consider the solutions of Eq. (12.42). At $u = 0$ we have $\phi_M(0) = \phi_E(0)$, the value at the center of the bounce. As u increases, ϕ_M approaches ϕ_{tv} and then oscillates about the true vacuum with an amplitude that falls as $u^{3/2}$. Let us define u_i as the value where ϕ_M approximates the true vacuum to some predetermined accuracy, thus defining $r_i = \sqrt{t^2 - u_i^2}$ to be the inner surface of the bubble wall. Similarly, we can define a value s_o where the field approximates the false vacuum, so that $r_o = \sqrt{s_o^2 + t^2}$ is the location of the outer surface of the bubble wall. The thickness of the bubble wall at large t is then

$$r_o - r_i \approx \frac{s_o^2 + u_i^2}{2t}.$$ (12.46)

Thus, even if the bubble does not initially have a thin wall, it develops one as time goes on. Although the interior is not precisely a true vacuum, as it is in the thin-wall approximation, the energy density from oscillations about the true vacuum decreases, eventually falling as t^{-3}. Since the walls are expanding outward at almost the speed of light, the total energy in the interior of the bubble remains essentially constant.

From an exterior point of view, it seems natural to use the coordinates r and t to describe the bubble, and to view it as a finite, but expanding, region

---

[5] The energy in the walls of this bubble is exactly canceled by the negative volume energy from the replacement of false vacuum by true in the bubble interior. Thus, energy is conserved, as it should be.

of approximate true vacuum. However, this is not the most natural description from the point of view of an interior observer. Inside the light cone of the origin we can define

$$\tilde{t} = \sqrt{t^2 - r^2},$$

$$\tilde{r} = \frac{r}{\sqrt{t^2 - r^2}}. \qquad (12.47)$$

Because ϕ_M depends only on $\tilde{t}$, these coordinates give a description in which the bubble interior (or at least the part within the light cone of the origin) is a time-dependent but spatially homogeneous region of infinite spatial extent. There is no privileged point that can be uniquely defined to be the "center" of the bubble. In terms of the new coordinates, the Minkowskian metric takes the form

$$ds^2 = d\tilde{t}^2 - \tilde{t}^2 \left[\frac{d\tilde{r}^2}{1 + \tilde{r}^2} + \tilde{r}^2 (d\theta^2 + \sin^2 \theta \, d\varphi^2) \right]$$
$$\equiv d\tilde{t}^2 - \tilde{t}^2 \, d\Omega_T^2. \qquad (12.48)$$

This describes an open Friedmann–Robertson–Walker universe with infinite negatively curved spacelike surfaces of constant time. These surfaces are spacelike hyperboloids (i.e., hyperboloids with timelike normals) and $d\Omega_T^2$ is the metric on the unit spacelike hyperboloid.

   This picture of an almost empty bubble interior is only valid to the extent that collisions with other bubbles can be ignored. One might expect that the continued nucleation and growth of bubbles would eventually cause them to coalesce, so that the bubble walls would disappear and the latent heat would be released. This is indeed what happens in cosmological first-order phase transitions that proceed rapidly with little supercooling. However, as will be discussed in Sec. 12.7, bubble nucleation in a quasi-de Sitter background can lead to a very different type of scenario [126].

12.4 Tunneling at finite temperature

Let us now consider the effects of a nonzero temperature $T = 1/\beta$. Specifically, let us assume that our system is initially in a quasi-equilibrium, with a thermal distribution of the states built upon the false vacuum, but with states built upon the true vacuum being unpopulated.

   The picture of bubble nucleation outlined in the previous sections must be modified in several ways [274, 275]. First, it is clear that in finding the bounce we must use the finite temperature effective potential $V_{\text{eff}}(\phi, T)$, which was discussed in Sec. 7.2, instead of the zero-temperature scalar field potential. Otherwise, we would find a nonzero rate for nucleating bubbles of the low-temperature phase even above the critical temperature of a first-order transition. To simplify the

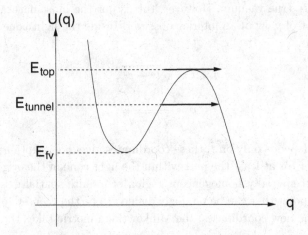

Fig. 12.8. The two modes for transitions at finite temperature. One is a fluctuation of the system to E_{top}, allowing it to pass over the potential energy barrier, the other thermally assisted tunneling at an energy E_{tunnel}.

notation, I will continue to simply write $V(\phi)$, but temperature dependence should be understood.

Second, there are new mechanisms available for the transition. The first of these might be termed thermally assisted quantum tunneling. Instead of tunneling directly from the false vacuum at the bottom of the potential energy well, the system can tunnel from a thermally excited higher energy state for which the barrier penetration integral is smaller. This is illustrated in Fig. 12.8 for a system with only a single degree of freedom. Taking a thermal average of the tunneling rate for this system and keeping track only of exponential factors gives

$$\Gamma_{\text{tunn}} \sim \int_{E_{\text{fv}}}^{E_{\text{top}}} dE \, e^{-\beta(E - E_{\text{fv}})} e^{-B(E)}$$

$$\sim e^{-\beta(E_* - E_{\text{fv}})} e^{-B(E_*)}, \tag{12.49}$$

where E_* is the value for which $\beta(E - E_{\text{fv}}) + B(E)$ is a minimum.

With many degrees of freedom and a Lagrangian with a quadratic kinetic energy, as in Eq. (9.7), we must minimize

$$\beta(E - E_{\text{fv}}) + 2 \int_{s_0}^{s_f} ds \, \sqrt{2[V(q(s)) - E]} \tag{12.50}$$

with respect to path, end points, and energy. We know that the minimization with respect to the first two of these is achieved by finding a solution of the Euclidean equations of motion. Minimization with respect to energy requires that

$$\beta = -2 \frac{d}{dE} \int_{s_0}^{s_f} ds \, \sqrt{2[V(q(s)) - E]}. \tag{12.51}$$

The integral has implicit dependences on E through the path and the end points. However, since we have already minimized with respect to path and end points, these implicit dependences do not contribute here, so we have

$$
\begin{aligned}
\beta &= 2 \int_{s_0}^{s_f} ds \, \frac{1}{\sqrt{2[V(q(s)) - E]}} \\
&= 2 \int_{s_0}^{s_f} ds \left[\sum_j \left(\frac{dq_j}{d\tau} \right)^2 \right]^{-1/2} \\
&= 2 \int_{s_0}^{s_f} ds \, \frac{d\tau}{ds} \\
&= 2(\tau_2 - \tau_1) \, .
\end{aligned}
\tag{12.52}
$$

Thus, the passage through the barrier must take a Euclidean time $\tau_2 - \tau_1 = \beta/2$; adding the τ-reversed passage doubles the time and tells us that we must look for a bounce solution that is periodic[6] in τ with period β. Repeating the manipulations of the zero-temperature case then gives

$$
\Gamma_{\text{tunn}} \sim e^{-[S(\phi_b) - S(\phi_{\text{fv}})]} \, ,
\tag{12.53}
$$

where in both actions the τ integration is understood to be over a single period β.

The second new possibility is for the system to be thermally excited all the way to the top of the barrier, with no quantum tunneling required at all, as also illustrated in Fig. 12.8. Again ignoring pre-exponential factors, we have

$$
\Gamma_{\text{therm}} \sim e^{-\beta(E_{\text{top}} - E_{\text{fv}})} \, .
\tag{12.54}
$$

When there is more than one degree of freedom, what appears as the top of the barrier in Fig. 12.8 is actually a saddle point, with the potential energy increasing in all of the remaining directions; i.e., it is the highest point on the lowest pass over the potential energy barrier. Because it is a stationary point of the potential energy, it is a static solution of the field equations.

In the field theory context this solution is a three-dimensional critical bubble, one where the outward pressure from the approximately true vacuum interior is precisely balanced by the surface tension; the density of such configurations in the thermal ensemble determines the bubble nucleation rate. In the thin-wall approximation the energy, relative to the false vacuum background, of a bubble of radius R is

$$
E = 4\pi R^2 \sigma - \frac{4\pi}{3} R^3 \epsilon \, .
\tag{12.55}
$$

---

[6] The fact that the bounce interpolates between a pair of classical turning points on opposite sides of the potential energy barrier means that, in addition to being periodic, the bounce must have two hypersurfaces of constant τ, separated by half a period, on which $\partial \phi / \partial \tau$ is identically zero.

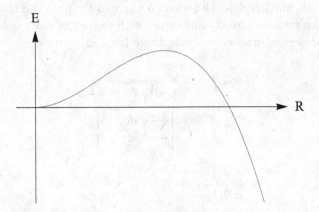

Fig. 12.9. Plot of energy as a function of bubble radius for thin-wall bubbles. The high-temperature critical bubble corresponds to the maximum of E, while the bubble created by quantum tunneling at $T = 0$ corresponds to the value of R where the curve crosses the horizontal axis.

As shown in Fig. 12.9, the critical bubble corresponds to the maximum of this function, while the bubble nucleated by purely quantum tunneling is one with $E = 0$.

We can view the three-dimensional saddle point configuration as a four-dimensional τ-independent bounce solution that extends for a period β. Equation (12.53) then applies equally well here, with the distinction between thermally assisted tunneling and a purely thermal transition being whether or not the bounce has a τ-dependence.

The problem can also be attacked by path integral methods. The partition function can be written as an integral over paths with periodic boundary conditions, which in the field theory case takes the form

$$Z = e^{-\beta F} = \sum_j e^{-\beta E_j} = \int [d\phi] e^{-S_E(\phi)} . \tag{12.56}$$

Because we are interested in a system that is a thermal distribution of false-vacuum-type states, the path integral is restricted to configurations on the false vacuum side of the potential barrier. The individual energies E_j and the free energy F then develop imaginary parts. Assuming these to be small, we have

$$\Gamma = -2 \operatorname{Im} F = -2Z^{-1} \sum_j e^{-\beta E_j} \operatorname{Im} E_j . \tag{12.57}$$

Thus, the imaginary part of the free energy gives us a thermally weighted average of the decay rate.

As before, the path integral can be evaluated as a sum of contributions from the stationary points and approximate stationary points of the action. When the

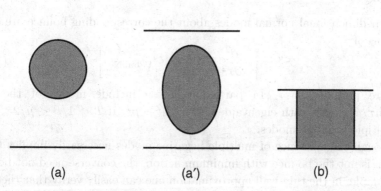

Fig. 12.10. Schematic illustration of the periodic bounces for finite temperature transitions. In all three cases the solid horizontal lines indicate the hypersurfaces $\tau = 0$ and $\tau = 1/\beta$, the bounds of one period. Bounces (a) and (a') describe thermally assisted tunneling and illustrate the deformation of the shaded true vacuum region with increasing temperature. Bounce (b) is τ-independent and describes thermal fluctuation over the potential energy barrier. Such bounces dominate at high temperature. At lower temperatures they remain solutions of the Euclidean field equation, but are unacceptable because they have multiple negative modes.

dilute-gas approximation is valid, these exponentiate and the imaginary part of the free energy can be read off from the coefficient of β in the exponent, leading to an expression for the pre-exponential factor analogous to that obtained in the zero-temperature case.

We have already seen that two distinct types of bounce can occur. At very low temperature (i.e., large β), the bounce corresponding to thermally assisted tunneling closely approximates the O(4)-symmetric zero-temperature bounce. As the temperature is increased this bounce retains an O(3) symmetry, but (except in the thin-wall approximation) is deformed in the τ-direction. Indeed, by considering such a bounce and its periodic recurrences, we see that the true vacuum region of the bounce is elongated, as shown in Fig. 12.10, eventually "colliding" with itself when the inverse temperature is comparable to the bounce radius.

The τ-independent bounce corresponding to a purely thermal transition is a solution of the Euclidean field equation at all temperatures.[7] However, at low temperatures it is not only subdominant, in that its action is greater than that of the bounce for thermally assisted tunneling, but it is also unacceptable because it has more than one negative mode. This can be seen by noting that the three-dimensional normal modes about the critical bubble solution $\phi_{\text{bubble}}(\mathbf{x})$ include a negative mode $\eta_-(\mathbf{x})$ for which

---

[7] There is no similar guarantee for the τ-dependent bounce. The absence of such a solution would indicate that the expression in Eq. (12.50) had no minimum and was instead a monotonically decreasing function of E in the range $E_{\text{fv}} < E < E_{\text{top}}$.

$$\left[-\boldsymbol{\nabla}^2 + V''(\phi_{\text{bubble}})\right]\eta_- = -\mu^2\eta_- . \tag{12.58}$$

The four-dimensional normal modes about the corresponding bounce are eigenfunctions of

$$S'' = \left[-\frac{\partial^2}{\partial\tau^2} - \boldsymbol{\nabla}^2 + V''(\phi_{\text{bubble}})\right] \tag{12.59}$$

that are periodic in τ. In particular, these include modes of the form $\cos(2\pi n\tau/\beta)\eta_-(\mathbf{x})$, with eigenvalues $(2\pi n/\beta)^2 - \mu^2$. If $T = 1/\beta < \mu/2\pi$, there are multiple negative modes.

Although the existence of multiple negative modes necessarily implies that a solution is not the bounce with minimum action, the converse need not be true. For example, in the thin-wall approximation one can easily verify that there is a temperature range in which the τ-independent bounce has only a single negative mode but nevertheless has a higher action than the τ-dependent bounce.

As a final point, note that the expansion of the bubble after nucleation is more complicated at nonzero temperature. In particular, its expansion within the thermal medium never approaches the speed of light.

12.5 Including gravity: bounce solutions

The effects of gravity on bubble nucleation have been neglected so far. While this is justified in a wide range of applications, there are situations where gravitational effects are clearly significant. The most obvious is when a transition involves mass scales close to the Planck scale, but gravitational effects can also come into play at lower mass scales if the bubble nucleates with a size large enough to be sensitive to the curvature of spacetime. Even when gravitational effects are unimportant for bubble nucleation itself they can be quite significant for the evolution of the bubble after nucleation.

With gravity ignored, the WKB and path integral approaches led us in a straightforward manner from first principles to expressions for the nucleation rate. The extension of these arguments to include gravity is a nontrivial matter, in large part because of the difficulties associated with globally defining energy in a curved spacetime, and to date no complete derivation of an expression for the nucleation rate has been produced. Nevertheless, one can argue from analogy, as was first done by Coleman and De Luccia [276]. As we will see, this leads to an expression that agrees with the flat spacetime result[8] in the weak gravity limit. However, some issues of interpretation arise when gravitational effects are large. I will show in Sec. 12.6 how these can be clarified by considering a limiting case in which the nucleation rate can be derived [277].

---

[8] Here, and in similar contexts, "flat spacetime" refers to the case where gravitational effects are ignored and the spacetime is treated as a fixed flat spacetime in both the true and false vacua.

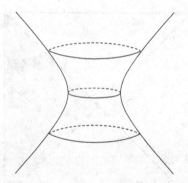

Fig. 12.11. De Sitter spacetime viewed as a hyperboloid. Time runs vertically. Horizontal spacelike slices through the hyperboloid are three-spheres.

It may be best to begin with a brief description of the initial false vacuum state, which requires some comment when gravity is included. The scalar field is of course spatially uniform, equal to ϕ_{fv} everywhere. The nature of the spacetime depends on the value of the potential in the false vacuum. If $V_{\mathrm{fv}} = 0$, the false vacuum spacetime is simply flat Minkowski space, and requires little further discussion. If it is nonzero, then V_{fv} is equivalent to a nonzero cosmological constant, and the false vacuum is either a de Sitter or an anti-de Sitter spacetime, according to whether V_{fv} is positive or negative, respectively.

As illustrated in Fig. 12.11, de Sitter spacetime can be viewed as the hyperboloid[9]

$$-v^2 + x^2 + y^2 + z^2 + w^2 = \Lambda^2 \tag{12.60}$$

embedded in a five-dimensional spacetime with metric

$$ds^2 = dv^2 - dx^2 - dy^2 - dz^2 - dw^2 \,. \tag{12.61}$$

Here Λ is related to the vacuum energy density V by

$$\Lambda = \sqrt{\frac{3}{\kappa V}} \,, \tag{12.62}$$

where

$$\kappa = 8\pi G = \frac{8\pi}{M_{\mathrm{Pl}}^2} \,. \tag{12.63}$$

There are several common choices of coordinates for this spacetime. The entire spacetime can be covered by a single coordinate patch in which the hypersurfaces of constant time are three-spheres and the metric takes the form

$$ds^2 = dt^2 - \cosh^2(t/\Lambda)\, d\Omega_3^2 \,, \tag{12.64}$$

---

[9] It may appear that the "equator" of the hyperboloid is somehow distinguished. This is an artifact of the visualization. In fact, by a symmetry transformation one can bring any point in the spacetime to the equator.

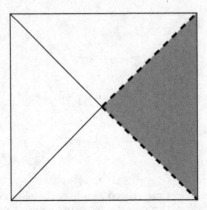

Fig. 12.12. The Penrose diagram of de Sitter spacetime. The shaded triangular region on the right represents the static patch. The vertical boundary on the right corresponds to $\tilde{r} = 0$. The dashed diagonal lines correspond to $\tilde{r} = \Lambda$, with the upper and lower ones being at $\tilde{t} = \infty$ and $\tilde{t} = -\infty$, respectively.

where $d\Omega_3$ is the line element on the unit three-sphere. This describes a closed Robertson–Walker spacetime[10] that expands exponentially fast at large time, with Hubble parameter $H = \Lambda^{-1}$. A consequence of this exponential expansion is the existence of an event horizon, beyond which an observer cannot see. This is nicely captured by a second set of coordinates for which the metric takes the static form

$$ds^2 = \left(1 - \frac{\tilde{r}^2}{\Lambda^2}\right) d\tilde{t}^2 - \left(1 - \frac{\tilde{r}^2}{\Lambda^2}\right)^{-1} d\tilde{r}^2 - \tilde{r}^2 d\Omega_2^2 . \qquad (12.65)$$

The "static patch" covered by these coordinates is bounded by the horizon at $\tilde{r} = \Lambda$ and includes half of the three-sphere equator of the de Sitter hyperboloid. It corresponds to the causal diamond in the Penrose diagram of Fig. 12.12.

The existence of a black hole horizon gives the black hole thermal properties characterized by the Hawking temperature. In a similar fashion, the presence of the de Sitter horizon leads to the Gibbons–Hawking temperature [278]

$$T_{\mathrm{dS}} = \frac{H}{2\pi} = \frac{1}{2\pi\Lambda} . \qquad (12.66)$$

This temperature will play an important role in our later discussion.

With a negative vacuum energy density V we obtain anti-de Sitter space. A convenient way of expressing its metric is

$$ds^2 = \left(1 + \frac{r^2}{L^2}\right) dt^2 - \left(1 + \frac{r^2}{L^2}\right)^{-1} dr^2 - r^2 d\Omega_2^2 , \qquad (12.67)$$

---

[10] One can also introduce foliations that give flat or open Robertson–Walker coordinates, but these do not cover the entire spacetime.

where

$$L = \sqrt{-\frac{3}{\kappa V}} \, . \tag{12.68}$$

There is no horizon, and therefore no temperature. Note, however, the behavior of the volume of a sphere as its radius increases. On a hypersurface of constant time, the two-sphere $r = R$ has area $4\pi R^2$, and encloses a volume

$$\mathcal{V}(R) = 4\pi \int_0^R dr \, \frac{r^2}{\sqrt{1 + (r^2/L^2)}} \, . \tag{12.69}$$

For $R \gg L$, this volume only grows as R^2, rather than as R^3, so the ratio of volume to area is bounded and never greater than $L/2$.

Let us turn now to our tunneling problem. Coleman and De Luccia argued that with gravity included the nucleation rate should still be based on a bounce solution, but with the Euclidean action now including a gravitational contribution. In other words,

$$\frac{\Gamma}{\mathcal{V}} = Ae^{-B} = Ae^{-[S(\phi_b) - S(\phi_{\text{fv}})]} \, , \tag{12.70}$$

where ϕ_b is a bounce solution of the Euler–Lagrange equations following from a Euclidean action

$$S = \int d^4x \sqrt{g} \left[-\frac{1}{2\kappa} R + \frac{1}{2} g^{ab} \, \partial_a \phi \, \partial_b \phi + V(\phi) \right] + S_{\text{bdy}} \tag{12.71}$$

that includes a Euclideanized Einstein–Hilbert term with the metric g_{ab} understood to have a positive definite Euclidean signature. The last term on the right-hand side, S_{bdy}, is the Euclidean version of the Gibbons–Hawking boundary action [279] that must be added to remove the second derivative terms from the action.

We know that in flat spacetime, at least for a theory with a single scalar field, the bounce of minimum action has an O(4) symmetry. Although the corresponding result in curved spacetime has not been proven, it is widely believed to be true. Let us assume that this is the case. Any O(4)-symmetric Euclidean metric can be written in the form

$$ds^2 = d\xi^2 + \rho(\xi)^2 d\Omega_3^2 \, . \tag{12.72}$$

The curvature scalar is then

$$R = -\frac{6}{\rho^2} \left(\rho \rho'' + \rho'^2 - 1 \right) , \tag{12.73}$$

with primes denoting differentiation with respect to ξ, and the action is

$$S = 2\pi^2 \int_{\xi_{min}}^{\xi_{max}} d\xi \left\{ \rho^3 \left[\frac{1}{2}\phi'^2 + V(\phi) \right] + \frac{3}{\kappa} \left(\rho^2 \rho'' + \rho\rho'^2 - \rho \right) \right\} + S_{bdy}$$

$$= 2\pi^2 \int_{\xi_{min}}^{\xi_{max}} d\xi \left\{ \rho^3 \left[\frac{1}{2}\phi'^2 + V(\phi) \right] - \frac{3}{\kappa}(\rho\rho'^2 + \rho) \right\}$$

$$+ \frac{6\pi^2}{\kappa} \left(\rho^2 \rho' \right) \Big|_{\xi=\xi_{min}}^{\xi=\xi_{max}} + S_{bdy} .$$

(12.74)

The boundary term from the integration by parts is exactly canceled by the Gibbons–Hawking term, and so we have

$$S = 2\pi^2 \int_{\xi_{min}}^{\xi_{max}} d\xi \left\{ \rho^3 \left[\frac{1}{2}\phi'^2 + V(\phi) \right] - \frac{3}{\kappa}(\rho\rho'^2 + \rho) \right\} .$$

(12.75)

The scalar field equation is

$$\phi'' + \frac{3\rho'}{\rho}\phi' = \frac{dV}{d\phi} ,$$

(12.76)

while the $\xi\xi$-component of the Euclidean Einstein equation gives

$$\rho'^2 = 1 + \frac{\kappa}{3}\rho^2 \left(\frac{1}{2}\phi'^2 - V \right) .$$

(12.77)

Another useful equation,

$$\rho'' = -\frac{\kappa}{3} \left(\frac{1}{2}\phi'^2 + V \right) \rho ,$$

(12.78)

is obtained by differentiating Eq. (12.77) and then using Eq. (12.76). By combining Eqs. (12.75) and (12.77), we can express the action for a solution of the field equations as

$$S = 4\pi^2 \int_{\xi_{min}}^{\xi_{max}} d\xi \left[\rho^3 V - \frac{3}{\kappa}\rho \right] .$$

(12.79)

The boundary conditions for the field equations depend on the topology of the solution which, as is often the case in general relativity, cannot be specified in advance. Subtracting Eqs. (12.77) and (12.78) shows that $\rho'^2 - \rho''\rho \geq 1$, from which we can conclude that ρ cannot have a nonzero minimum and must have at least one zero. The invariance of the metric under translation of ξ allows us to put one such zero at $\xi = \xi_{min} = 0$. If this is the only zero of ρ, then ξ ranges from 0 to ∞ and the solution has the topology of R^4. The boundary conditions for Eq. (12.76) are then that $\phi'(0) = 0$, to avoid a singularity at the origin, and that $\phi(\infty) = \phi_{fv}$, just as in flat spacetime. However, it is possible that ρ has a second zero at some finite ξ_{max}. In fact, this is always the case if $V(\phi) \geq 0$ everywhere and is not identically zero [126]. The solution then has the topology of a four-sphere, and the boundary conditions for Eq. (12.76) are that $\phi'(0)$ and $\phi'(\xi_{max})$ both vanish. In this case there is no requirement that there be any point where $\phi = \phi_{fv}$; in fact, although in the bounce solution $\phi - \phi_{fv}$ can become exponentially small, there is no point on the four-sphere where it precisely vanishes.

With these preliminaries completed, let us study the Euclidean solutions, beginning with the case of a de Sitter false vacuum. We need both the bounce solution and the Euclidean version of the false vacuum. The latter is easily obtained. With ϕ constant and V_{fv} positive, Eq. (12.77) gives

$$\rho = \Lambda_{fv} \sin(\xi/\Lambda_{fv}) \tag{12.80}$$

and $\xi_{max} = \pi \Lambda_{fv}$. With this form for ρ, Eq. (12.72) gives the standard round metric on a four-sphere of radius Λ_{fv}, which is the Euclidean counterpart of the Lorentzian hyperboloid. Using Eq. (12.79), we find that its action is

$$S(\phi_{fv}) = -\frac{24\pi^2}{\kappa^2 V_{fv}} = -\frac{8\pi^2}{\kappa} \Lambda_{fv}^2 . \tag{12.81}$$

Let us turn now to the bounce itself, beginning with a limit where gravitational effects might be expected to be small. Suppose that $V(\phi)$ is such that (1) the variation in V in the range between ϕ_{fv} and ϕ_{tv} is much smaller than M_{Pl}^4 and (2) the radius of the flat-spacetime bounce (i.e., of the region where ϕ is appreciably different from ϕ_{fv}) is much less than Λ_{fv}. Using Eq. (12.77), the first condition tells us that $\rho' \approx 1$, and hence $\rho \approx \xi$, in the region $0 \le \rho \ll \Lambda_{fv}$. Equation (12.76) then reduces to the flat spacetime Eq. (12.14), so that in this region the bounce is essentially the same as in flat spacetime. Our second condition then tells us that ϕ becomes exponentially close to ϕ_{fv} (although never quite equal to it) before leaving this region, so that outside this region ρ is well approximated by Eq. (12.80), except for a small shift $\xi \to \xi - \Delta$. Here Δ is of the order of $\mathcal{M}/M_{Pl}^2$, where $\mathcal{M}$ is a mass scale characterizing the size and shape of the barrier in $V(\phi)$ that separates the two vacua. The Euclidean action is finite and negative, but subtracting the false vacuum action gives a positive value for B that differs from the flat spacetime result only by a fractional amount of order $\mathcal{M}^2/M_{Pl}^2$.

We can also consider the thin-wall limit in which the bounce solution consists of a region of pure true vacuum, a region of pure false vacuum, and a relatively thin wall separating the two. The action of such a thin-wall configuration can be written as the sum of three terms, one from the wall and one each from the true vacuum and false vacuum regions.

If the $3\rho'\phi'/\rho$ term in Eq. (12.76) can be neglected, the matter field contribution to the wall action density σ is given by the flat-space expression, Eq. (12.30). The contribution from the second half of the integrand in Eq. (12.75) vanishes in this limit, so if $\rho = \bar{\rho}$ at the wall, the wall action is

$$S_{wall} = 2\pi^2 \bar{\rho}^3 \sigma , \tag{12.82}$$

In calculating the contribution from the vacuum regions, we need to keep in mind that the region bounded by a wall with curvature radius ρ can be either more or less than half of a four-sphere. Using Eq. (12.79), we obtain

$$S_{\text{vol}} = -\frac{4\pi^2}{\kappa}\Lambda^2\left[1 \mp \left(1 - \frac{\bar{\rho}^2}{\Lambda^2}\right)^{3/2}\right], \tag{12.83}$$

where the upper (lower) sign applies if the region is less (more) than a hemisphere.

It might seem that there are four cases to consider, with the true vacuum and false vacuum regions each being either more or less than a hemisphere. However, the two cases in which the true vacuum is more than a hemisphere are ruled out because ρ'' would have to be positive at the wall, in contradiction with Eq. (12.78). Thus, we only need to consider two cases: Case A, with more than a hemisphere of false vacuum, and case B, with less than a hemisphere of false vacuum. The weak gravity limit is a limiting case of case A.

After subtracting the action of the pure false vacuum we obtain

$$B = 2\pi^2\bar{\rho}^3\sigma + \frac{4\pi^2}{\kappa}\Lambda_{\text{fv}}^2\left[1 \mp \left(1 - \frac{\bar{\rho}^2}{\Lambda_{\text{fv}}^2}\right)^{3/2}\right] - \frac{4\pi^2}{\kappa}\Lambda_{\text{tv}}^2\left[1 - \left(1 - \frac{\bar{\rho}^2}{\Lambda_{\text{tv}}^2}\right)^{3/2}\right], \tag{12.84}$$

where the upper and lower signs correspond to case A and case B, respectively. The bounce corresponds to a stationary point of this function with respect to $\bar{\rho}$. It turns out that for any value of the parameters such a stationary point occurs for only one of the two cases, with the bounce being of type A or of type B according to whether

$$\frac{\epsilon}{3\sigma} - \frac{\kappa\sigma}{4}$$

is positive or negative, respectively. In either case, the stationary point occurs when

$$\frac{1}{\bar{\rho}^2} = \frac{1}{\Lambda_{\text{fv}}^2} + \left(\frac{\epsilon}{3\sigma} - \frac{\kappa\sigma}{4}\right)^2$$

$$= \frac{1}{\Lambda_{\text{tv}}^2} + \left(\frac{\epsilon}{3\sigma} + \frac{\kappa\sigma}{4}\right)^2. \tag{12.85}$$

Substituting this value for $\bar{\rho}$ into Eq. (12.81) gives [280]

$$B = \bar{\rho}^3\left[2\pi^2\sigma - \frac{4\pi^2}{\kappa}\Lambda_{\text{fv}}^2\left(\frac{\epsilon}{3\sigma} - \frac{\kappa\sigma}{4}\right)^3 + \frac{4\pi^2}{\kappa}\Lambda_{\text{tv}}^2\left(\frac{\epsilon}{3\sigma} + \frac{\kappa\sigma}{4}\right)^3\right]$$

$$+ \frac{4\pi^2}{\kappa}\left(\Lambda_{\text{fv}}^2 - \Lambda_{\text{tv}}^2\right). \tag{12.86}$$

The signs in this expression are the same for both the type A and the type B solutions, because the explicit sign change in Eq. (12.84) is compensated by an

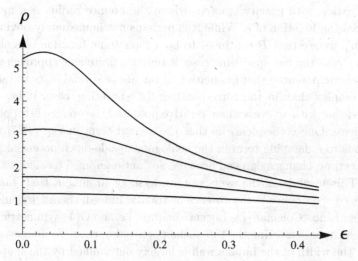

Fig. 12.13. The behavior of $\bar{\rho}$ as a function of ϵ in the thin-wall regime for three values of $\Lambda_{\rm fv}$. In units with $\kappa = 1$, these have $\sigma = 0.2$ and $V_{\rm fv} = 0.1$, 0.3, 1.0, and 2.0 (reading from the top). Note that at large ϵ the true vacuum is anti-de Sitter.

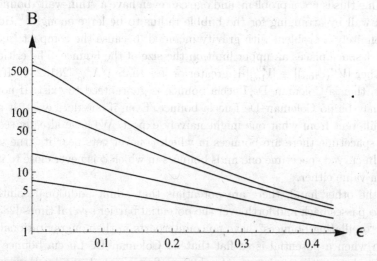

Fig. 12.14. The behavior of B as a function of ϵ in the thin-wall regime for three values of $\Lambda_{\rm fv}$. In units with $\kappa = 1$, these have $\sigma = 0.2$ and $V_{\rm fv} = 0.1$, 0.3, 1.0 and 2.0 (reading from the top). At large ϵ the true vacuum is anti-de Sitter.

implicit sign change coming from the fact that the second term in the square brackets has a different sign in the two cases.

The behaviors of $\bar{\rho}$ and B as functions of ϵ are illustrated in Figs. 12.13 and 12.14. Note that $\bar{\rho}$ reaches its maximum value, $\Lambda_{\rm fv}$, at a nonzero value of ϵ and then decreases as ϵ is reduced below this critical value. This should be compared to the situation with gravity ignored, where the bounce radius is a monotonically decreasing function of ϵ. While it is perhaps not immediately obvious from Eq. (12.86), we see that B continues to be a monotonic function of ϵ although, in contrast with the flat spacetime case, it remains finite as ϵ approaches zero.

For a generic potential that fits neither of the above special cases the situation is more complex than in the corresponding flat spacetime case. It has already been noted that with an everywhere positive potential the bounce is topologically a four-sphere. One consequence is that the second term in Eq. (12.76), which corresponds to a damping term in the analogous upside-down potential problem in flat spacetime, changes sign and becomes an "antidamping" term for part of the interval. This invalidates the overshoot–undershoot argument that guaranteed the existence of a bounce in the absence of gravity. Indeed, there are situations in which there is no "Coleman–De Luccia" bounce; i.e., no O(4)-symmetric bounce with nonconstant ϕ crossing the top of the potential barrier. To understand this, note that the width of the bubble wall is largely determined by the shape of the barrier in $V(\phi)$, with a natural scale for this width being set by the curvature of V at $\phi = \phi_{\rm top}$, the top of the barrier. A very flat potential barrier corresponds to a low mass scale, or a large distance scale, and thus to a thick wall. In flat spacetime this is not a problem, and one can even have a "thin-wall" bounce with a thick wall by arranging for the bubble radius to be large enough.[11] However, it is definitely a problem with gravity included, because the compact Euclidean de Sitter space places an upper limit on the size of the bounce. The critical test is whether $|V''(\phi_{\rm top})| \equiv |V''_{\rm top}|$ is greater or less than $4/\Lambda^2_{\rm top}$ [281–284]. If $V''_{\rm top}$ is greater, then a Coleman–De Luccia bounce is guaranteed to exist. If not, then there may be no Coleman–De Luccia bounce. Even if one does exist, it may be quite different from what one might naïvely expect. We have already seen that in flat spacetime there are bounces in which ϕ never gets near its true vacuum value. In curved spacetime one finds bounces in which ϕ is never close to its false vacuum value either.

On the other hand, there are potentials that admit additional solutions in which ϕ passes back and forth over the potential barrier several times [283, 285]. These "oscillating bounces" have no counterparts in the nongravitational case.

Even when a potential is so flat that no Coleman–De Luccia bounce exists, there is always a second type of solution, first pointed out by Hawking and Moss [286]. This is a homogeneous solution with $\phi = \phi_{\rm top}$ everywhere and

$$\rho = \Lambda_{\rm top} \sin(\xi/\Lambda_{\rm top}) \,. \tag{12.87}$$

Its Euclidean action is

$$S_{\rm HM} = -\frac{24\pi^2}{\kappa^2 V_{\rm top}} \,, \tag{12.88}$$

---

[11] Indeed, one might well have termed it the "large-bubble approximation".

leading to a tunneling exponent

$$
\begin{aligned}
B_{HM} &= S_{HM} - S(\phi_{fv}) \\
&= \frac{24\pi^2}{\kappa^2}\left(\frac{1}{V_{fv}} - \frac{1}{V_{top}}\right).
\end{aligned}
\tag{12.89}
$$

The Hawking–Moss solution is much more akin to the high-temperature thermal production of a critical bubble than to a quantum tunneling bounce, a point that will elaborated on more fully in Sec. 12.6. Such a process becomes physically plausible when we recall that de Sitter space has a nonzero temperature T_{dS}. In fact, the thermal nature of de Sitter space makes another sort of tunneling process possible.

The discussion of the Coleman–De Luccia bounces so far has been phrased in terms of the production of a true vacuum bubble, centered at the point $\xi = 0$, in a false vacuum background represented by the region centered at $\xi = \xi_{max}$. However, one side of a sphere is not intrinsically different from the other. We could just as well interchange the roles of $\xi = \xi_{max}$ and $\xi = 0$ and view the bounce as describing the production of a false vacuum bubble in a true vacuum background [287]. In the absence of gravity such a process, even if it could occur, would have negligible impact because the volume pressure and the surface tension would both act to cause the false vacuum bubble to collapse. In de Sitter spacetime, on the other hand, the cosmic expansion can protect such a bubble from collapse if it nucleates with a sufficiently large radius.

For the nucleation of a true vacuum bubble, the exponent B was the difference between the bounce action and the Euclidean action of the initial false vacuum. We should therefore expect B for the nucleation of a false vacuum bubble to be given by the difference between the bounce action and the Euclidean action of the true vacuum. The ratio of these rates would then be [287]

$$
\begin{aligned}
\frac{\Gamma_{fv\to tv}}{\Gamma_{tv\to fv}} &\approx \frac{e^{-[S(\phi_b)-S(\phi_{fv})]}}{e^{-[S(\phi_b)-S(\phi_{tv})]}} \\
&= e^{S(\phi_{fv})-S(\phi_{tv})}.
\end{aligned}
\tag{12.90}
$$

Let us suppose that the true and false vacua are almost degenerate, with $V_{fv} - V_{tv} \ll V_{fv}$. To leading approximation, the two vacua can be treated as having a common horizon length Λ and temperature $T_{dS} = 1/(2\pi\Lambda)$, so that the exponent in the last line above becomes

$$
\begin{aligned}
S(\phi_{fv}) - S(\phi_{tv}) &= \frac{24\pi^2}{\kappa^2}\left(\frac{1}{V_{tv}} - \frac{1}{V_{fv}}\right) \\
&\approx -\frac{\frac{4\pi}{3}\Lambda^3(V_{fv} - V_{tv})}{T_{dS}} \\
&\approx \frac{\Delta U_{hor}}{T_{dS}},
\end{aligned}
\tag{12.91}
$$

where ΔU_{hor} is the potential energy difference between a horizon volume of false vacuum and one of true vacuum.[12] A false vacuum bubble will expand after nucleation, raising the possibility that true vacuum bubbles might nucleate and expand within it, and then false vacuum bubbles within them, and so forth. One might expect to eventually reach an equilibrium in which the relative volumes of true and false vacua were related by a Boltzmann factor, with

$$\frac{\mathcal{V}_{\mathrm{fv}}}{\mathcal{V}_{\mathrm{tv}}} \approx \frac{\Gamma_{\mathrm{fv}\to\mathrm{tv}}}{\Gamma_{\mathrm{tv}\to\mathrm{fv}}} \approx e^{-\Delta U_{\mathrm{hor}}/T_{\mathrm{dS}}} . \tag{12.92}$$

In actuality, matters are more complicated. One has to specify the spacelike hypersurface on which $\mathcal{V}_{\mathrm{fv}}$ and $\mathcal{V}_{\mathrm{tv}}$ are being calculated. With a uniform de Sitter metric one would naturally take this to be a hypersurface of constant t. However, with differing vacuum energies the global spacetime is not purely de Sitter and the choice of hypersurface is hardly self-evident. Finding the correct way of evaluating this and similar ratios has become known as the measure problem, and is a major issue in theories that predict eternal inflation.

The discussion thus far has focused on the case of an initial de Sitter vacuum and implicitly assumed that the final vacuum was also de Sitter. The situation is rather similar for decay from de Sitter to anti-de Sitter or de Sitter to Minkowski. In both cases the bounce retains its four-sphere topology. If the thin-wall approximation is appropriate, the true vacuum region is described by

$$\rho = \begin{cases} \xi, & \text{Minkowski}, \\[2mm] L_{\mathrm{tv}} \sinh(\xi/L_{\mathrm{tv}}), & \text{anti-de Sitter}, \end{cases} \tag{12.93}$$

with the anti-de Sitter length scale L defined as in Eq. (12.68). Equations (12.84), (12.85), and (12.86) are still valid, provided that one makes the substitutions $\Lambda_{\mathrm{tv}}^2 \to -L_{\mathrm{tv}}^2$ or $\Lambda_{\mathrm{tv}} \to \infty$ for anti-de Sitter or Minkowski true vacua, respectively. In the case of a Minkowski true vacuum these results can also be written as [276]

$$\bar{\rho} = \frac{\bar{\rho}_0}{1 + (\bar{\rho}_0/2\Lambda_{\mathrm{fv}})^2} \tag{12.94}$$

and

$$B = \frac{B_0}{\left[1 + (\bar{\rho}_0/2\Lambda_{\mathrm{fv}})^2\right]^2} , \tag{12.95}$$

where $\bar{\rho}_0$ and B_0 are the flat spacetime expressions given in Eqs (12.34) and (12.35).

---

[12] There are subtleties in going from the second line to the third in this equation. Because the space is curved, the horizon volume is not given simply by the flat space formula $(4\pi/3)\Lambda^3$. However, when redshift factors are properly taken into account, the curvature effects cancel out in the calculation of the horizon energy difference. This will become clearer from the discussion in the next section.

Because the bounces for these transitions, like those that describe de Sitter to de Sitter transitions, are topologically four-spheres, one could imagine interchanging the initial and final states and viewing these as describing transitions upward from an initial Minkowski or anti-de Sitter vacuum to a de Sitter one. However, to obtain the tunneling exponent B we must subtract the action for the initial state from the action of the bounce. For both the Minkowski and anti-de Sitter cases the vacuum action is equal to $-\infty$. Subtracting this from a finite bounce action gives $B = \infty$, and thus a vanishing bubble nucleation rate.[13] This result is of course not surprising, since there is no Gibbons–Hawking temperature in a spacetime with negative or vanishing cosmological constant.

We can also consider the nucleation of bubbles of a lower vacuum from an initial anti-de Sitter or Minkowski vacuum. Both the Euclideanized initial vacuum and the bounce solution itself are infinite in extent, with the topology of R^4 and with the scalar field equal to its false vacuum value at Euclidean infinity.

In the thin-wall limit the equations from the de Sitter case can again be used, with the replacements $\Lambda^2 \to -L^2$ or $\Lambda \to \infty$ as appropriate. With the upper sign in Eqs.(12.83) and (12.84) both the bounce and the initial false vacuum are infinite in extent. B is the (finite) difference between the two formally infinite actions. For the particular case of anti-de Sitter bubbles nucleating in Minkowski space, the analogues of Eqs. (12.94) and (12.95) are

$$\bar{\rho} = \frac{\bar{\rho}_0}{1 - (\bar{\rho}_0/2L_{\text{tv}})^2} \tag{12.96}$$

and

$$B = \frac{B_0}{[1 - (\bar{\rho}_0/2L_{\text{tv}})^2]^2}. \tag{12.97}$$

The analogues of the type B bounces that correspond to the lower signs in Eqs. (12.83) and (12.84) have finite false vacuum regions and thus finite actions. Subtracting the negatively infinite initial state actions from these gives $B = \infty$. Hence, they do not lead to bubble nucleation. Thus, in the thin-wall approximation there is no tunneling if [276]

$$\epsilon \leq \frac{3}{4}\kappa\sigma^2. \tag{12.98}$$

This might seem surprising, since one might expect to always be able to tunnel from a vacuum of higher energy to one of lower energy. However, it must be remembered that the bounce does not describe tunneling from vacuum to vacuum, but rather from a vacuum to a state with a bubble surrounded by vacuum. There is a conserved energy if the spacetime is asymptotically flat, so a

---

[13] Note that this result depends crucially on the proper treatment of the boundary terms. If we had omitted the Gibbons–Hawking term in Eq. (12.71), we would have obtained either zero (in the Minkowski case) or $+\infty$ (in the anti-de Sitter case) for the initial state action.

bubble that nucleates in Minkowski spacetime must have zero total energy; i.e., it must be large enough that the negative vacuum energy in the interior compensates for the positive energy in the bubble wall. With gravity neglected one can always make the bubble large enough that this is the case. When gravitational effects are included this is not necessarily so, because of the peculiar feature of anti-de Sitter spacetime that the volume enclosed by a three-sphere only grows as R^2 at large radius. If the false vacuum is anti-de Sitter rather than Minkowski, the Poincaré symmetry generator corresponding to the total energy must be replaced by a generator of the O(3,2) anti-de Sitter symmetry, but a similar result is obtained [288].

12.6 Interpretation of the bounce solutions

As derived for flat spacetime, the bounce solution plays a dual role. It not only determines the exponent in the bubble nucleation rate, but it also describes the optimal tunneling path through configuration space and gives the most likely configuration of the bubble at nucleation. We saw in Sec. 12.1 that a sequence of spacelike slices through the bounce gives a family of configurations along this optimal path.

A similar understanding of the bounce seems natural for the bounces with R^4 topology that govern decay from a Minkowski or anti-de Sitter false vacuum. However, the situation is not so straightforward for the bounces associated with bubble nucleation in a de Sitter vacuum. Because these bounces are finite in extent, there is no hypersurface at $\tau = -\infty$ to be identified with the initial false vacuum configuration. Indeed, ϕ never quite reaches its false vacuum value anywhere on the bounce. For "small bounces", such as that illustrated in Fig. 12.15, these may seem minor points. The slice through the lower part of the four-sphere in that figure gives a close approximation to the initial vacuum. The equatorial three-sphere contains a nucleated bubble very similar to that obtained without gravity. When mapped onto the equator of the de Sitter hyperboloid, this seems to be a plausible initial configuration for the Lorentzian evolution after nucleation. However, the identification of the initial configuration is more problematic for the large-bounce solution shown in Fig. 12.16. The difficulty is even greater for the Hawking–Moss solution. Not only is there no reference to the initial state, but identification of the equatorial slice with the configuration after the transition would imply an implausible process in which the field over all of de Sitter space fluctuated to the top of the potential barrier.

The difficulties here can be attributed in part to the fact that the formalism we have been using was obtained by arguing from analogy, rather than by an explicit derivation of the answer to a precisely posed question. Some insight can be obtained by considering a limiting case in which the Coleman–De Luccia prescription for the bounce follows straightforwardly from the results that we have already obtained [277]. To this end, consider the limit where the variation

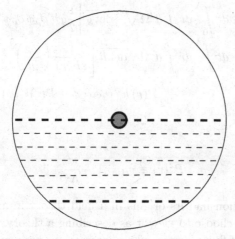

Fig. 12.15. A plausible (but incorrect) curved spaced analogue of Fig. 12.3 for the case when the radius of the shaded true vacuum region is much less than the de Sitter horizon length.

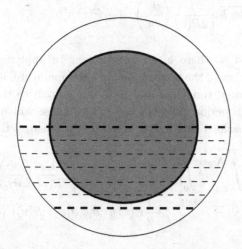

Fig. 12.16. The extension of Fig. 12.15 to the case when the radius of the true vacuum region is comparable to the de Sitter horizon length.

of $V(\phi)$ in the interval between ϕ_{fv} and ϕ_{tv} is small compared to its actual value. To a first approximation, the tunneling can then be viewed as taking place within a fixed de Sitter background. Working in this fixed background approximation, let us focus on the fields within the static patch, the region covered by the metric of Eq. (12.65). In terms of the static coordinates (with the tildes omitted) the matter part of the real-time Lorentzian action on this patch is

$$S_L = \int_{-\infty}^{\infty} dt \int_0^{\Lambda} dr \int d^2\Omega \, \sqrt{-\det g} \left[\frac{1}{2} g^{\mu\nu} \partial_\mu \phi \, \partial_\nu \phi - V(\phi) \right]$$

$$= \int_{-\infty}^{\infty} dt \int_0^{\Lambda} dr \int d^2\Omega \, \sqrt{\det h} \left[\frac{1}{2P(r)} \left(\frac{\partial \phi}{\partial t} \right)^2 \right.$$

$$\left. - \frac{1}{2} P(r) \, h^{ij} \partial_i \phi \, \partial_j \phi - P(r) V(\phi) \right], \qquad (12.99)$$

where $h_{ij} = g_{ij}$ and

$$P^2(r) \equiv g_{tt} = 1 - \frac{r^2}{\Lambda^2}. \qquad (12.100)$$

Although this action had its origins in a field theory in a curved spacetime, we can equally well choose to view it as describing a theory, on a curved three-dimensional space of finite extent, whose interactions happen to have a peculiar position dependence that is described by the various factors of $P(r)$. Because the background spacetime is static, there is a well-defined energy functional

$$E = \int d^3x \, \sqrt{\det h} \left[\frac{1}{2P(r)} \left(\frac{\partial \phi}{\partial t} \right)^2 + \frac{1}{2} P(r) \, h^{ij} \partial_i \phi \, \partial_j \phi + P(r) V(\phi) \right], \qquad (12.101)$$

for this theory, and so there is no impediment to applying the methods of Secs. 12.1–12.4. In view of the nonzero de Sitter temperature, it seems reasonable to view this as thermal tunneling and to look for bounces that are periodic in Euclidean time with period $\beta = 1/T_{\mathrm{dS}} = 2\pi\Lambda$. These will be solutions of the Euler–Lagrange equations that follow from the Euclidean action

$$S_E = \int_{-\pi\Lambda}^{\pi\Lambda} d\tau \int_0^{\Lambda} dr \int d^2\Omega \, \sqrt{\det h} \left[\frac{1}{2P(r)} \left(\frac{\partial \phi}{\partial \tau} \right)^2 \right.$$

$$\left. + \frac{1}{2} P(r) \, h^{ij} \partial_i \phi \, \partial_j \phi + P(r) V(\phi) \right]. \qquad (12.102)$$

As illustrated in Fig. 12.17, these bounces can be understood as a sequences of field configurations, each of which is on a three-dimensional ball bounded by the horizon two-sphere $r = \Lambda$. The configurations are labeled by a Euclidean time τ that runs from $-\pi\Lambda$ to $\pi\Lambda$, with the hypersurfaces $\tau = -\pi\Lambda$ and $\tau = \pi\Lambda$ identified.

As in flat spacetime, there are two modes of finite temperature decay. For the case of thermally assisted tunneling, which is shown in Fig. 12.17, we can take the hypersurface at $\tau = -\pi\Lambda$ to give the approximately false vacuum initial configuration, and the hypersurface $\tau = 0$ to give the configuration immediately after tunneling. This bounce appears to bear little resemblance to that of Coleman and De Luccia. However, we can define a Euclidean metric

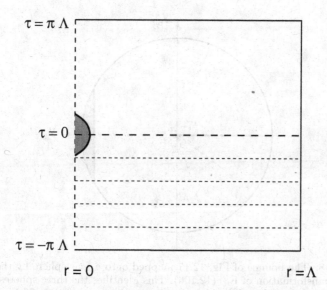

Fig. 12.17. The Coleman–De Luccia bounce, as it appears in static de Sitter coordinates. The horizontal lines represent hypersurfaces of constant τ. These are three-dimensional balls with their centers at $r = 0$ (the dotted line on the left) and their outer edges at $r = \Lambda$ (the solid line on the right). The shaded region is the true vacuum part of the bounce.

$$\tilde{g}_{ab}\, dx^a dx^b = P^2 d\tau^2 + h_{ij} dx^i dx^j$$
$$= \left(1 - \frac{r^2}{\Lambda^2}\right) d\tau^2 + \left(1 - \frac{r^2}{\Lambda^2}\right)^{-1} dr^2 + r^2 (d\theta^2 + \sin^2\theta d\phi^2). \tag{12.103}$$

With r ranging from 0 to Λ and $\tau = -\pi\Lambda$ and $\tau = \pi\Lambda$ identified, this is just the round metric on a four-sphere. The Euclidean action of Eq. (12.102) can now be rewritten as

$$S_E = \int d^4x \sqrt{\det \tilde{g}} \left[\frac{1}{2}\tilde{g}_{ab}\, \partial_a\phi\, \partial_b\phi + V(\phi)\right] \tag{12.104}$$

and the action for the decay rate is

$$\frac{\Gamma}{V} \sim e^{-[S_E(\phi_b) - S_E(\phi_{\rm fv})]}. \tag{12.105}$$

The last two equations differ from the corresponding ones in the Coleman–De Luccia formulation in that the Einstein–Hilbert curvature term has been omitted from the Euclidean action. However, in the fixed background approximation that we are using this term is the same for both the bounce and the pure false vacuum, and so cancels between the two terms in the exponent.

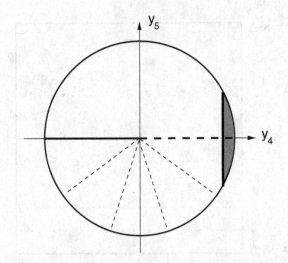

Fig. 12.18. The bounce of Fig. 12.17 mapped onto a four-sphere by the coordinate transformation of Eq. (12.106). This identifies the three-spheres $r = \Lambda$ for all values of τ. The solid and dashed lines correspond to those in Fig. 12.17.

The interpretation of the metric in terms of a four-sphere can be seen more explicitly by defining

$$y_1 = r \sin\theta \cos\phi$$
$$y_2 = r \sin\theta \sin\phi$$
$$y_3 = r \cos\theta$$
$$y_4 = \sqrt{\Lambda^2 - r^2} \, \cos(\tau/\Lambda)$$
$$y_5 = \sqrt{\Lambda^2 - r^2} \, \sin(\tau/\Lambda) \qquad (12.106)$$

so that the Euclidean space of the bounce is just the four-sphere $\sum y_a^2 = \Lambda^2$ in a five-dimensional Euclidean space. In terms of these variables the constant τ slices of Fig. 12.17 become the "radial" slices shown in Fig. 12.18. Each of these slices represents a three-dimensional ball, with the center of the ball, $r = 0$, lying on the outer edge of the diagram (where $y_4^2 + y_5^2 = \Lambda^2$), and the outer boundary of the ball, the horizon two-sphere at $r = \Lambda$, being mapped to the two-sphere[14] defined by $y_4 = y_5 = 0$. Contrary to what one might have guessed—and to what has often been assumed in the literature—the configuration after tunneling does not correspond to the full equatorial slice of the bounce. Instead, half of this slice gives the configuration before tunneling and the other half the configuration after

---

[14] Notice that this mapping takes the horizon two-spheres at all values of τ to a single two-sphere. Although this could have led to a singularity in either the metric or the scalar field as viewed on the four-sphere, this turns out to not be the case for the O(4)-symmetric bounce.

tunneling. In neither case are we given a full spatial hypersurface, but instead only a constant time slice through the static patch that is bounded by the horizon. The nature of the spacetime beyond the horizon never enters the problem, except insofar as it provides the thermal background represented by T_{dS}.

The second mode of decay, fluctuation to a saddle point over the barrier, is represented by a τ-independent bounce. One such example is the Hawking–Moss solution in which the field throughout the entire horizon volume—not the entire universe—is thermally excited to the top of the potential energy barrier.[15] This manifestly τ-independent bounce is completely analogous to the high-temperature τ-independent bounce in flat spacetime. This thermal interpretation is further supported by the expression for the tunneling exponent given in Eq. (12.89). In the fixed background approximation, with $V_{\mathrm{top}} - V_{\mathrm{fv}} \ll V_{\mathrm{fv}}$, we have [see also Eq. (12.91)]

$$B_{\mathrm{HM}} \approx \frac{\left(\frac{4\pi}{3}\Lambda^3\right)(V_{\mathrm{top}} - V_{\mathrm{fv}})}{T_{\mathrm{dS}}}. \tag{12.107}$$

The numerator is precisely the jump in the energy of Eq. (12.101) needed to take a uniform false vacuum horizon volume to a configuration with the field everywhere at ϕ_{top}. [Note that the metric factors in the integrand of Eq. (12.101) cancel in such a way that one ends up with the result that would be obtained by a naïve calculation that used a flat space volume calculation and ignored redshift factors.]

Perhaps surprisingly, the Hawking–Moss solution is not the only τ-independent bounce. By reorienting the Coleman–De Luccia bounce of Fig. 12.18 we can obtain the τ-independent bounce solution illustrated in Fig. 12.19 [277]. This corresponds to the purely thermal creation of a true vacuum region that intersects the horizon. Thus, the same Euclidean solution can correspond to either thermally assisted quantum tunneling or to a purely thermal process. This is perhaps not so surprising if we recall that the de Sitter temperature itself is a quantum effect.

There is another class of solutions whose form emphasizes the significance of inferring the configuration after bubble nucleation from just half of the equatorial slice. As was noted in Sec. 12.4, in flat spacetime the dominant bounces at high temperature are independent of the periodic Euclidean time. Hence, assuming O(3) symmetry of the spatial slices, these bounces have an O(2)×O(3) symmetry. In a de Sitter context, there is a natural generalization of these bounces, with the temperature given by T_{dS}. The O(2)×O(3) symmetry is retained, although now the constant τ slices only extend out to the horizon. The scalar field is the same on all of these slices, which correspond to the horizontal dotted lines

---

[15] The stochastic interpretation [296–298] of the Hawking–Moss bounce may be viewed as describing a mechanism by which the thermal distribution of field configurations is established.

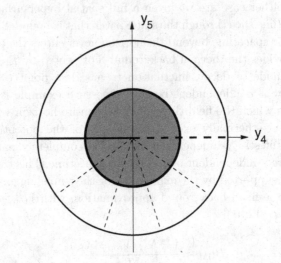

Fig. 12.19. The bounce of Fig. 12.18 rotated to give a τ-independent bounce.

in Fig. 12.17 or to the radial dotted lines in Fig. 12.18. A full equatorial slice would then contain two copies of a true vacuum bubble. Interpreting this full slice as giving the configuration after the transition would lead to the surprising conclusion that the bounce described the production of a pair of bubbles [301]. On the other hand, restricting to the horizon region described by the half-slice yields the more plausible production of a single bubble [302].

In contrast with the flat spacetime case where there is a demonstration that the bounce of lowest action has $O(4)$ symmetry, there is no similar result when gravity is included. These $O(2) \times O(3)$-symmetric bounces thus provide a useful test of whether the $O(4)$-symmetric bounces remain dominant in curved spacetime. For all the cases that have been examined, the $O(4)$-symmetric bounce has the lower action [301, 302], although there are parameter ranges where the two types of bounces both merge into the Hawking–Moss solution and thus have equal action in the limit.

The gravitational effects on the pre-exponential factor have not been discussed thus far. When the fixed-background approximation is appropriate one can simply follow the methods used in flat spacetime and obtain an expression involving a functional determinant of the second variation of the matter action. As before, one must seek a bounce with a single negative mode. Multiple negative modes indicate that the bounce is not even a local minimum of B and must therefore be discarded. This turns out to be the case both for the Hawking–Moss solution when $V''_{\text{top}} > 4/\Lambda^2_{\text{top}}$ (i.e., precisely when an ordinary Coleman–De Luccia bounce is guaranteed to exist), as well as for all of the oscillating bounce solutions [289, 290].

Outside this approximation the situation is less clear. First, there is as yet no derivation leading to a precise expression for the decay rate. Second, even if one assumes a similar form for the pre-exponential factor, matters are complicated by the fact that the dynamics of the gravitational and matter fields are no longer decoupled and must both contribute nontrivially to the functional determinant. However, gravity, like any gauge theory, is a constrained system. If these constraints are not properly taken into account when determining the normal modes, spurious negative modes can appear. Some partial results have been obtained, with demonstrations of the existence of a negative mode under certain conditions [291–294], but also arguments suggesting the absence of a negative mode in other regimes [295]. More work is needed to clarify matters here.

12.7 Curved spacetime evolution after bubble nucleation

The discussion of decay from a de Sitter vacuum in Sec. 12.6 was framed in the context of a fixed de Sitter background metric. We saw that the configuration of the field within the static patch after nucleation was given by one-half of an equatorial slice of the bounce. This suggests that in the more general case the initial data for classical evolution after bubble nucleation is given by that part of an equatorial slice that lies within the horizon of some given point P.

Because the initial configuration outside the horizon is not specified, the future evolution is only determined within the causal diamond of P. It might then seem to be best to work with coordinates that were naturally restricted to this region; e.g., the static de Sitter coordinates in the fixed background approximation. However, for the case of an O(4)-symmetric Coleman–De Luccia bounce it turns out to be more convenient to work with coordinates that extend beyond the causal diamond, keeping in mind that only a portion of the resulting classical evolution is reliable.

An O(4)-symmetric bounce has a Euclidean metric of the form

$$ds^2 = d\xi^2 + \rho(\xi)^2 d\Omega_3^2$$
$$= d\xi^2 + \rho(\xi)^2 \left(d\tau'^2 + \cos^2 \tau' \, d\Omega_2^2 \right) . \tag{12.108}$$

The equatorial hypersurface corresponds to $\tau' = 0$. For the case of an exactly de Sitter Euclidean metric, one-half of this hypersurface coincides with the $\tau = 0$ hypersurface of the metric of Eq. (12.103), although for nonzero values the hypersurfaces of constant τ' and of constant τ do not coincide. On the hypersurface $\tau' = \tau = 0$, we have $d\phi/d\tau' = d\phi/d\tau = 0$.

As was the case for flat spacetime, the regions inside and outside the light cone of the origin (i.e., the point $\xi = 0$) must be treated separately. Outside the light cone the metric can be written as

$$ds^2 = -d\xi^2 - \rho(\xi)^2 d\Omega_S^2$$
$$= -d\xi^2 - \rho(\xi)^2 \left(-d\tau'^2 + \cosh^2 \tau' \, d\Omega_2^2 \right) \tag{12.109}$$

with the Euclidean $d\Omega_3^2$ replaced by the Lorentzian $d\Omega_S^2$, which is the metric on a unit hyperboloid with a spacelike normal vector. As a result of this change, $\rho = 0$ specifies not just a point, but rather the full light cone of the origin. The metric function ρ and the scalar field obey

$$\rho'^2 = 1 + \frac{\kappa}{3}\rho^2 \left(\frac{1}{2}\phi'^2 - V\right) \tag{12.110}$$

and

$$\phi'' + \frac{3\rho'}{\rho}\phi' = \frac{dV}{d\phi}, \tag{12.111}$$

with primes denoting differentiation with respect to ξ. These are the same as the equations for the bounce. Therefore, in complete parallel with the flat spacetime results, $\phi(\xi)$ and $\rho(\xi)$ have the same functional form in this region as they did in the bounce.

The metric inside the light cone of the origin can be written in the open Robertson–Walker form

$$ds^2 = dt^2 - a(t)^2 d\Omega_T^2, \tag{12.112}$$

where $d\Omega_T^2$, the metric on a unit hyperboloid with a timelike normal vector, was given in Eq. (12.48). The scale factor obeys the Friedmann equation

$$\dot{a}^2 = 1 + \frac{\kappa}{3}a^2 \left(\frac{1}{2}\dot{\phi}^2 + V\right) \tag{12.113}$$

(with overdots denoting differentiation with respect to t) and the scalar field satisfies

$$\ddot{\phi} + \frac{3\dot{a}}{a}\dot{\phi} = -\frac{dV}{d\phi}. \tag{12.114}$$

The zero of a, which can be taken to occur at $t = 0$, occurs on the light cone. The interior and exterior metrics join smoothly across this boundary.[16]

Just as in the absence of gravity, in the thin-wall approximation the bubble wall lies in the region outside the light cone. If there is not a well-defined thin wall at the time of nucleation, one develops over time and becomes thinner as the bubble expands. Also as before, the trajectory of the bubble wall asymptotically approaches a light cone. In Minkowski spacetime, any pair of light cones eventually intersects. This is not so in de Sitter spacetime, because of the presence of horizons. As a result, a given pair of bubbles expanding in a de Sitter false vacuum will never collide if they are nucleated at spacetime points that are too far apart. Consequently, if the bubble nucleation rate is too low compared

---

[16] For decay from a de Sitter vacuum there is a second zero of ρ at ξ_{max}. If the continuation from Euclidean to Lorentzian signature were carried out over the entire equatorial slice, there would be a second light cone boundary at $\xi = \xi_{max}$, with a second open Robertson–Walker region on the other side. However, as we have seen, there is no physical justification for extending the continuation to the region near this second zero.

to the cosmic expansion, the bubbles do not coalesce to form a uniform region of the equilibrium true vacuum phase. In particular, the regions of true vacuum do not even percolate if the dimensionless ratio $\gamma = (\Gamma/\mathcal{V})H^{-4} = (\Gamma/\mathcal{V})\Lambda^4$ is less than a critical value γ_{cr} that has been shown [126] to lie in the range $0.24 > \gamma_{\mathrm{cr}} > 1.1 \times 10^{-6}$.

If γ is less than this critical value, the true vacuum regions remain confined to isolated bubbles or clusters of bubbles. We saw in Sec. 12.3 that the interior of an uncollided flat spacetime bubble was essentially pure vacuum, with almost all of the energy from the latent heat of transition going into the acceleration of the bubble walls. A similar effect occurs in curved spacetime. It can be quantified by considering the entropy density inside the bubble. As in flat spacetime, the scalar field in the interior of the bubble eventually begins to oscillate about its true vacuum value. After some time the energy in these oscillations is thermalized, generating an entropy density S. An adiabatically expanding Robertson–Walker universe is characterized by the dimensionless quantity $a^3 S$; in our universe, observations indicate that this is greater than 10^{90}. By contrast, it was shown in Ref. [126] that for a generic potential with a de Sitter or Minkowski true vacuum $a^3 S$ in the interior of the bubble would be of order unity, implying that a slow first-order transition would not lead to an acceptable cosmology. However, Linde [299] and Albrecht and Steinhardt [300] realized that a non-generic potential could give very different results. If $V(\phi)$ has a suitably flat region between the potential barrier and the true vacuum minimum at ϕ_{tv}, the scale factor can grow exponentially before ϕ enters the oscillatory regime. As a result, this "new inflation" scenario can yield values of $a^3 S$ that easily satisfy the observational constraint.

Although this scenario would allow the observed universe to evolve inside a single bubble, bubbles nucleated near this reference bubble would collide with it, producing a bubble cluster. The exponentially expanding de Sitter spacetime in the exterior false vacuum region leads to a very unusual distribution of bubble sizes in this cluster [126], with almost all bubbles being exponentially smaller than the largest bubble.[17] The extent to which the distribution of bubbles in the cluster reveals any indication of the onset of inflation in the exterior region, and indeed whether a typical observer would see any observable traces of the bubble collisions, have been the subject of a number of recent papers [304–309]

One can also consider the case where the true vacuum has a negative vacuum energy and so is an anti-de Sitter spacetime. In this case the interior region has a very different fate [276]. If the thin-wall approximation were exact, ϕ would be precisely equal to ϕ_{tv} and Eq. (12.113) would take the form

$$\dot{a}^2 = 1 + \frac{\kappa}{3}a^2 V_{\mathrm{tv}} = 1 - \frac{a^2}{L^2} \tag{12.115}$$

---

[17] By the de Sitter analogue of a Lorentz transformation, one can choose a frame in which the three largest bubble in the cluster are of equal size.

with the solution

$$a = L \sin(t/L).\tag{12.116}$$

Like the zero on the light cone at $t = 0$, the zero of a at $t = \pi L$ is a coordinate singularity, not a physical one. However, Eq. (12.114) shows that a real singularity in ϕ, and hence a physical singularity in the metric, will develop unless $\dot{\phi}$ is precisely zero whenever $a = 0$. The corrections to the thin-wall limit, even though exponentially small, give ϕ a time-dependence such that for generic initial conditions $\dot{\phi}$ is nonzero at the zero of a, and a singularity ensues.

This difficulty persists even when the thin-wall approximation is not applicable [310]. Equation (12.113) implies that

$$\dot{a}^2 \geq 1 - \frac{a^2}{L^2},\tag{12.117}$$

so that a second zero of a is inevitable once a starts decreasing and becomes small compared to L. Except for exceptional values of $\phi(0)$, $\dot{\phi}$ is nonvanishing at this second zero, and one is inevitably driven to a singularity within a finite time after the nucleation of the bubble.

Appendix A
Roots and weights

In this appendix I review some facts concerning compact Lie groups and their Lie algebras that are used in this book. I will focus in particular on the roots and weights. As we will see, these can be understood by generalizing concepts that are familiar from the simplest non-Abelian Lie group, SU(2).

A compact Lie group and its algebra may be simple, semisimple, or neither. If it is simple, the generators of the algebra cannot be divided into two mutually commuting sets and the group cannot be written, even locally, as the product of two groups. If the group can be written, at least locally, as a product of groups, but without any U(1) factors, it is semisimple, but not simple. Thus, SU(3) is simple and semisimple, SO(4) = [SU(2)×SU(2)]/Z_2 is semisimple but not simple, and SU(2) × U(1) is neither. Because all compact Lie groups can be built up from the simple Lie groups, I will focus on these.

A.1 Root systems

The usual choice of Hermitian generators for SU(2) is such that in the fundamental two-dimensional representation they are given by $T_a = \sigma_a/2$. They then satisfy the standard angular momentum commutation relations, and in any irreducible representation obey the relation

$$\operatorname{tr}(T_a T_b) = \delta_{ab} T(R), \tag{A.1}$$

where $T(R)$ is a constant that depends on the representation; for the fundamental doublet representation, $T = 1/2$. From these Hermitian generators one can construct two ladder operators $T_\pm = T_1 \pm iT_2$ that raise and lower the value of the third generator, T_3, by one unit.

This construction generalizes in a natural way to larger Lie groups. If the group is compact, it is always possible to choose its generators so that they obey Eq. (A.1). From these, one can choose a maximal set of mutually commuting generators H_j $(j = 1, 2, \ldots, r)$. These generalize the role of T_3, in that states of an irreducible representation can be taken to be simultaneous eigenvectors of the H_j; their eigenvalues are the weights that will be discussed in the next section. The r-dimensional vector space spanned by the H_j is called the Cartan subalgebra, and r is known as the rank of the group. The remaining generators can be combined to form a set of ladder operators E_α that obey

$$[H_j, E_\alpha] = \alpha_j E_\alpha \,,$$
$$[E_\alpha, E_{-\alpha}] = \alpha_j H_j \equiv \boldsymbol{\alpha} \cdot \mathbf{H} \,, \tag{A.2}$$

with $E_{-\alpha} = E_\alpha^\dagger$. The r-component vectors $\boldsymbol{\alpha}$ are known as roots and are naturally viewed as living in an r-dimensional Euclidean space.

This can be illustrated with the case of SU(3). A standard choice of the generators in the fundamental three-dimensional representation is $T_a = \lambda_a/2$, where the Gell-Mann matrices λ_a generalize the Pauli matrices and are given by

$$\lambda_1 = \begin{pmatrix} 0 & 1 & 0 \\ 1 & 0 & 0 \\ 0 & 0 & 0 \end{pmatrix} \qquad \lambda_2 = \begin{pmatrix} 0 & -i & 0 \\ i & 0 & 0 \\ 0 & 0 & 0 \end{pmatrix} \qquad \lambda_3 = \begin{pmatrix} 1 & 0 & 0 \\ 0 & -1 & 0 \\ 0 & 0 & 0 \end{pmatrix}$$

$$\lambda_4 = \begin{pmatrix} 0 & 0 & 1 \\ 0 & 0 & 0 \\ 1 & 0 & 0 \end{pmatrix} \qquad \lambda_5 = \begin{pmatrix} 0 & 0 & -i \\ 0 & 0 & 0 \\ i & 0 & 0 \end{pmatrix} \qquad \lambda_6 = \begin{pmatrix} 0 & 0 & 0 \\ 0 & 0 & 1 \\ 0 & 1 & 0 \end{pmatrix}$$

$$\lambda_7 = \begin{pmatrix} 0 & 0 & 0 \\ 0 & 0 & -i \\ 0 & i & 0 \end{pmatrix} \qquad \lambda_8 = \sqrt{\frac{1}{3}} \begin{pmatrix} 1 & 0 & 0 \\ 0 & 1 & 0 \\ 0 & 0 & -2 \end{pmatrix} . \tag{A.3}$$

The generators of the Cartan subalgebra can be take to be $H_1 = T_3$ and $H_2 = T_8$. We can then define ladder operators

$$E_{\pm\alpha} = \frac{1}{\sqrt{2}} \left(T_1 \pm i T_2 \right) ,$$
$$E_{\pm\beta} = \frac{1}{\sqrt{2}} \left(T_4 \pm i T_5 \right) ,$$
$$E_{\pm\gamma} = \frac{1}{\sqrt{2}} \left(T_6 \pm i T_7 \right) . \tag{A.4}$$

Evaluating the commutators of these with the H_i, we find that E_α raises H_1 by one unit and leaves H_2 unchanged, E_β raises H_1 by $1/2$ and H_2 by $\sqrt{3}/2$, and E_γ lowers H_1 by $1/2$ but raises H_2 by $\sqrt{3}/2$. These relations can be summarized by a root diagram, such as that shown in Fig. A.1a.

As a second example, consider the rotation group in N dimensions, SO(N). Its generators can be taken to be the $J_{ij} = -J_{ji}$ that generate rotations in the i-j plane. These obey

$$[J_{ij}, J_{kl}] = i \left(\delta_{ik} J_{jl} - \delta_{il} J_{jk} - \delta_{jk} J_{il} + \delta_{jl} J_{ik} \right) . \tag{A.5}$$

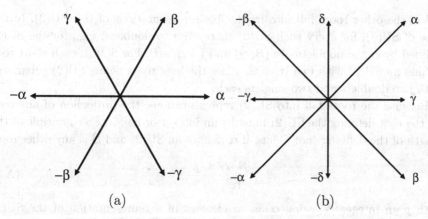

Fig. A.1. The root diagrams for SU(3) and SO(5). The horizontal and vertical directions correspond to H_1 and H_2, respectively.

If $N = 2k$ is even, the generators of the Cartan subalgebra can be taken to be $J_{12}, J_{34}, \ldots, J_{2k-1,2k}$. The same choice suffices for $N = 2k + 1$. Thus, SO($2k$) and SO($2k + 1$) both have rank k.

Let us focus on the case of SO(5), and define

$$E_\alpha = \frac{1}{2} \left(J_{13} + i J_{23} \right) + \frac{i}{2} \left(J_{14} + i J_{24} \right),$$

$$E_\beta = \frac{1}{2} \left(J_{13} + i J_{23} \right) - \frac{i}{2} \left(J_{14} + i J_{24} \right),$$

$$E_\gamma = \frac{1}{\sqrt{2}} \left(J_{15} + i J_{25} \right),$$

$$E_\delta = \frac{1}{\sqrt{2}} \left(J_{35} + i J_{45} \right). \tag{A.6}$$

Using Eq. (A.5) we find that E_α raises both $H_1 = J_{12}$ and $H_2 = J_{34}$ by one unit, while E_β raises H_1 but lowers H_2, each by one unit. Further, E_γ raises H_1 and commutes with H_2, while E_δ commutes with H_1 and raises H_2. The corresponding root diagram is shown in Fig. A.1b. In contrast with SU(3), there are two different root lengths.

Note that removing the roots $\pm\gamma$ and $\pm\delta$, formed from the generators J_{j5}, gives the root diagram for the semisimple SO(4).

There are strong constraints on the possible root systems. These can be obtained by noting that any root α can be used to define an SU(2) subgroup with generators

$$t_1(\alpha) = \frac{1}{\sqrt{2\alpha^2}} \left(E_\alpha + E_{-\alpha} \right),$$

$$t_2(\alpha) = -\frac{i}{\sqrt{2\alpha^2}} \left(E_\alpha - E_{-\alpha} \right),$$

$$t_3(\alpha) = \frac{1}{\alpha^2} \alpha_j H_j. \tag{A.7}$$

All of the other roots fall into irreducible representations of this SU(2). In the case of SU(3), for every such SU(2) there are two doublets; e.g., for the SU(2) defined by α the doublets are (β, γ) and $(-\gamma, -\beta)$. For SO(5), each short root defines an SU(2) with two triplets, while the long roots define SU(2) subgroups with two doublets and two singlets each.

Because the roots fall into SU(2) representations, the projection of any root on the root defining the SU(2) must be an integer or half-integer multiple of the length of the defining root. Thus, if α defines an SU(2) and β is any other root,

$$\frac{\beta \cdot \alpha}{\alpha^2} = \frac{p}{2}, \tag{A.8}$$

with p an integer. Viewing α as an element of a representation of the SU(2) defined by β gives a dual relation,

$$\frac{\alpha \cdot \beta}{\beta^2} = \frac{q}{2}, \tag{A.9}$$

with q also an integer. Combining these two equations yields the two relations

$$4\cos^2\theta = \frac{4(\alpha \cdot \beta)^2}{\alpha^2 \beta^2} = pq, \tag{A.10}$$

$$\frac{\beta^2}{\alpha^2} = \frac{p}{q}, \tag{A.11}$$

where θ is the angle between α and β.

Since $\cos\theta$ must lie between -1 and 1, we see that $0 \leq pq \leq 3$. The only possibilities are

- $p = q = 0$: The two roots are orthogonal.
- $p = q = \pm 1$: The roots are either $60°$ or $120°$ apart and of equal length.
- $p = 2q = \pm 2$ or $q = 2p = \pm 2$: The angle between the roots is either $45°$ or $135°$ and one root is $\sqrt{2}$ times the length of the other.
- $p = 3q = \pm 3$ or $q = 3p = \pm 3$: The angle between the roots is either $30°$ or $150°$ and one root is $\sqrt{3}$ times the length of the other.

A further consequence follows from the fact that the roots form complete SU(2) representations. Reflection of a root β in the hyperplane through the origin orthogonal to α gives a vector $\beta - 2(\beta \cdot \alpha)/|\alpha|$ that is also a root; the group generated by such reflections is known as the Weyl group.

It is a straightforward geometrical exercise to determine all possible root systems satisfying these constraints; the details can be found, e.g., in [311] or [312]. There are four infinite series, A_N, B_N, C_N, and D_N (the classical Lie algebras) and five exceptional algebras, G_2, F_4, E_6, E_7, and E_8; in each case the subscript denotes the rank. The roots of these algebras (with an arbitrary overall normalization) can be given in terms of orthogonal unit vectors $\mathbf{e}_j$.

The roots and dimensions[1] for the infinite series are

- A_N, with dimension $N(N+2)$:
  $$\mathbf{e}_i - \mathbf{e}_j \ (1 \leq i \neq j \leq N+1).$$
- B_N, with dimension $N(2N+1)$:
  $$\pm\mathbf{e}_i \pm \mathbf{e}_j \ (1 \leq i \neq j \leq N);$$
  $$\pm\mathbf{e}_i \ (1 \leq i \leq N).$$
- C_N, with dimension $N(2N+1)$:
  $$\pm\mathbf{e}_i \pm \mathbf{e}_j \ (1 \leq i \neq j \leq N);$$
  $$\pm 2\mathbf{e}_i \ (1 \leq i \leq N).$$
- D_N, with dimension $N(2N-1)$:
  $$\pm\mathbf{e}_i \pm \mathbf{e}_j \ (1 \leq i \neq j \leq N).$$

Two of these series generalize our results for SO(4) and SO(5): the B_N are the algebras for the odd orthogonal groups SO($2N+1$), and the D_N are the algebras for the even orthogonal groups SO($2N$).

A_N is the algebra of SU($N+1$). Although its roots are written in terms of an $(N+1)$-dimensional space, they all lie in the N-dimensional subspace orthogonal to $\mathbf{e}_1 + \mathbf{e}_2 + \cdots + \mathbf{e}_{N+1}$. Writing them in this form makes it clear that the roots of SU(N) are a subset of those of SO($2N$), corresponding to the fact the former is a subgroup of the latter.

Given any root one can define a dual root

$$\boldsymbol{\alpha}^* = \frac{\boldsymbol{\alpha}}{\alpha^2}. \tag{A.12}$$

It is easy to see that the duals of a root system satisfying our constraints also form an acceptable root system. A_N and D_N are self-dual (after an irrelevant rescaling), but taking the dual of B_N leads to the last of the infinite series, C_N, which is the Lie algebra of the symplectic group Sp($2N$).

The dimensions and roots for the exceptional groups are

- G_2 (dimension 14):
  $$\mathbf{e}_i - \mathbf{e}_j \ (1 \leq i \neq j \leq 3);$$
  $$\pm(3\mathbf{e}_j - \mathbf{e}_1 - \mathbf{e}_2 - \mathbf{e}_3) \ (j = 1, 2, 3).$$
- F_4 (dimension 52):
  $$\pm(\mathbf{e}_i \pm \mathbf{e}_j) \ (1 \leq i \neq j \leq 4);$$
  $$\pm\mathbf{e}_i \ (i = 1, 2, 3, 4);$$
  $$\tfrac{1}{2}(\pm\mathbf{e}_1 \pm \mathbf{e}_2 \pm \mathbf{e}_3 \pm \mathbf{e}_4).$$
- E_8 (dimension 248):
  $$\pm\mathbf{e}_i \pm \mathbf{e}_j \ (1 \leq i \neq j \leq 8);$$
  $$\tfrac{1}{2}(\pm\mathbf{e}_1 \pm \mathbf{e}_2 \pm \mathbf{e}_3 \pm \mathbf{e}_4 \pm \mathbf{e}_5 \pm \mathbf{e}_6 \pm \mathbf{e}_7 \pm \mathbf{e}_8), \text{ with an even number of plus}$$
  signs.

---

[1] Recall that the dimension of a Lie group and of its algebra is the dimension of the group as a manifold and of the algebra as a vector space. It is equal to the number of generators.

- E_7 (dimension 133):

  $\pm\mathbf{e}_i \pm \mathbf{e}_j$ $(1 \le i \ne j \le 6)$;

  $\pm(\mathbf{e}_7 + \mathbf{e}_8)$;

  $\frac{1}{2}[(\pm\mathbf{e}_1 \pm \mathbf{e}_2 \pm \mathbf{e}_3 \pm \mathbf{e}_4 \pm \mathbf{e}_5 \pm \mathbf{e}_6) \pm (\mathbf{e}_7 + \mathbf{e}_8)]$, with an even number of plus signs in the first pair of parentheses.

- E_6 (dimension 78):

  $\pm\mathbf{e}_i \pm \mathbf{e}_j$ $(1 \le i \ne j \le 5)$;

  $\frac{1}{2}[\pm\mathbf{e}_1 \pm \mathbf{e}_2 \pm \mathbf{e}_3 \pm \mathbf{e}_4 \pm \mathbf{e}_5 \pm (\mathbf{e}_6 + \mathbf{e}_7 + \mathbf{e}_8)]$, with an even number of plus signs.

Note that the roots of E_7 are those roots of E_8 that are orthogonal to the root $\mathbf{e}_7 - \mathbf{e}_8$, and the roots of E_6 are the roots orthogonal to both $\mathbf{e}_6 - \mathbf{e}_7$ and $\mathbf{e}_7 - \mathbf{e}_8$. One could continue this pattern to obtain smaller algebras. However, the algebras obtained in this manner are not new: $E_5 = D_5 = \mathrm{SO}(10)$, $E_4 = A_4 = \mathrm{SU}(5)$, and $E_3 = A_2 \times A_1 = \mathrm{SU}(3) \times \mathrm{SU}(2)$.

In fact, the full root system of any of these algebras can be reconstructed from a special subset of r roots $\boldsymbol{\beta}_a$, known as simple roots. These form a basis for the root system and are chosen so that the inner product of any two simple roots is less than or equal to zero. Once these are known, all of the other roots can be obtained by applying successive Weyl reflections. Furthermore, all other roots are linear combinations of the simple roots with integral coefficients that are all of the same sign.

The essential properties of the simple roots can be summarized compactly by the Dynkin diagrams, which are displayed in Fig. A.2. Each node in these diagrams corresponds to a simple root. Two roots $\boldsymbol{\beta}_a$ and $\boldsymbol{\beta}_b$ are linked by

$$m_{ab} = \frac{4(\boldsymbol{\beta}_a \cdot \boldsymbol{\beta}_b)^2}{\beta_a^2 \beta_b^2} \tag{A.13}$$

lines. If $m_{ab} = 1$, then $\boldsymbol{\beta}_a$ and $\boldsymbol{\beta}_b$ are of equal length. If not, there is an arrow pointing toward the longer root.

Note that the Dynkin diagrams for low ranks are not all distinct. This reflects the equivalences $A_1 = B_1 = C_1$, $B_2 = C_2$, and $A_3 = D_3$. Also, note that D_2 is not simple.

The choice of the set of simple roots is not unique, although the m_{ab} are. Given any r-component vector $\mathbf{h}$, it is always possible to choose the simple roots so that their inner products with $\mathbf{h}$ are all positive. If these inner products are all nonzero, this picks out a unique set of simple roots.

With the full root systems specified as previously, the simple roots can be taken to be

- A_N: $\mathbf{e}_j - \mathbf{e}_{j+1}$ $(j = 1, 2, \ldots, N)$.
- B_N: $\mathbf{e}_j - \mathbf{e}_{j+1}$ $(j = 1, 2, \ldots, N-1)$;

  $\mathbf{e}_N$.

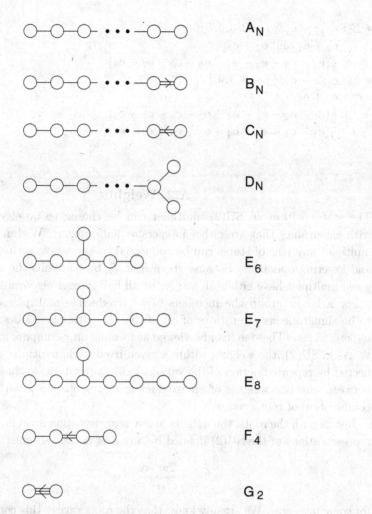

Fig. A.2. Dynkin diagrams for the simple Lie algebras.

- C_N: $e_j - e_{j+1}$ $(j = 1, 2, \ldots, N-1)$;
  $2e_N$.
- D_N: $e_j - e_{j+1}$ $(j = 1, 2, \ldots, N-1)$;
  $e_{N-1} + e_N$.
- G_2: $e_2 - e_3$ and $e_1 - 2e_2 + e_3$.
- F_4: $e_2 - e_3$ and $e_3 - e_4$;
  e_4;
  $\frac{1}{2}(e_1 - e_2 - e_3 - e_4)$.
- E_8: $e_{j+1} - e_j$ $(j = 2, 3, \ldots, 7)$;
  $e_2 + e_3$;
  $\frac{1}{2}(e_1 + e_2 - e_3 - e_4 - e_5 - e_6 - e_7 - e_8)$.

- E_7: $\mathbf{e}_{j+1} - \mathbf{e}_j$, $(j = 2, 3, 4, 5)$;

    $\mathbf{e}_2 + \mathbf{e}_3$ and $\mathbf{e}_7 + \mathbf{e}_8$;

    $\frac{1}{2}(\mathbf{e}_1 + \mathbf{e}_2 - \mathbf{e}_3 - \mathbf{e}_4 - \mathbf{e}_5 - \mathbf{e}_6 - \mathbf{e}_7 - \mathbf{e}_8)$.

- E_6: $\mathbf{e}_{j+1} - \mathbf{e}_j$, $(j = 2, 3, 4)$;

    $\mathbf{e}_2 + \mathbf{e}_3$;

    $\frac{1}{2}(\mathbf{e}_1 + \mathbf{e}_2 - \mathbf{e}_3 - \mathbf{e}_4 - \mathbf{e}_5 - \mathbf{e}_6 - \mathbf{e}_7 - \mathbf{e}_8)$;

    $\frac{1}{2}(\mathbf{e}_1 - \mathbf{e}_2 - \mathbf{e}_3 - \mathbf{e}_4 - \mathbf{e}_5 + \mathbf{e}_6 + \mathbf{e}_7 + \mathbf{e}_8)$.

A.2 Weights

The states within an SU(2) multiplet can be chosen to be eigenstates of T_3, with eigenvalues that are either integers or half-integers. Within an irreducible multiplet any pair of states can be connected by successive action of the raising and lowering operators. Because these shift T_3 by one unit, the states within a given multiplet have either all integer or all half-integer eigenvalues.

For a larger group the members of an irreducible multiplet can be chosen to be simultaneous eigenstates of all of the H_j, with eigenvalues, or "weights", $w_1, w_2, \ldots, w_r$. These w_j can be viewed as forming an r-component weight vector $\mathbf{w}$. As in SU(2), the weights within a given irreducible multiplet can all be connected by repeated action of the various ladder operators, so that the difference between any two weights of an irreducible representation is an integral linear combination of root vectors.

Just as with the roots, the weights in any representation must fall into complete representations of the SU(2) defined by any root $\boldsymbol{\alpha}$, and so must obey

$$\frac{2\mathbf{w} \cdot \boldsymbol{\alpha}}{\boldsymbol{\alpha}^2} = n \tag{A.14}$$

for some integer n. We already know that the roots satisfy this constraint, so the roots themselves are weights; more specifically, they are the nonzero weights of the adjoint representation.

The set of all possible weights for a simple Lie group can be displayed as a lattice of discrete points in an r-dimensional Euclidean space. For SU(2), this is just the set of integer and half-integer points along the real line. The weight lattices for the rank 2 groups SU(3) and Spin(5), the covering group of SO(5), are shown in Fig. A.3. As indicated in the figure, these lattices can be divided into sublattices such that any two points within a sublattice can be connected by an integral sum of roots, and that two points within different sublattices cannot be. This is completely analogous to the distinction between integer and half-integer weights for SU(2).

The existence of these sublattices is closely related to the fact that several Lie groups can share the same Lie algebra. Let G be the universal covering group of a simple Lie algebra. The center K of G is defined to be the set of elements that

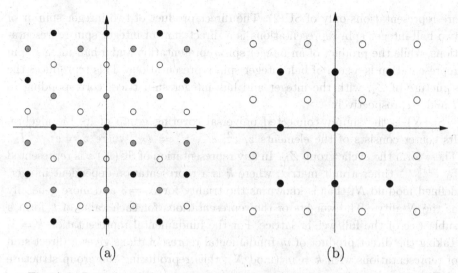

(a)   (b)

Fig. A.3. Portions of the weight lattices for (a) SU(3) and (b) Spin(5). In
both cases the root lattice is indicated by solid black circles. For SU(3) the
sublattices with triality 1 and triality 2 are indicated by white and gray circles,
respectively. For Spin(5) the white circles indicate the weights of the spinor
representations. The horizontal and vertical directions correspond to H_1 and
H_2, respectively.

commute with all elements of G; these form a finite group. A new group G/K,
also known as the adjoint group, can be defined by identifying the elements g
and kg of G, where k is any element of the center. A similar construction can be
carried out with K replaced by any of its subgroups.

For a representation of G to also be a true (i.e., single-valued) representation
of G/K, the elements of K must be represented by a unit matrix. The weights
of these latter representations form a sublattice of the full weight lattice of G.
This sublattice contains the root vectors, and may be termed the root lattice.[2]
The remaining representations of G are not true representations of G/K. Their
weights fall into sublattices that can be put in a one-to-one correspondence with
the nontrivial elements of K, with the properties of the direct products of these
representations reflecting the group structure of K. (The groups obtained by
factoring by a subgroup of the center have correspondingly fewer sublattices,
and the adjoint group has only the root lattice.)

For example, SU(2) is the universal covering group of its Lie algebra, with
a center Z_2 composed of the two elements I and $-I$. Identifying the elements
u and $-u$ yields the group SO(3). The integer spin representations are representations of both SU(2) and SO(3), but the half-integer spin representations

---

[2] The root lattice is the only sublattice that is itself a lattice. The others are not, because
they are not closed under addition.

are representations only of SU(2). The direct product of two integer spin or of two half-integer spin representations is a direct sum of integer spin representations, while the product of an integer spin representation and a half-integer spin representation is a sum of half-integer spin representations. This reproduces the structure of Z_2, with the integer and half-integer sublattices corresponding to I and $-I$, respectively.

SU(N) is the simply connected universal covering group of its Lie algebra. Its center consists of the elements z, z^2, ..., $z^N = I_N$, where $z = e^{2\pi i/N} I_N$. These form the cyclic group Z_N. In any representation of SU(N) z is represented by $e^{2\pi ki/N}$ times a unit matrix, where k is a representation-dependent integer, defined modulo N, that is known as the triality for $N = 3$ and more generally as the N-ality. The weights of the representations for each value of k form a sublattice of the full weight lattice. For the fundamental representation $k = 1$. Taking the direct product of m fundamental representations gives a direct sum of representations with $k = m \pmod N$, thus reproducing the group structure of Z_N.

Spin(N) is the universal covering group of SO(N). For odd $N = 2r + 1$, its center is Z_2 and the weight lattice consists of two sublattices. One contains the weights of the vector and tensor representations that are also representations of SO($2r + 1$), while the other contains the weights of the spinor representations that are representations only of Spin($2r + 1$). With the roots normalized as in the previous section, the weights of the former are of the form $\sum p_a \mathbf{e}_a$, with the p_a all integers, while the spinor sublattice has weights $\sum (p_a + \frac{1}{2}) \mathbf{e}_a$.

The structure for even N is more complicated. The center of Spin($2r$) is Z_4 if r is odd and $Z_2 \times Z_2$ if r is even. [This extra structure can be traced to the fact that SO($2r$) has a nontrivial center composed of the elements $\pm I_{2r}$.] The full weight lattice of Spin($2r$) is the same as that of Spin($2r + 1$), but the two sublattices of the latter are each divided further into two sublattices, according to whether $\sum p_a$ is even or odd. For further discussion of these and the quotient groups of Spin($2r$), see [98].

With the normalization used previously, the weights of Sp($2N$) are of the form $\sum p_a \mathbf{e}_a$, with the p_a again all integers. There are two sublattices, with $\sum p_a$ even or odd, respectively, corresponding to the fact that the center of Sp($2N$) is Z_2.

Of the exceptional groups, E_6 and E_7 have centers Z_3 and Z_2, respectively, while the other three have trivial centers.

Appendix B
Index theorems for BPS solitons

We saw in Sec. 10.8 how index theory methods could be used to count the parameters entering a general Yang–Mills instanton solution. In this appendix I will discuss the application of similar methods to BPS solitons.

To this end, let us suppose that we have a BPS soliton that is the solution of a Bogomolny-type first-order differential equation. We are interested in counting the zero modes that are perturbations about this solution that obey both the linearized Bogomolny equation and an appropriate condition that eliminates purely gauge modes. Taken together, these equations can be written as a matrix differential equation

$$\mathcal{D}\delta\Psi = 0, \tag{B.1}$$

where $\mathcal{D}$ is a matrix differential operator. The number of linearly independent solutions of this equation gives the dimension of the moduli space and, after the subtraction of any global gauge modes, the number of parameters entering the general BPS solution.

The index of $\mathcal{D}$ is defined to be

$$\mathcal{I}(\mathcal{D}) = \dim \ker(\mathcal{D}) - \dim \ker(\mathcal{D}^\dagger), \tag{B.2}$$

i.e., the difference in the numbers of zero modes of $\mathcal{D}$ and of $\mathcal{D}^\dagger$. Often, as in the case of the instantons, there is a vanishing theorem showing that $\mathcal{D}^\dagger$ has no zero modes, in which case $\mathcal{I}(\mathcal{D})$ is precisely the number we seek. Otherwise, it only gives a lower bound.

It is useful to define an auxiliary quantity

$$\mathcal{I}(\mathcal{D}, M^2) = \text{Tr} \left(\frac{M^2}{\mathcal{D}^\dagger \mathcal{D} + M^2} - \frac{M^2}{\mathcal{D}\mathcal{D}^\dagger + M^2} \right). \tag{B.3}$$

where the trace is to be understood in the functional sense. In the $M^2 \to 0$ limit only modes with zero eigenvalues contribute, and so we expect that

$$\lim_{M^2 \to 0} \mathcal{I}(\mathcal{D}, M^2) = \mathcal{I}(\mathcal{D}). \tag{B.4}$$

An exception can occur if $\mathcal{D}$ and $\mathcal{D}^\dagger$ have continuum spectra extending down to zero. If the density of eigenvalues is sufficiently singular near 0, there is a possibility of a continuum contribution that survives in this limit. Since low eigenvalues are associated with large distance, this possibility can be analyzed by considering the large r behavior of the operators and their eigenfunctions.

If the space is compact, so that $\mathcal{D}$ and $\mathcal{D}^\dagger$ have discrete spectra, then, as we saw in Sec. 10.8, the nonzero eigenvalues of these two operators are equal, so their contributions to Eq. (B.3) cancel. Hence $\mathcal{I}(\mathcal{D}, M^2)$ simply counts the zero eigenvalues and is therefore equal to the index for all values of M^2. The same is true for a system on an open space if the falloff at large distance is sufficiently rapid, as we saw for the case of the instanton in Sec. 10.8. In such cases, we can evaluate the index using any value of M^2. As in the instanton case, the limit $M^2 \to \infty$ is often particularly convenient.

A general strategy is to expand the two terms in Eq. (B.3) in power series. In cases where the index can be obtained from the $M^2 \to \infty$ limit, the expansion is chosen so that all but the first few terms are proportional to negative powers of M^2 and so can be ignored in the limit, while of the remaining terms all but one vanish by direct calculation.

In some other cases, with a slower falloff of the fields at large distance, the topological charges are given by surface integrals at spatial infinity. The arguments for M^2 independence fail and the index must be evaluated by taking the $M^2 \to 0$ limit. A suitable expansion is then one where successive terms involve progressively higher powers of a quantity proportional to an inverse power of r. If the volume integral implied by the trace can be converted to an integral over the surface at spatial infinity, only a finite number of terms can contribute to the surface integral with, again, all but one of these vanishing by direct calculation.

B.1 Vortices

An example of the first type is provided by the two-dimensional U(1) model with equal scalar and vector masses [41]. The BPS vortices in this theory obey Eq. (8.25). Let us require that the perturbations about a solution obey the background gauge condition of Eq. (3.60). This, together with the linearization of the Bogomolny equations, implies that the zero modes satisfy

$$\mathcal{D}\eta = 0, \tag{B.5}$$

where $\eta = (\delta\phi_1, \delta\phi_2, \delta A_1, \delta_2)^t$,

$$\mathcal{D} = \begin{pmatrix} (\partial_1 + eA_2) & (\partial_2 - eA_1) & -e\phi_2 & e\phi_1 \\ (-\partial_2 + eA_1) & (\partial_1 + eA_2) & e\phi_1 & e\phi_2 \\ e\phi_1 & e\phi_2 & -\partial_2 & \partial_1 \\ -e\phi_2 & e\phi_1 & \partial_1 & \partial_2 \end{pmatrix}, \tag{B.6}$$

and ϕ_1 and ϕ_2 are the real and imaginary parts of the scalar field,

A calculation then shows that

$$\mathcal{D}^\dagger \mathcal{D} = -I\nabla^2 - L_1,$$
$$\mathcal{D}\mathcal{D}^\dagger = -I\nabla^2 - L_2, \tag{B.7}$$

where I is the unit matrix and the L_j are composed of terms with at most one derivative. We can write

$$\frac{1}{\mathcal{D}^\dagger\mathcal{D} + M^2} = \frac{1}{(-\nabla^2 + M^2)} + \frac{1}{(-\nabla^2 + M^2)}L_1\frac{1}{(-\nabla^2 + M^2)}$$
$$+ \frac{1}{(-\nabla^2 + M^2)}L_1\frac{1}{(-\nabla^2 + M^2)}L_1\frac{1}{(-\nabla^2 + M^2)} + \cdots,$$

$$\frac{1}{\mathcal{D}\mathcal{D}^\dagger + M^2} = \frac{1}{(-\nabla^2 + M^2)} + \frac{1}{(-\nabla^2 + M^2)}L_2\frac{1}{(-\nabla^2 + M^2)}$$
$$+ \frac{1}{(-\nabla^2 + M^2)}L_2\frac{1}{(-\nabla^2 + M^2)}L_2\frac{1}{(-\nabla^2 + M^2)} + \cdots. \tag{B.8}$$

The traces of these series can now be evaluated term by term, with successive terms proportional to lower and lower powers of M^2.

The first terms in the two expansions are identical and cancel. The second terms give

$$\mathcal{I}^{(2)}(\mathcal{D}, M) = \operatorname{Tr} \frac{M^2}{(-\nabla^2 + M^2)}(L_1 - L_2)\frac{1}{(-\nabla^2 + M^2)}$$
$$= \operatorname{Tr} \frac{M^2}{(-\nabla^2 + M^2)^2}(L_1 - L_2)$$
$$= \int d^2x \left\langle x \left| \frac{M^2}{(-\nabla^2 + M^2)^2} \operatorname{tr}(L_1 - L_2) \right| x \right\rangle. \tag{B.9}$$

The second line follows from the cyclic property of the trace. In the third line, the functional trace has been decomposed into a spatial integration and an ordinary matrix trace. We can see from this line that only the diagonal elements of the L_j are needed at this stage. Evaluating the products of $\mathcal{D}$ and its adjoint, we find that

$$\operatorname{tr}(L_1 - L_2) = 4eF_{12} \tag{B.10}$$

and so

$$\mathcal{I}^{(2)}(\mathcal{D}, M^2) = 4e \int d^2x\, F_{12}(x) \int \frac{d^2k}{(2\pi)^2} \frac{M^2}{(k^2 + M^2)^2}$$
$$= 4e \left(\frac{1}{4\pi}\right) \int d^2x\, F_{12}(x). \tag{B.11}$$

Recalling that the magnetic flux is quantized, with

$$\int d^2x\, F_{12}(x) = \frac{2\pi n}{e}, \tag{B.12}$$

we see that $\mathcal{I}^{(2)}(\mathcal{D}, M^2) = 2n$, where n is the vorticity. The higher-order terms in the expansion of $\mathcal{I}(\mathcal{D}, M^2)$ are all proportional to inverse powers[1] of M^2. Taking the limit $M^2 \to \infty$, we have

$$\mathcal{I}(\mathcal{D}) = 2n. \tag{B.13}$$

It is not hard to check that $\mathcal{D}\mathcal{D}^\dagger$ is positive definite, so that $\mathcal{D}^\dagger$ has no zero modes [41]. It then follows that the index is indeed equal to the dimension of the moduli space, and that the latter is twice the vorticity, just as asserted in Sec. 8.4. For the special case of axially symmetric solutions [313], this counting of the zero modes can be verified by direct analysis of the differential equations [41].

B.2 Monopoles

The second type of calculation is exemplified by the case of BPS monopoles. Considerable simplification is obtained by noting that the Bogomolny equation (8.19) is identical to the instanton self-duality condition, Eq. (10.80), if we make the correspondence $\Phi \to -A_4$ and require that all quantities be independent of x_4. Following precisely the same steps as in Sec. 10.8, we perturb about this equation to obtain the zero-mode equation and impose the background gauge condition. After writing the perturbations as fermion fields, we eventually arrive at

$$\mathcal{I}(\mathcal{D}, M^2) = 2\,\mathrm{Tr}\left[-\gamma_5 \frac{M^2}{-(\gamma \cdot D)^2 + M^2}\right], \tag{B.14}$$

where the γ_p are Euclidean Dirac matrices. The factor of two difference compared to Eq. (10.129) is the same factor as in Eq. (10.123) and is present because we are interested in the zero modes of the bosonic operator.

In the instanton case this was evaluated by performing a series expansion of the quantity on the right-hand side and noting that all but one of the terms vanished in the $M^2 \to \infty$ limit, in a fashion similar to that of the vortex calculation in the previous section. This strategy is not available to us for the monopole case, where $\mathcal{I}(\mathcal{D}, M^2)$ is not independent of M^2 and so must be evaluated in the $M^2 \to 0$ limit. Instead, we will proceed by showing that the right-hand side of Eq. (B.14) is the volume integral of a total divergence and so can be converted to an integral over the surface at spatial infinity. The integrand of this surface integral can then be expanded in an infinite series in which all but a finite number of terms vanish at large distance.

---

[1] This is not completely trivial. The L_j contain terms with first derivatives that could, from dimensional analysis, give order unity contributions from the third terms in the expansions. However, one can check that these contributions cancel.

In detail, we start by writing

$$
\begin{aligned}
\mathcal{I}(\mathcal{D}, M^2) &= -\int d^3x \operatorname{tr} \left\langle x \left| \gamma_5 \frac{2M^2}{-(\gamma \cdot D)^2 + M^2} \right| x \right\rangle \\
&= -\int d^3x \operatorname{tr} \left\langle x \left| \gamma_5 \frac{2M(-\gamma \cdot D + M)}{-(\gamma \cdot D)^2 + M^2} \right| x \right\rangle \\
&= -\int d^3x \operatorname{tr} \left\langle x \left| \gamma_5 \frac{2M}{\gamma \cdot D + M} \right| x \right\rangle .
\end{aligned}
\tag{B.15}
$$

The second equality here was obtained by using the cyclic property of the trace and noting that γ^5 anticommutes with an odd number of gamma matrices. Next, we define a nonlocal current

$$
J^i(x, y) = \operatorname{tr} \left\langle x \left| \gamma_5 \gamma_i \frac{1}{\gamma \cdot D + M} \right| y \right\rangle .
\tag{B.16}
$$

By noting that

$$
\begin{aligned}
\delta(x - y) &= \left(\gamma_i \frac{\partial}{\partial x^i} - ie\gamma \cdot A + M \right) \left\langle x \left| \frac{1}{\gamma \cdot D + M} \right| y \right\rangle \\
&= \left\langle x \left| \frac{1}{\gamma \cdot D + M} \right| y \right\rangle \left(-\gamma_i \frac{\overleftarrow{\partial}}{\partial y^i} - ie\gamma \cdot A + M \right) ,
\end{aligned}
\tag{B.17}
$$

and using the cyclic property of the trace, we find that

$$
\begin{aligned}
&\left(\frac{\partial}{\partial x^i} + \frac{\partial}{\partial y^i} \right) J^i(x, y) \\
&= -\operatorname{tr} \left\{ [2M + ie\gamma \cdot A(x) - ie\gamma \cdot A(y)] \left\langle x \left| \gamma_5 \frac{1}{\gamma \cdot D + M} \right| y \right\rangle \right\} .
\end{aligned}
\tag{B.18}
$$

Taking the limit $x \to y$, we obtain

$$
\mathcal{I}(\mathcal{D}, M^2) = \int d^3x \, \partial_i J^i(x, x) = \lim_{R \to \infty} \int_R dS_i \, J^i(x, x) ,
\tag{B.19}
$$

with the surface integral being over a sphere of radius R.

We now rewrite the current as

$$
J^i(x, x) = -\operatorname{tr} \left\langle x \left| \gamma_5 \gamma_i \gamma \cdot D \frac{1}{-(\gamma \cdot D)^2 + M^2} \right| x \right\rangle .
\tag{B.20}
$$

Because

$$
-(\gamma \cdot D)^2 + M^2 = -D_j^2 + e^2 \Phi^2 + M^2 + \frac{ie}{4} [\gamma_p, \gamma_q] F_{pq}
\tag{B.21}
$$

we can expand the last factor in J^i as

$$\frac{1}{-(\gamma \cdot D)^2 + M^2} = \frac{1}{-D_j^2 + e^2\Phi^2 + M^2}$$
$$-\frac{1}{-D_j^2 + e^2\Phi^2 + M^2}\left(\frac{ie}{4}[\gamma_p, \gamma_q]F_{pq}\right)\frac{1}{-D_j^2 + e^2\Phi^2 + M^2}$$
$$+\cdots. \tag{B.22}$$

When this is inserted into Eq. (B.20), the first term vanishes after the trace over Dirac indices is taken and the terms represented by the ellipsis can be neglected because of the $1/r^2$ falloff of F_{pq}. This leaves only the terms linear in F_{pq}.

At large distance we have

$$F_{ij} = \epsilon^{ijk}F_{k4} = -\epsilon^{ijk}\frac{\hat{r}^k}{4\pi r^2}Q_M + O(1/r^3), \tag{B.23}$$

where Q_M is the magnetic charge. Substituting this into our expression for the current and evaluating the Dirac traces, we find that

$$\hat{r}^i J^i = \frac{e^2}{\pi r^2}\text{tr}\left\langle x \middle| \Phi \frac{1}{-\nabla^2 + e^2\Phi^2 + M^2}Q_M\frac{1}{-\nabla^2 + e^2\Phi^2 + M^2}\middle| x\right\rangle$$
$$+O(1/r^3), \tag{B.24}$$

where now the trace is only over group indices and the $1/r^3$ piece includes the terms from replacing the covariant derivatives with ordinary derivatives in the matrix element. It is convenient to work in a gauge where Φ and Q_M are both diagonal. They can then be taken outside the matrix element, which is also then diagonal, and whose nonzero components can be obtained with the aid of the integral

$$\left\langle x \middle| \frac{1}{(-\nabla^2 + \mu^2)^2}\middle| x\right\rangle = \int \frac{d^3k}{(2\pi)^3}\frac{1}{(k^2 + \mu^2)^2} = \frac{1}{8\pi\mu}. \tag{B.25}$$

Assembling all the pieces we find that for an SU(2) BPS solution with n units of magnetic charge [83]

$$\lim_{M^2 \to 0} \mathcal{I}(\mathcal{D}, M^2) = 4n. \tag{B.26}$$

We are not quite done. Because massless fields are present, both $\mathcal{D}^\dagger\mathcal{D}$ and $\mathcal{D}\mathcal{D}^\dagger$ have continuum spectra running down to zero. This raises the possibility of a continuum contribution of the form

$$\mathcal{I}_{\text{cont}} = \lim_{M^2 \to 0}\int \frac{d^3k}{(2\pi^3)}\frac{M^2}{k^2 + M^2}\left[\rho_{\mathcal{D}^\dagger\mathcal{D}}(k^2) - \rho_{\mathcal{D}\mathcal{D}^\dagger}(k^2)\right], \tag{B.27}$$

where $\rho_{\mathcal{O}}$ is the density of continuum eigenvalues of the operator $\mathcal{O}$.

The small eigenvalues are sensitive to the large-distance behavior of the operators. This is most easily analyzed using the string gauge formalism of Sec. 5.3. At large r the charged fields are exponentially small while the $1/r$ tails of the

Higgs and gauge fields do not interact. It is then straightforward to show that neither of the densities in Eq. (B.27) is singular [83].

For larger gauge groups, with maximal symmetry breaking, substituting Eqs. (8.64) and (8.65) into Eq. (B.24) leads to Eq. (8.74). As in the SU(2) case, the fields with $1/r$ tails do not interact with each other, while the remaining fields fall exponentially, so $\mathcal{I}_{cont}$ again vanishes. If instead there is an unbroken non-Abelian subgroup, there are interactions between the fields with long-range tails. If the total magnetic charge is Abelian, so that the non-Abelian part of the magnetic field falls at least as fast as $1/r^3$, one can show that there is no continuum contribution. These arguments fail if the magnetic charge has a non-Abelian component. In fact, one can find examples in which the zero-mode equations can be solved explicitly to confirm the presence of a nonzero continuum term [199, 208]. For another example of an index calculation where a continuum contribution arises, see [314].

References

[1] R. F. Dashen, B. Hasslacher, and A. Neveu, "Nonperturbative methods and extended hadron models in field theory. 2. Two-dimensional models and extended hadrons," *Phys. Rev. D* **10**, 4130 (1974).

[2] A. M. Polyakov, "Particle spectrum in the quantum field theory," *JETP Lett.* **20**, 194 (1974).

[3] T. H. R. Skyrme, "A nonlinear theory of strong interactions," *Proc. Roy. Soc. Lond. A* **247**, 260 (1958).

[4] T. H. R. Skyrme, "Particle states of a quantized meson field," *Proc. Roy. Soc. Lond. A* **262**, 237 (1961).

[5] P. M. Morse and H. Feshbach, *Methods of Theoretical Physics* (New York: McGraw-Hill, 1953), p. 734.

[6] J. Goldstone and R. Jackiw, "Quantization of nonlinear waves," *Phys. Rev. D* **11**, 1486 (1975).

[7] J.-L. Gervais and B. Sakita, "Extended particles in quantum field theories," *Phys. Rev. D* **11**, 2943 (1975).

[8] J.-L. Gervais, A. Jevicki, and B. Sakita, "Perturbation expansion around extended particle states in quantum field theory," *Phys. Rev. D* **12**, 1038 (1975).

[9] C. G. Callan, Jr. and D. J. Gross, "Quantum perturbation theory of solitons," *Nucl. Phys. B* **93**, 29 (1975).

[10] N. H. Christ and T. D. Lee, "Quantum expansion of soliton solutions," *Phys. Rev. D* **12**, 1606 (1975).

[11] E. Tomboulis, "Canonical quantization of nonlinear waves," *Phys. Rev. D* **12**, 1678 (1975).

[12] M. Creutz, "Quantum mechanics of extended objects in relativistic field theory," *Phys. Rev. D* **12**, 3126 (1975).

[13] R. Rajaraman and E. J. Weinberg, "Internal symmetry and the semiclassical method in quantum field theory," *Phys. Rev. D* **11**, 2950 (1975).

[14] R. Jackiw and C. Rebbi, "Solitons with fermion number 1/2," *Phys. Rev. D* **13**, 3398 (1976).

[15] R. Jackiw and J. R. Schrieffer, "Solitons with fermion number 1/2 in condensed matter and relativistic field theories," *Nucl. Phys. B* **190**, 253 (1981).

[16] R. Rajaraman, "Intersoliton forces in weak coupling quantum field theories," *Phys. Rev. D* **15**, 2866 (1977).

[17] N. S. Manton, "The force between 't Hooft–Polyakov monopoles," *Nucl. Phys. B* **126**, 525 (1977).

[18] N. S. Manton, "An effective Lagrangian for solitons," *Nucl. Phys. B* **150**, 397 (1979).

[19] J. K. Perring and T. H. R. Skyrme, "A model unified field equation," *Nucl. Phys.* **31**, 550 (1962).

[20] R. F. Dashen, B. Hasslacher, and A. Neveu, "Nonperturbative methods and extended hadron models in field theory. I. Semiclassical functional methods," *Phys. Rev. D* **10**, 4114 (1974).

[21] R. F. Dashen, B. Hasslacher, and A. Neveu, "The particle spectrum in model field theories from semiclassical functional integral techniques," *Phys. Rev. D* **11**, 3424 (1975).

[22] R. Easther, J. T. Giblin, Jr, L. Hui, and E. A. Lim, "New mechanism for bubble nucleation: Classical transitions," *Phys. Rev. D* **80**, 123519 (2009).

[23] J. T. Giblin, Jr, L. Hui, E. A. Lim, and I.-S. Yang, "How to run through walls: Dynamics of bubble and soliton collisions," *Phys. Rev. D* **82**, 045019 (2010).

[24] I. L. Bogolyubsky and V. G. Makhankov, "On the pulsed soliton lifetime in two classical relativistic theory models," *JETP Lett.* **24**, 12 (1976).

[25] M. Gleiser, "Pseudostable bubbles," *Phys. Rev. D* **49**, 2978 (1994).

[26] E. J. Copeland, M. Gleiser, and H.-R. Muller, "Oscillons: Resonant configurations during bubble collapse," *Phys. Rev. D* **52**, 1920 (1995).

[27] M. Gleiser and D. Sicilia, "General theory of oscillon dynamics," *Phys. Rev. D* **80**, 125037 (2009).

[28] M. A. Amin and D. Shirokoff, "Flat-top oscillons in an expanding universe," *Phys. Rev. D* **81**, 085045 (2010).

[29] M. P. Hertzberg, "Quantum radiation of oscillons," *Phys. Rev. D* **82**, 045022 (2010).

[30] A. B. Zamolodchikov and A. B. Zamolodchikov, "Relativistic factorized S-matrix in two dimensions having O(N) isotopic symmetry," *Nucl. Phys. B* **133**, 525 (1978).

[31] A. B. Zamolodchikov and A. B. Zamolodchikov, "Factorized S-matrices in two dimensions as the exact solutions of certain relativistic quantum field models," *Annals Phys.* **120**, 253 (1979).

[32] W. E. Thirring, "A soluble relativistic field theory?," *Annals Phys.* **3**, 91 (1958).

[33] S. Coleman, "The quantum sine-Gordon equation as the massive Thirring model," *Phys. Rev. D* **11**, 2088 (1975).

[34] S. Mandelstam, "Soliton operators for the quantized sine-Gordon equation," *Phys. Rev. D* **11**, 3026 (1975).

[35] G. H. Derrick, "Comments on nonlinear wave equations as models for elementary particles," *J. Math. Phys.* **5**, 1252 (1964).

[36] A. M. Polyakov and A. A. Belavin, "Metastable states of two-dimensional isotropic ferromagnets," *JETP Lett.* **22**, 245 (1975).

[37] H. B. Nielsen and P. Olesen, "Vortex line models for dual strings," *Nucl. Phys. B* **61**, 45 (1973).

[38] B. Plohr, "The behavior at infinity of isotropic vortices and monopoles," *J. Math. Phys.* **22**, 2184 (1981).

[39] L. Perivolaropoulos, "Asymptotics of Nielsen–Olesen vortices," *Phys. Rev. D* **48**, 5961 (1993).

[40] L. Jacobs and C. Rebbi, "Interaction energy of superconducting vortices," *Phys. Rev. B* **19**, 4486 (1979).

[41] E. J. Weinberg, "Multivortex solutions of the Ginzburg–Landau equations," *Phys. Rev. D* **19**, 3008 (1979).

[42] C. H. Taubes, "Arbitrary *N*-vortex solutions to the first order Landau–Ginzburg equations," *Commun. Math. Phys.* **72**, 277 (1980).

[43] R. Jackiw and P. Rossi, "Zero modes of the vortex–fermion system," *Nucl. Phys. B* **190**, 681 (1981).

[44] E. J. Weinberg, "Index calculations for the fermion–vortex system," *Phys. Rev. D* **24**, 2669 (1981).

[45] E. Witten, "Superconducting strings," *Nucl. Phys. B* **249**, 557 (1985).

[46] C. G. Callan, Jr. and J. A. Harvey, "Anomalies and fermion zero modes on strings and domain walls," *Nucl. Phys. B* **250**, 427 (1985).

[47] G. Lazarides and Q. Shafi, "Superconducting strings in axion models," *Phys. Lett.* **151B**, 123 (1985).

[48] A. Vilenkin and E. P. S. Shellard, *Cosmic Strings and other Topological Defects* (Cambridge University Press, 1994).

[49] N. D. Mermin, "The topological theory of defects in ordered media," *Rev. Mod. Phys.* **51**, 591 (1979).

[50] S. Coleman, "Classical lumps and their quantum descendants." In *Aspects of Symmetry*, S. Coleman (Cambridge University Press, 1985).

[51] T. Vachaspati and A. Achucarro, "Semilocal cosmic strings," *Phys. Rev. D* **44**, 3067 (1991).

[52] M. Hindmarsh, "Existence and stability of semilocal strings," *Phys. Rev. Lett.* **68**, 1263 (1992).

[53] M. Hindmarsh, "Semilocal topological defects," *Nucl. Phys. B* **392**, 461 (1993).

[54] A. Achucarro, K. Kuijken, L. Perivolaropoulos, and T. Vachaspati, "Dynamical simulations of semilocal strings," *Nucl. Phys. B* **388**, 435 (1992).

[55] T. Vachaspati, "Vortex solutions in the Weinberg–Salam model," *Phys. Rev. Lett.* **68**, 1977 (1992).

[56] T. Vachaspati, "Electroweak strings," *Nucl. Phys. B* **397**, 648 (1993).

[57] M. James, L. Perivolaropoulos, and T. Vachaspati, "Detailed stability analysis of electroweak strings," *Nucl. Phys. B* **395**, 534 (1993).

[58] A. S. Schwarz, "Field theories with no local conservation of the electric charge," *Nucl. Phys. B* **208**, 141 (1982).

[59] M. G. Alford, K. Benson, S. Coleman, J. March-Russell, and F. Wilczek, "The interactions and excitations of non-Abelian vortices," *Phys. Rev. Lett.* **64**, 1632 (1990).

[60] M. G. Alford, K. Benson, S. Coleman, J. March-Russell, and F. Wilczek, "Zero modes of non-Abelian vortices," *Nucl. Phys. B* **349**, 414 (1991).

[61] M. Bucher, H.-K. Lo, and J. Preskill, "Topological approach to Alice electrodynamics," *Nucl. Phys. B* **386**, 3 (1992).

[62] M. Bucher, K.-M. Lee, and J. Preskill, "On detecting discrete Cheshire charge," *Nucl. Phys. B* **386**, 27 (1992).

[63] J. Preskill and L. M. Krauss, "Local discrete symmetry and quantum mechanical hair," *Nucl. Phys. B* **341**, 50 (1990).

[64] E. Cartan, "La topologie des espaces représentatifs des groupes de Lie," *Œuvres complètes I, 2* (Paris: Éditions du CNRS, 1984), p. 1307.

[65] G. 't Hooft, "Magnetic monopoles in unified gauge theories," *Nucl. Phys. B* **79**, 276 (1974).

[66] P. A. M. Dirac, "Quantized singularities in the electromagnetic field," *Proc. Roy. Soc. Lond. A* **133**, 60 (1931).

[67] D. Zwanziger, "Quantum field theory of particles with both electric and magnetic charges," *Phys. Rev.* **176**, 1489 (1968).

[68] J. S. Schwinger, "Sources and magnetic charge," *Phys. Rev.* **173**, 1536 (1968).

[69] E. Witten, "Dyons of charge $e\theta/2\pi$," *Phys. Lett.* **86B**, 283 (1979).

[70] T. T. Wu and C. N. Yang, "Concept of nonintegrable phase factors and global formulation of gauge fields," *Phys. Rev. D* **12**, 3845 (1975).

[71] J. J. Thomson, "On momentum in the electric field," *Philos. Mag.* **8**, 331 (1904).

[72] I. Tamm, "Die verallgemeinerten Kugelfunktionen und die Wellenfunktionen eines Elektrons im Felde eines Magnetpoles," *Z. Phys.* **71**, 141 (1931).

[73] T. T. Wu and C. N. Yang, "Dirac monopole without strings: Monopole harmonics," *Nucl. Phys. B* **107**, 365 (1976).

[74] H. A. Olsen, P. Osland, and T. T. Wu, "On the existence of bound states for a massive spin-one particle and a magnetic monopole," *Phys. Rev. D* **42**, 665 (1990).

[75] E. J. Weinberg, "Monopole vector spherical harmonics," *Phys. Rev. D* **49**, 1086 (1994).

[76] M. I. Monastyrsky and A. M. Perelomov, "Concerning the existence of monopoles in gauge field theories," *JETP Lett.* **21**, 43 (1975).

[77] H. Georgi and S. L. Glashow, "Unified weak and electromagnetic interactions without neutral currents," *Phys. Rev. Lett.* **28**, 1494 (1972).

[78] T. W. Kirkman and C. K. Zachos, "Asymptotic analysis of the monopole structure," *Phys. Rev. D* **24**, 999 (1981).

[79] K. Lee and E. J. Weinberg, "Nontopological magnetic monopoles and new magnetically charged black holes," *Phys. Rev. Lett.* **73**, 1203 (1994).

[80] E. J. Weinberg and A. H. Guth, "Nonexistence of spherically symmetric monopoles with multiple magnetic charge," *Phys. Rev. D* **14**, 1660 (1976).

[81] E. B. Bogomolny, "Stability of classical solutions," *Sov. J. Nucl. Phys.* **24**, 449 (1976).

[82] M. K. Prasad and C. M. Sommerfield, "An exact classical solution for the 't Hooft monopole and the Julia–Zee dyon," *Phys. Rev. Lett.* **35**, 760 (1975).

[83] E. J. Weinberg, "Parameter counting for multimonopole solutions," *Phys. Rev. D* **20** (1979) 936.

[84] A. Jaffe and C. Taubes, *Vortices and Monopoles* (Boston: Birkhäuser, 1980).

[85] N. H. Christ, A. H. Guth, and E. J. Weinberg, "Canonical formalism for gauge theories with application to monopole solutions," *Nucl. Phys. B* **114**, 61 (1976).

[86] B. Julia and A. Zee, "Poles with both magnetic and electric charges in non-Abelian gauge theory," *Phys. Rev. D* **11**, 2227 (1975).

[87] R. Jackiw and C. Rebbi, "Spin from isospin in a gauge theory," *Phys. Rev. Lett.* **36**, 1116 (1976).

[88] P. Hasenfratz and G. 't Hooft, "Fermion–boson puzzle in a gauge theory," *Phys. Rev. Lett.* **36**, 1119 (1976).

[89] A. S. Goldhaber, "Spin and statistics connection for charge–monopole composites," *Phys. Rev. Lett.* **36**, 1122 (1976).

[90] C. Callias, "Index theorems on open spaces," *Commun. Math. Phys.* **62**, 213 (1978).

[91] V. A. Rubakov, "Adler–Bell–Jackiw anomaly and fermion number breaking in the presence of a magnetic monopole," *Nucl. Phys. B* **203**, 311 (1982).

[92] C. G. Callan, Jr., "Disappearing dyons," *Phys. Rev. D* **25**, 2141 (1982).

[93] C. G. Callan, Jr., "Dyon–fermion dynamics," *Phys. Rev. D* **26**, 2058 (1982).

[94] A. S. Blaer, N. H. Christ, and J.-F. Tang, "Anomalous fermion production by a Julia–Zee dyon," *Phys. Rev. Lett.* **47**, 1364 (1981).

[95] A. S. Blaer, N. H. Christ, and J.-F. Tang, "Fermion emission from a Julia–Zee dyon," *Phys. Rev. D* **25**, 2128 (1982).

[96] P. Klimo and J. S. Dowker, "Dirac monopoles for general gauge theories," *Int. J. Theor. Phys.* **8**, 409 (1973).

[97] F. Englert and P. Windey, "Quantization condition for 't Hooft monopoles in compact simple Lie groups," *Phys. Rev. D* **14**, 2728 (1976).

[98] P. Goddard, J. Nuyts, and D. I. Olive, "Gauge theories and magnetic charge," *Nucl. Phys. B* **125**, 1 (1977).

[99] E. Lubkin, "Geometric definition of gauge invariance," *Annals Phys.* **23**, 233 (1963).

[100] R. A. Brandt and F. Neri, "Stability analysis for singular non-Abelian magnetic monopoles," *Nucl. Phys. B* **161**, 253 (1979).

[101] S. Coleman, "The magnetic monopole fifty years later." In *The Unity of Fundamental Interactions*, ed. A. Zichichi (New York: Plenum, 1983).

[102] A. Sinha, "SU(3) magnetic monopoles," *Phys. Rev. D* **14**, 2016 (1976).

[103] Yu. S. Tyupkin, V. A. Fateev, and A. S. Shvarts, "Existence of heavy particles in gauge field theories," *JETP Lett.* **21**, 41 (1975).

[104] E. J. Weinberg, D. London, and J. L. Rosner, "Magnetic monopoles with Z_n charges," *Nucl. Phys. B* **236**, 90 (1984).

[105] C. P. Dokos and T. N. Tomaras, "Monopoles and dyons in the SU(5) model," *Phys. Rev. D* **21**, 2940 (1980).

[106] C. L. Gardner and J. A. Harvey, "Stable grand unified monopoles with multiple Dirac charge," *Phys. Rev. Lett.* **52**, 879 (1984).

[107] G. Lazarides and Q. Shafi, "The fate of primordial magnetic monopoles," *Phys. Lett.* **94B**, 149 (1980).

[108] A. Abouelsaood, "Are there chromodyons?," *Nucl. Phys. B* **226**, 309 (1983).

[109] P. C. Nelson and A. Manohar, "Global color is not always defined," *Phys. Rev. Lett.* **50**, 943 (1983).

[110] A. P. Balachandran, G. Marmo, N. Mukunda, J. S. Nilsson, E. C. G. Sudarshan, and F. Zaccaria, "Monopole topology and the problem of color," *Phys. Rev. Lett.* **50**, 1553 (1983).

[111] A. P. Balachandran, G. Marmo, N. Mukunda, J. S. Nilsson, E. C. G. Sudarshan, and F. Zaccaria, "Nonabelian monopoles break color. I. Classical mechanics," *Phys. Rev. D* **29**, 2919 (1984).

[112] A. P. Balachandran, G. Marmo, N. Mukunda, J. S. Nilsson, E. C. G. Sudarshan, and F. Zaccaria, "Nonabelian monopoles break color. II. Field theory and quantum mechanics," *Phys. Rev. D* **29**, 2936 (1984).

[113] P. A. Horvathy and J. H. Rawnsley, "Internal symmetries of nonabelian gauge field configurations," *Phys. Rev. D* **32**, 968 (1985).

[114] P. A. Horvathy and J. H. Rawnsley, "The problem of 'global color' in gauge theories," *J. Math. Phys.* **27**, 982 (1986).

[115] H. Guo and E. J. Weinberg, "Instabilities of chromodyons in SO(5) gauge theory," *Phys. Rev. D* **77**, 105026 (2008).

[116] V. A. Rubakov, "Superheavy magnetic monopoles and proton decay," *JETP Lett.* **33**, 644 (1981).

[117] C. G. Callan, Jr., "Monopole catalysis of baryon decay," *Nucl. Phys. B* **212**, 391 (1983).

[118] F. Wilczek, "Remarks on dyons," *Phys. Rev. Lett.* **48**, 1146 (1982).

[119] S. Dawson and A. N. Schellekens, "Monopole catalysis of proton decay in SO(10) grand unified models," *Phys. Rev. D* **27**, 2119 (1983).

[120] A. H. Guth, "The inflationary universe: A possible solution to the horizon and flatness problems," *Phys. Rev. D* **23**, 347 (1981).

[121] D. A. Kirzhnits and A. D. Linde, "Macroscopic consequences of the Weinberg model," *Phys. Lett.* **42B**, 471 (1972).

[122] L. A. Dolan and R. Jackiw, "Symmetry behavior at finite temperature," *Phys. Rev. D* **9**, 3320 (1974).

[123] S. Weinberg, "Gauge and global symmetries at high temperature," *Phys. Rev. D* **9**, 3357 (1974).

[124] D. A. Kirzhnits and A. D. Linde, "Symmetry behavior in gauge theories," *Annals Phys.* **101**, 195 (1976).

[125] S. Coleman and E. J. Weinberg, "Radiative corrections as the origin of spontaneous symmetry breaking," *Phys. Rev. D* **7**, 1888 (1973).

[126] A. H. Guth and E. J. Weinberg, "Could the universe have recovered from a slow first-order phase transition?," *Nucl. Phys. B* **212**, 321 (1983).

[127] A. H. Guth and E. J. Weinberg, "Cosmological consequences of a first-order phase transition in the SU(5) grand unified model," *Phys. Rev. D* **23**, 876 (1981).

[128] T. W. B. Kibble, "Topology of cosmic domains and strings," *J. Phys. A* **9**, 1387 (1976).

[129] M. B. Einhorn, D. L. Stein, and D. Toussaint, "Are grand unified theories compatible with standard cosmology?," *Phys. Rev. D* **21**, 3295 (1980).

[130] A. Vilenkin, "Gravitational field of vacuum domain walls and strings," *Phys. Rev. D* **23**, 852 (1981).

[131] A. Vilenkin, "Gravitational field of vacuum domain walls," *Phys. Lett.* **133B**, 177 (1983).

[132] J. Ipser and P. Sikivie, "Gravitationally repulsive domain wall," *Phys. Rev. D* **30**, 712 (1984).

[133] Ya. B. Zeldovich, I. Yu. Kobzarev, and L. B. Okun, "Cosmological consequences of a spontaneous breakdown of a discrete symmetry," *JETP* **40**, 1 (1975).

[134] T. Vachaspati, *Kinks and Domain Walls: An Introduction to Classical and Quantum Solitons* (Cambridge University Press, 2006).

[135] N. Bevis, M. Hindmarsh, M. Kunz, and J. Urrestilla, "Fitting CMB data with cosmic strings and inflation," *Phys. Rev. Lett.* **100**, 021301 (2008).

[136] R. Battye and A. Moss, "Updated constraints on the cosmic string tension," *Phys. Rev. D* **82**, 023521 (2010).

[137] T. W. B. Kibble, "Cosmic strings reborn?," [astro-ph/0410073].

[138] J. Polchinski, "Introduction to cosmic F- and D-strings," [hep-th/0412244].

[139] Ya. B. Zeldovich and M. Y. Khlopov, "On the concentration of relic magnetic monopoles in the universe," *Phys. Lett.* **79B**, 239 (1978).

[140] J. Preskill, "Cosmological production of superheavy magnetic monopoles," *Phys. Rev. Lett.* **43**, 1365 (1979).

[141] E. N. Parker, "The origin of magnetic fields," *Astrophys. J.* **160**, 383 (1970).

[142] M. S. Turner, E. N. Parker, and T. J. Bogdan, "Magnetic monopoles and the survival of galactic magnetic fields," *Phys. Rev. D* **26**, 1296 (1982).

[143] F. C. Adams, M. Fatuzzo, K. Freese, G. Tarle, R. Watkins, and M. S. Turner, "Extension of the Parker bound on the flux of magnetic monopoles," *Phys. Rev. Lett.* **70**, 2511 (1993).

[144] Y. Rephaeli and M. S. Turner, "The magnetic monopole flux and the survival of intracluster magnetic fields," *Phys. Lett.* **121B**, 115 (1983).

[145] M. Ambrosio *et al.* [MACRO Collaboration], "Final results of magnetic monopole searches with the MACRO experiment," *Eur. Phys. J. C* **25**, 511 (2002).

[146] E. W. Kolb, S. A. Colgate, and J. A. Harvey, "Monopole catalysis of nucleon decay in neutron stars," *Phys. Rev. Lett.* **49**, 1373 (1982).

[147] S. Dimopoulos, J. Preskill, and F. Wilczek, "Catalyzed nucleon decay in neutron stars," *Phys. Lett.* **119B**, 320 (1982).

[148] K. Freese, M. S. Turner, and D. N. Schramm, "Monopole catalysis of nucleon decay in old pulsars," *Phys. Rev. Lett.* **51**, 1625 (1983).

[149] E. W. Kolb and M. S. Turner, "Limits from the soft X-ray background on the temperature of old neutron stars and on the flux of superheavy magnetic monopoles," *Astrophys. J.* **286**, 702 (1984).

[150] J. A. Harvey, "Monopoles in neutron stars," *Nucl. Phys. B* **236**, 255 (1984).

[151] K. Freese and E. Krasteva, "Bound on the flux of magnetic monopoles from catalysis of nucleon decay in white dwarfs," *Phys. Rev. D* **59**, 063007 (1999).

[152] J. Arafune, M. Fukugita, and S. Yanagita, "Monopole abundance in the Solar System and the intrinsic heat in the Jovian planets," *Phys. Rev. D* **32**, 2586 (1985).

[153] P. Langacker and S.-Y. Pi, "Magnetic monopoles in grand unified theories," *Phys. Rev. Lett.* **45**, 1 (1980).

[154] T. W. B. Kibble and E. J. Weinberg, "When does causality constrain the monopole abundance?," *Phys. Rev. D* **43**, 3188 (1991).

[155] E. J. Weinberg and P. Yi, "Magnetic monopole dynamics, supersymmetry, and duality," *Phys. Rept.* **438**, 65 (2007).

[156] S. Coleman, S. J. Parke, A. Neveu, and C. M. Sommerfield, "Can one dent a dyon?," *Phys. Rev. D* **15**, 544 (1977).

[157] C. H. Taubes, "The existence of a nonminimal solution to the SU(2) Yang–Mills–Higgs equations on R^3. Part I," *Commun. Math. Phys.* **86**, 257 (1982).

[158] C. H. Taubes, "The existence of a nonminimal solution to the SU(2) Yang–Mills–Higgs equations on R^3. Part II," *Commun. Math. Phys.* **86**, 299 (1982).

[159] J. Hong, Y. Kim, and P. Y. Pac, "On the multivortex solutions of the Abelian Chern–Simons–Higgs theory," *Phys. Rev. Lett.* **64**, 2230 (1990).

[160] R. Jackiw and E. J. Weinberg, "Self-dual Chern–Simons vortices," *Phys. Rev. Lett.* **64**, 2234 (1990).

[161] R. Jackiw, K. Lee, and E. J. Weinberg, "Self-dual Chern–Simons solitons," *Phys. Rev. D* **42**, 3488 (1990).

[162] C. Lee, K. Lee, and H. Min, "Self-dual Maxwell–Chern–Simons solitons," *Phys. Lett. B* **252**, 79 (1990).

[163] L. Brink, J. H. Schwarz, and J. Scherk, "Supersymmetric Yang–Mills theories," *Nucl. Phys. B* **121**, 77 (1977).

[164] E. Witten and D. I. Olive, "Supersymmetry algebras that include topological charges," *Phys. Lett.* **78B**, 97 (1978).

[165] C. M. Miller, K. Schalm, and E. J. Weinberg, "Nonextremal black holes are BPS," *Phys. Rev. D* **76**, 044001 (2007).

[166] H. Nastase, M. A. Stephanov, P. van Nieuwenhuizen, and A. Rebhan, "Topological boundary conditions, the BPS bound, and elimination of ambiguities in the quantum mass of solitons," *Nucl. Phys. B* **542**, 471 (1999).

[167] N. Graham and R. L. Jaffe, "Energy, central charge, and the BPS bound for (1+1)-dimensional supersymmetric solitons," *Nucl. Phys. B* **544**, 432 (1999).

[168] M. A. Shifman, A. I. Vainshtein, and M. B. Voloshin, "Anomaly and quantum corrections to solitons in two-dimensional theories with minimal supersymmetry," *Phys. Rev. D* **59**, 045016 (1999).

[169] O. Bergman, "Three-pronged strings and 1/4 BPS states in $N = 4$ super-Yang–Mills theory," *Nucl. Phys. B* **525**, 104 (1998).

[170] O. Bergman and B. Kol, "String webs and 1/4 BPS monopoles," *Nucl. Phys. B* **536**, 149 (1998).

[171] K. Lee and P. Yi, "Dyons in $N = 4$ supersymmetric theories and three-pronged strings," *Phys. Rev. D* **58**, 066005 (1998).

[172] W. Nahm, "The construction of all self-dual multimonopoles by the ADHM method." In *Monopoles in Quantum Field Theory*, eds. N. S. Craigie *et al.* (Singapore: World Scientific, 1982).

[173] W. Nahm, "Multimonopoles in the ADHM construction." In *Gauge Theories and Lepton Hadron Interactions*, eds. Z. Horvath *et al.* (Budapest: Central Research Institute for Physics, 1982).

[174] W. Nahm, "All self-dual multimonopoles for arbitrary gauge groups." In *Structural Elements in Particle Physics and Statistical Mechanics*, eds. J. Honerkamp *et al.* (New York: Plenum, 1983).

[175] W. Nahm, "Self-dual monopoles and calorons." In *Group Theoretical Methods in Physics*, eds. G. Denardo *et al.* (Berlin: Springer-Verlag, 1984).

[176] N. Manton and P. Sutcliffe, *Topological Solitons* (Cambridge University Press, 2004).

[177] S. A. Brown, H. Panagopoulos, and M. K. Prasad, "Two separated SU(2) Yang–Mills–Higgs monopoles in the ADHMN Construction," *Phys. Rev. D* **26**, 854 (1982).

[178] P. Houston and L. O'Raifeartaigh, "On the charge distribution of static axial and mirror symmetric monopole systems," *Phys. Lett.* **94B**, 153 (1980).

[179] R. S. Ward, "A Yang–Mills–Higgs monopole of charge 2," *Commun. Math. Phys.* **79**, 317 (1981).

[180] P. Forgacs, Z. Horvath, and L. Palla, "Exact multimonopole solutions in the Bogomolny–Prasad–Sommerfield limit," *Phys. Lett.* **99B**, 232 (1981) [Erratum-ibid. **101**, 457 (1981)].

[181] M. K. Prasad and P. Rossi, "Construction of exact Yang–Mills–Higgs multimonopoles of arbitrary charge," *Phys. Rev. Lett.* **46**, 806 (1981).

[182] C. Rebbi and P. Rossi, "Multimonopole solutions in the Prasad–Sommerfield limit," *Phys. Rev. D* **22**, 2010 (1980).

[183] N. J. Hitchin, N. S. Manton, and M. K. Murray, "Symmetric monopoles," *Nonlinearity* **8**, 661 (1995).

[184] C. J. Houghton and P. M. Sutcliffe, "Tetrahedral and cubic monopoles," *Commun. Math. Phys.* **180**, 343 (1996).

[185] C. J. Houghton and P. M. Sutcliffe, "Monopole scattering with a twist," *Nucl. Phys. B* **464**, 59 (1996).

[186] P. M. Sutcliffe, "Monopole zeros," *Phys. Lett. B* **376**, 103 (1996).

[187] C. J. Houghton and P. M. Sutcliffe, "Octahedral and dodecahedral monopoles," *Nonlinearity* **9**, 385 (1996).

[188] C. J. Houghton, N. S. Manton, and P. M. Sutcliffe, "Rational maps, monopoles and skyrmions," *Nucl. Phys. B* **510**, 507 (1998).

[189] N. S. Manton, "A remark on the scattering of BPS monopoles," *Phys. Lett.* **110B**, 54 (1982).

[190] P. J. Ruback, "Vortex string motion in the Abelian Higgs model," *Nucl. Phys. B* **296**, 669 (1988).

[191] N. S. Manton and T. M. Samols, "Radiation from monopole scattering," *Phys. Lett. B* **215**, 559 (1988).

[192] D. Stuart, "The geodesic approximation for the Yang–Mills–Higgs equations," *Commun. Math. Phys.* **166**, 149 (1994).

[193] N. S. Manton, "Monopole interactions at long range," *Phys. Lett.* **154B**, 397 (1985).

[194] G. W. Gibbons and N. S. Manton, "The moduli space metric for well separated BPS monopoles," *Phys. Lett. B* **356**, 32 (1995).

[195] M. F. Atiyah and N. J. Hitchin, "Low-energy scattering of non-Abelian magnetic monopoles," *Phil. Trans. Roy. Soc. Lond. A* **315**, 459 (1985).

[196] M. F. Atiyah and N. J. Hitchin, "Low-energy scattering of non-Abelian monopoles," *Phys. Lett.* **107A**, 21 (1985).

[197] M. F. Atiyah and N. J. Hitchin, *The Geometry and Dynamics of Magnetic Monopoles* (Princeton University Press, 1988).

[198] G. W. Gibbons and N. S. Manton, "Classical and quantum dynamics of BPS monopoles," *Nucl. Phys. B* **274**, 183 (1986).

[199] E. J. Weinberg, "Fundamental monopoles and multimonopole solutions for arbitrary simple gauge groups," *Nucl. Phys. B* **167**, 500 (1980).

[200] E. J. Weinberg and P. Yi, "Explicit multimonopole solutions in SU(N) gauge theory," *Phys. Rev. D* **58**, 046001 (1998).

[201] S. A. Connell, "The dynamics of the SU(3) charge (1,1) magnetic monopole," University of South Australia preprint (1994).

[202] J. P. Gauntlett and D. A. Lowe, "Dyons and S-duality in $N = 4$ supersymmetric gauge theory," *Nucl. Phys. B* **472**, 194 (1996).

[203] K. Lee, E. J. Weinberg, and P. Yi, "Electromagnetic duality and SU(3) monopoles," *Phys. Lett. B* **376**, 97 (1996).

[204] K. Lee, E. J. Weinberg, and P. Yi, "The moduli space of many BPS monopoles for arbitrary gauge groups," *Phys. Rev. D* **54**, 1633 (1996).

[205] M. K. Murray, "A note on the (1, 1, ..., 1) monopole metric," *J. Geom. Phys.* **23**, 31 (1997).

[206] G. Chalmers, "Multimonopole moduli spaces for SU(N) gauge group," hep-th/9605182 (1996).

[207] C. Lu, "Two monopole systems and the formation of non-Abelian clouds," *Phys. Rev. D* **58**, 125010 (1998).

[208] E. J. Weinberg, "Fundamental monopoles in theories with arbitrary symmetry breaking," *Nucl. Phys. B* **203**, 445 (1982).

[209] K. Lee, E. J. Weinberg, and P. Yi, "Massive and massless monopoles with non-Abelian magnetic charges," *Phys. Rev. D* **54**, 6351 (1996).

[210] E. J. Weinberg, "A continuous family of magnetic monopole solutions," *Phys. Lett.* **119B**, 151 (1982).

[211] R. S. Ward, "Magnetic monopoles with gauge group SU(3) broken to U(2)," *Phys. Lett.* **107B**, 281 (1981).

[212] A. S. Dancer and R. A. Leese, "A numerical study of SU(3) charge-two monopoles with minimal symmetry breaking," *Phys. Lett. B* **390**, 252 (1997).

[213] A. S. Dancer, "Nahm data and SU(3) monopoles," *Nonlinearity* **5**, 1355 (1992).

[214] P. Irwin, "SU(3) monopoles and their fields," *Phys. Rev. D* **56**, 5200 (1997).

[215] C. J. Houghton and E. J. Weinberg, "Multicloud solutions with massless and massive monopoles," *Phys. Rev. D* **66**, 125002 (2002).

[216] A. S. Dancer, "Nahm's equations and hyper-Kähler geometry," *Commun. Math. Phys.* **158**, 545 (1993).

[217] A. Dancer and R. Leese, "Dynamics of SU(3) monopoles," *Proc. Roy. Soc. Lond. A* **440**, 421 (1993).

[218] X. Chen and E. J. Weinberg, "Scattering of massless and massive monopoles in an SU(N) theory," *Phys. Rev. D* **64**, 065010 (2001).

[219] C. M. Miller and E. J. Weinberg, "Interactions of massless monopole clouds," *Phys. Rev. D* **80**, 065025 (2009).

[220] X. Chen, H. Guo, and E. J. Weinberg, "Massless monopoles and the moduli space approximation," *Phys. Rev. D* **64**, 125004 (2001).

[221] C. Montonen and D. I. Olive, "Magnetic monopoles as gauge particles?," *Phys. Lett.* **72B**, 117 (1977).

[222] H. Osborn, "Topological charges for $N = 4$ supersymmetric gauge theories and monopoles of spin 1," *Phys. Lett.* **83B**, 321 (1979).

[223] A. Sen, "Dyon–monopole bound states, self-dual harmonic forms on the multimonopole moduli space, and SL(2,Z) invariance in string theory," *Phys. Lett. B* **329**, 217 (1994).

[224] T. Banks, C. M. Bender, and T. T. Wu, "Coupled anharmonic oscillators. I. Equal-mass case," *Phys. Rev. D* **8**, 3346 (1973).

[225] T. Banks and C. M. Bender, "Coupled anharmonic oscillators. II. Unequal-mass case," *Phys. Rev. D* **8**, 3366 (1973).

[226] S. Coleman, "Fate of the false vacuum: Semiclassical theory," *Phys. Rev. D* **15**, 2929 (1977).

[227] A. A. Belavin, A. M. Polyakov, A. S. Shvarts, and Y. S. Tyupkin, "Pseudoparticle solutions of the Yang–Mills equations," *Phys. Lett.* **59B**, 85 (1975).

[228] S. Coleman, "The uses of instantons." In *Aspects of Symmetry*, S. Coleman (Cambridge University Press, 1985).

[229] J. S. Langer, "Theory of the condensation point," *Annals Phys.* **41**, 108 (1967).

[230] C. G. Callan, Jr. and S. Coleman, "Fate of the false vacuum. II. First quantum corrections," *Phys. Rev. D* **16**, 1762 (1977).

[231] S. Coleman, "Quantum tunneling and negative eigenvalues," *Nucl. Phys. B* **298**, 178 (1988).

[232] R. P. Feynman and A. R. Hibbs, *Quantum Mechanics and Path Integrals* (New York: McGraw-Hill, 1965).

[233] R. Jackiw and C. Rebbi, "Vacuum periodicity in a Yang–Mills quantum theory," *Phys. Rev. Lett.* **37**, 172 (1976).

[234] C. G. Callan, Jr., R. F. Dashen, and D. J. Gross, "The structure of the gauge theory vacuum," *Phys. Lett.* **63B**, 334 (1976).

[235] C. W. Bernard and E. J. Weinberg, "The interpretation of pseudoparticles in physical gauges," *Phys. Rev. D* **15**, 3656 (1977).

[236] V. N. Gribov, "Quantization of non-Abelian gauge theories," *Nucl. Phys. B* **139**, 1 (1978).

[237] R. Jackiw and C. Rebbi, "Conformal properties of a Yang–Mills pseudoparticle," *Phys. Rev. D* **14**, 517 (1976).

[238] G. 't Hooft, "Computation of the quantum effects due to a four-dimensional pseudoparticle," *Phys. Rev. D* **14**, 3432 (1976).

[239] G. 't Hooft, unpublished

[240] R. Jackiw, C. Nohl, and C. Rebbi, "Conformal properties of pseudoparticle configurations," *Phys. Rev. D* **15**, 1642 (1977).

[241] A. S. Schwarz, "On regular solutions of Euclidean Yang–Mills equations," *Phys. Lett.* **67B**, 172 (1977).

[242] R. Jackiw and C. Rebbi, "Degrees of freedom in pseudoparticle systems," *Phys. Lett.* **67B**, 189 (1977).

[243] M. F. Atiyah, N. J. Hitchin, and I. M. Singer, "Deformations of instantons," *Proc. Nat. Acad. Sci.* **74**, 2662 (1977).

[244] L. S. Brown, R. D. Carlitz, and C. Lee, "Massless excitations in instanton fields," *Phys. Rev. D* **16**, 417 (1977).

[245] M. F. Atiyah and I. M. Singer, "The index of elliptic operators. 1," *Annals Math.* **87**, 484 (1968).

[246] M. F. Atiyah, N. J. Hitchin, V. G. Drinfeld, and Y. I. Manin, "Construction of instantons," *Phys. Lett.* **65A**, 185 (1978).

[247] V. G. Drinfeld and Y. I. Manin, "A description of instantons," *Commun. Math. Phys.* **63**, 177 (1978).

[248] N. H. Christ, E. J. Weinberg, and N. K. Stanton, "General self-dual Yang–Mills solutions," *Phys. Rev. D* **18**, 2013 (1978).

[249] E. Corrigan, D. B. Fairlie, S. Templeton, and P. Goddard, "A Green's function for the general self-dual gauge field," *Nucl. Phys. B* **140**, 31 (1978).

[250] E. Corrigan and P. Goddard, "Construction of instanton and monopole solutions and reciprocity," *Annals Phys.* **154**, 253 (1984).

[251] E. Witten, "Small instantons in string theory," *Nucl. Phys. B* **460**, 541 (1996).

[252] M. R. Douglas, "Gauge fields and D-branes," *J. Geom. Phys.* **28**, 255 (1998).

[253] A. A. Belavin and A. M. Polyakov, "Quantum fluctuations of pseudoparticles," *Nucl. Phys. B* **123**, 429 (1977).

[254] C. W. Bernard, N. H. Christ, A. H. Guth, and E. J. Weinberg, "Pseudoparticle parameters for arbitrary gauge groups," *Phys. Rev. D* **16**, 2967 (1977).

[255] J. S. Bell and R. Jackiw, "A PCAC puzzle: $\pi_0 \to \gamma\gamma$ in the σ-model," *Nuovo Cim.* **A60**, 47 (1969).

[256] S. L. Adler, "Axial vector vertex in spinor electrodynamics," *Phys. Rev.* **177**, 2426 (1969).

[257] W. A. Bardeen, "Anomalous Ward identities in spinor field theories," *Phys. Rev.* **184**, 1848 (1969).

[258] K. Fujikawa, "Path integral measure for gauge invariant fermion theories," *Phys. Rev. Lett.* **42**, 1195 (1979).

[259] C. G. Callan, Jr., R. F. Dashen, and D. J. Gross, "Toward a theory of the strong interactions," *Phys. Rev. D* **17**, 2717 (1978).

[260] S. Weinberg, "The U(1) problem," *Phys. Rev. D* **11**, 3583 (1975).

[261] G. 't Hooft, "Symmetry breaking through Bell–Jackiw anomalies," *Phys. Rev. Lett.* **37**, 8 (1976).

[262] N. S. Manton, "Topology in the Weinberg–Salam theory," *Phys. Rev. D* **28**, 2019 (1983).

[263] F. R. Klinkhamer and N. S. Manton, "A saddle point solution in the Weinberg–Salam theory," *Phys. Rev. D* **30**, 2212 (1984).

[264] V. A. Rubakov and M. E. Shaposhnikov, "Electroweak baryon number nonconservation in the early universe and in high-energy collisions," *Usp. Fiz. Nauk* **166**, 493 (1996).

[265] K. Nakamura *et al.* [Particle Data Group Collaboration], "Review of particle physics," *J. Phys. G* **37**, 075021 (2010).

[266] R. J. Crewther, P. Di Vecchia, G. Veneziano, and E. Witten, "Chiral estimate of the electric dipole moment of the neutron in quantum chromodynamics," *Phys. Lett.* **88B**, 123 (1979).

[267] R. D. Peccei and H. R. Quinn, "CP conservation in the presence of instantons," *Phys. Rev. Lett.* **38**, 1440 (1977).

[268] R. D. Peccei and H. R. Quinn, "Constraints imposed by CP conservation in the presence of instantons," *Phys. Rev. D* **16**, 1791 (1977).

[269] S. Weinberg, "A new light boson?," *Phys. Rev. Lett.* **40**, 223 (1978).

[270] F. Wilczek, "Problem of strong P and T invariance in the presence of instantons," *Phys. Rev. Lett.* **40**, 279 (1978).

[271] S. Coleman, V. Glaser, and A. Martin, "Action minima among solutions to a class of Euclidean scalar field equations," *Commun. Math. Phys.* **58**, 211 (1978).

[272] A. Kusenko, K. Lee, and E. J. Weinberg, "Vacuum decay and internal symmetries," *Phys. Rev. D* **55**, 4903 (1997).

[273] E. J. Weinberg, "Vacuum decay in theories with symmetry breaking by radiative corrections," *Phys. Rev. D* **47**, 4614 (1993).

[274] I. Affleck, "Quantum statistical metastability," *Phys. Rev. Lett.* **46**, 388 (1981).

[275] A. D. Linde, "Decay of the false vacuum at finite temperature," *Nucl. Phys. B* **216**, 421 (1983).

[276] S. Coleman and F. De Luccia, "Gravitational effects on and of vacuum decay," *Phys. Rev. D* **21**, 3305 (1980).

[277] A. R. Brown and E. J. Weinberg, "Thermal derivation of the Coleman–De Luccia tunneling prescription," *Phys. Rev. D* **76**, 064003 (2007).

[278] G. W. Gibbons and S. W. Hawking, "Cosmological event horizons, thermodynamics, and particle creation," *Phys. Rev. D* **15**, 2738 (1977).

[279] G. W. Gibbons and S. W. Hawking, "Action integrals and partition functions in quantum gravity," *Phys. Rev. D* **15**, 2752 (1977).

[280] S. J. Parke, "Gravity, the decay of the false vacuum and the new inflationary universe scenario," *Phys. Lett.* **121B**, 313 (1983).

[281] L. G. Jensen and P. J. Steinhardt, "Bubble nucleation and the Coleman–Weinberg model," *Nucl. Phys. B* **237**, 176 (1984).

[282] L. G. Jensen and P. J. Steinhardt, "Bubble nucleation for flat potential barriers," *Nucl. Phys. B* **317**, 693 (1989).

[283] J. C. Hackworth and E. J. Weinberg, "Oscillating bounce solutions and vacuum tunneling in de Sitter spacetime," *Phys. Rev. D* **71**, 044014 (2005).

[284] P. Batra and M. Kleban, "Transitions between de Sitter minima," *Phys. Rev. D* **76**, 103510 (2007).

[285] T. Banks, "Heretics of the false vacuum: Gravitational effects on and of vacuum decay. 2," hep-th/0211160 (2002).

[286] S. W. Hawking and I. G. Moss, "Supercooled phase transitions in the very early universe," *Phys. Lett.* **110B**, 35 (1982).

[287] K. Lee and E. J. Weinberg, "Decay of the true vacuum in curved space-time," *Phys. Rev. D* **36**, 1088 (1987).

[288] L. F. Abbott and S. Deser, "Stability of gravity with a cosmological constant," *Nucl. Phys. B* **195**, 76 (1982).

[289] J. C. Hackworth, "Vacuum decay in de Sitter spacetime," Ph. D. thesis, Columbia University (2006).

[290] G. Lavrelashvili, "The number of negative modes of the oscillating bounces," *Phys. Rev. D* **73**, 083513 (2006).

[291] T. Tanaka, "The no-negative mode theorem in false vacuum decay with gravity," *Nucl. Phys. B* **556**, 373 (1999).

[292] A. Khvedelidze, G. V. Lavrelashvili, and T. Tanaka, "On cosmological perturbations in closed FRW model with scalar field and false vacuum decay," *Phys. Rev. D* **62**, 083501 (2000).

[293] G. V. Lavrelashvili, "Negative mode problem in false vacuum decay with gravity," Nucl. Phys. Proc. Suppl. **88**, 75 (2000).

[294] S. Gratton and N. Turok, "Homogeneous modes of cosmological instantons," *Phys. Rev. D* **63**, 123514 (2001).

[295] S. Coleman and P. J. Steinhardt, unpublished.

[296] A. A. Starobinsky, "Stochastic de Sitter (inflationary) stage in the early universe." In *Field Theory, Quantum Gravity and Strings*, eds. H. J. De Vega and N. Sanchez (New York: Springer-Verlag, 1986).

[297] A. S. Goncharov, A. D. Linde, and V. F. Mukhanov, "The global structure of the inflationary universe," *Int. J. Mod. Phys. A* **2**, 561 (1987).

[298] A. D. Linde, "Hard art of the universe creation (stochastic approach to tunneling and baby universe formation)," *Nucl. Phys. B* **372**, 421 (1992).

[299] A. D. Linde, "A new inflationary universe scenario: A possible solution of the horizon, flatness, homogeneity, isotropy and primordial monopole problems," *Phys. Lett.* **108B**, 389 (1982).

[300] A. Albrecht and P. J. Steinhardt, "Cosmology for grand unified theories with radiatively induced symmetry breaking," *Phys. Rev. Lett.* **48**, 1220 (1982).

[301] J. Garriga and A. Megevand, "Coincident brane nucleation and the neutralization of Λ," *Phys. Rev. D* **69**, 083510 (2004).

[302] A. Masoumi and E. J. Weinberg, "Bounces with $O(3) \times O(2)$ symmetry." (2012).

[303] S. W. Hawking, I. G. Moss, and J. M. Stewart, "Bubble collisions in the very early universe," *Phys. Rev. D* **26**, 2681 (1982).

[304] J. Garriga, A. H. Guth, and A. Vilenkin, "Eternal inflation, bubble collisions, and the persistence of memory," *Phys. Rev. D* **76**, 123512 (2007).

[305] S. Chang, M. Kleban, and T. S. Levi, "When worlds collide," *JCAP* **0804**, 034 (2008).

[306] S. Chang, M. Kleban, and T. S. Levi, "Watching worlds collide: Effects on the CMB from cosmological bubble collisions," *JCAP* **0904**, 025 (2009).

[307] A. Aguirre, M. C. Johnson, and M. Tysanner, "Surviving the crash: Assessing the aftermath of cosmic bubble collisions," *Phys. Rev. D* **79**, 123514 (2009).

[308] B. Freivogel, M. Kleban, A. Nicolis, and K. Sigurdson, "Eternal inflation, bubble collisions, and the disintegration of the persistence of memory," *JCAP* **0908**, 036 (2009).

[309] J. J. Blanco-Pillado and M. P. Salem, "Observable effects of anisotropic bubble nucleation," *JCAP* **1007**, 007 (2010).

[310] L. F. Abbott and S. Coleman, "The collapse of an anti-de Sitter bubble," *Nucl. Phys. B* **259**, 170 (1985).

[311] J. E. Humphreys, *Introduction to Lie Algebras and Representation Theory* (New York: Springer-Verlag, 1972).

[312] P. Ramond, *Group Theory* (Cambridge University Press, 2010).

[313] H. J. de Vega and F. A. Schaposnik, "Classical vortex solution of the Abelian Higgs model," *Phys. Rev. D* **14**, 1100 (1976).

[314] J. E. Kiskis, "Fermions in a pseudoparticle field," *Phys. Rev. D* **15**, 2329 (1977).

Index